LANCIA DELTA
Gold Portfolio
1979-1994

Compiled by
R.M.Clarke

ISBN 1 85520 257 3

BROOKLANDS BOOKS LTD.
P.O. BOX 146, COBHAM,
SURREY, KT11 1LG. UK

Printed in Hong Kong

BROOKLANDS BOOKS

BROOKLANDS ROAD TEST SERIES

Abarth Gold Portfolio 1950-1971
AC Ace & Aceca 1953-1983
Alfa Romeo Giulietta Gold Portfolio 1954-1965
Alfa Romeo Giulia Berlinas 1962-1976
Alfa Romeo Giulia Coupés 1963-1976
Alfa Romeo Giulia Coupés Gold P. 1963-1976
Alfa Romeo Spider 1966-1990
Alfa Romeo Spider Gold Portfolio 1966-1991
Alfa Romeo Alfasud 1972-1984
Alfa Romeo Alfetta Gold Portfolio 1972-1987
Alfa Romeo Alfetta GTV6 1980-1987
Allard Gold Portfolio 1937-1959
Alvis Gold Portfolio 1919-1967
AMX & Javelin Muscle Portfolio1968-1974
Armstrong Siddeley Gold Portfolio 1945-1960
Austin A30 & A35 1951-1962
Austin Healey 100 & 100/6 Gold P. 1952-1959
Austin Healey 3000 Gold Portfolio 1959-1967
Austin Healey Sprite 1958-1971
Barracuda Muscle Portfolio 1964-1974
BMW Six Cyl. Coupés 1969-1975
BMW 1600 Collection No.1 1966-1981
BMW 2002 Gold Portfolio1968-1976
BMW 316, 318, 320 (4 cyl.) Gold P. 1975-1990
BMW 320, 323, 325 (6 cyl.) Gold P .1977-1990
BMW M Series Performance Portfolio1976-1993
BMW 5 Series Gold Portfolio1981-1987
Bristol Cars Gold Portfolio 1946-1992
Buick Automobiles 1947-1960
Buick Muscle Cars 1965-1970
Cadillac Automobiles 1949-1959
Cadillac Automobiles 1960-1969
Charger Muscle Portfolio1966-1974
Chevrolet 1955-1957
Chevrolet Impala & SS 1958-1971
Chevrolet Corvair 1959-1969
Chevy II & Nova SS Muscle Portfolio 1962-1974
Chevy El Camino & SS 1959-1987
Chevelle & SS Muscle Portfolio 1964-1972
Chevrolet Muscle Cars 1966-1971
Chevy Blazer 1969-1981
Chevrolet Corvette Gold Portfolio 1953-1962
Chevrolet Corvette Sting Ray Gold P. 1963-1967
Chevrolet Corvette Gold Portfolio 1968-1977
High Performance Corvettes 1983-1989
Camaro Muscle Portfolio 1967-1973
Chevrolet Camaro Z28 & SS 1966-1973
Chevrolet Camaro & Z28 1973-1981
High Performance Camaros 1982-1988
Chrysler 300 Gold Portfolio 1955-1970
Chrysler Valiant 1960-1962
Citroen Traction Avant Gold Portfolio 1934-1957
Citroen 2CV Gold Portfolio 1948-1989
Citroen DS & ID 1955-1975
Citroen DS & ID Gold Portfolio 1955-1975
Citroen SM 1970-1975
Cobras & Replicas 1962-1983
Shelby Cobra Gold Portfolio 1962-1969
Cobras & Cobra Replicas Gold P. 1962-1989
Cunningham Automobiles 1951-1955
Daimler SP250 Sports & V-8 250 Saloon Gold Portfolio 1959-1969
Datsun Roadsters 1962-1971
Datsun 240Z 1970-1973
Datsun 280Z & ZX 1975-1983
The De Lorean 1977-1993
De Tomaso Collection No. 1 1962-1981
Dodge Muscle Cars 1967-1970
Dodge Viper on the Road
ERA Gold Portfolio 1934-1994
Excalibur Collection No. 1 1952-1981
Facel Vega 1954-1964
Ferrari Dino 1965-1974
Ferrari Dino 308 1974-1979
Ferrari 308 & Mondial 1980-1984
Fiat 500 Gold Portfolio 1936-1972
Fiat 600 & 850 Gold Portfolio 1955-1972
Fiat Pininfarina 124 & 2000 Spider 1968-1985
Fiat-Bertone X1/9 1973-1988
Ford Consul, Zephyr, Zodiac Mk.I & II 1950-1962
Ford Zephyr, Zodiac, Executive, Mk.III & Mk.IV 1962-1971
Ford Cortina 1600E & GT 1967-1970
High Performance Capris Gold P. 1969-1987
Capri Muscle Portfolio 1974-1987
High Performance Fiestas 1979-1991
High Performance Escorts Mk.I 1968-1974
High Performance Escorts Mk.II 1975-1980
High Performance Escorts 1980-1985
High Performance Escorts 1985-1990
High Performance Sierras & Merkurs Gold Portfolio 1983-1990
Ford Automobiles 1949-1959
Ford Fairlane 1955-1970
Ford Ranchero 1957-1959
Ford Thunderbird 1955-1957
Ford Thunderbird 1958-1963
Ford Thunderbird 1964-1976
Ford GT40 Gold Portfolio 1964-1987
Ford Bronco 1966-1977
Ford Bronco 1978-1988
Holden 1948-1962
Honda CRX 1983-1987
Isetta 1953-1964
ISO & Bizzarrini Gold Portfolio 1962-1974
Jaguar and SS Gold Portfolio 1931-1951
Jaguar XK120, 140, 150 Gold P. 1948-1960
Jaguar Mk.VII, VIII, IX, X, 420 Gold P.1950-1970
Jaguar Mk.1 & Mk.2 Gold Portfolio 1959-1969
Jaguar E-Type Gold Portfolio 1961-1971
Jaguar E-Type V-12 1971-1975
Jaguar XJ12, XJ5.3, V12 Gold P. 1972-1990
Jaguar XJ6 Series I & II Gold P. 1968-1979
Jaguar XJ6 Series II 1973-1979
Jaguar XJ6 Series III 1979-1986
Jaguar XJS Gold Portfolio 1975-1990
Jeep CJ5 & CJ6 1960-1976
Jeep CJ5 & CJ7 1976-1986
Jensen Cars 1946-1967
Jensen Cars 1967-1979
Jensen Interceptor Gold Portfolio 1966-1986
Jensen Healey 1972-1976
Lagonda Gold Portfolio 1919-1964
Lamborghini Cars 1964-1970
Lamborghini Countach & Urraco 1974-1980
Lamborghini Countach & Jalpa 1980-1985
Lancia Fulvia Gold Portfolio 1963-1976
Lancia Beta Gold Portfolio 1972-1984
Lancia Delta Gold Portfolio 1979-1994
Lancia Stratos 1972-1985
Land Rover Series I 1948-1958
Land Rover Series II & IIa 1958-1971
Land Rover Series III 1971-1985
Land Rover 90 & 110 Defender Gold Portfolio 1983-1994
Land Rover Discovery 1989-1994
Lincoln Gold Portfolio 1949-1960
Lincoln Continental 1961-1969
Lincoln Continental 1969-1976
Lotus Sports Racers Gold Portfolio 1953-1965
Lotus & Caterham Seven Gold P. 1957-1989
Lotus Elite 1957-1964
Lotus Elite & Eclat 1974-1982
Lotus Elan Gold Portfolio 1962-1974
Lotus Elan Collection No. 2 1963-1972
Lotus Elan & SE 1989-1992
Lotus Cortina Gold Portfolio 1963-1970
Lotus Europa Gold Portfolio 1966-1975
Lotus Elite & Eclat 1974-1982
Lotus Turbo Esprit 1980-1986
Marcos Cars 1960-1988
Maserati 1965-1970
Maserati 1970-1975
Mercedes 190 & 300 SL 1954-1963
Mercedes 230/250/280SL 1963-1971
Mercedes Benz SLs & SLCs Gold P. 1971-1989
Mercedes S & 600 1965-1972
Mercedes S Class 1972-1979
Mercedes SLs Performance Portfolio 1989-1994
Mercury Muscle Cars 1966-1971
Messerschmitt Gold Portfolio1954-1964
Metropolitan 1954-1962
MG Gold Portfolio 1929-1939
MG TC 1945-1949
MG TD 1949-1953
MG TF 1953-1955
MGA & Twin Cam Gold Portfolio 1955-1962
MG Midget Gold Portfolio1961-1979
MGB Roadsters 1962-1980
MGB MGC & V8 Gold Portfolio 1962-1980
MGB GT 1965-1980
Mini Cooper Gold Portfolio 1961-1971
Mini Muscle Cars 1961-1979
Mini Moke Gold Portfolio1964-1994
Mopar Muscle Cars 1964-1967
Morgan Three-Wheeler Gold Portfolio 1910-1952
Morgan Plus 4 & Four 4 Gold P. 1936-1967
Morgan Cars 1960-1970
Morgan Cars Gold Portfolio 1968-1989
Morris Minor Collection No. 1 1948-1980
Shelby Mustang Muscle Portfolio 1965-1970
High Performance Mustang IIs 1974-1978
High Performance Mustangs 1982-1988
Oldsmobile Automobiles 1955-1963
Oldsmobile Muscle Cars 1964-1971
Oldsmobile Toronado 1966-1978
Opel GT 1968-1973
Packard Gold Portfolio 1946-1958
Pantera Gold Portfolio 1970-1989
Panther Gold Portfolio 1972-1990
Plymouth Muscle Cars 1966-1971
Pontiac Tempest & GTO 1961-1965
Pontiac Muscle Cars 1966-1972
Pontiac Firebird & Trans-Am 1973-1981
High Performance Firebirds 1982-1988
Pontiac Fiero 1984-1988
Porsche 356 Gold Portfolio1953-1965
Porsche 911 1965-1969
Porsche 911 1970-1972
Porsche 911 1973-1977
Porsche 911 Carrera 1973-1977
Porsche 911 Turbo 1975-1984
Porsche 911 SC 1978-1983
Porsche 914 Collection No. 1 1969-1983
Porsche 914 Gold Portfolio 1969-1976
Porsche 924 Gold Portfolio 1975-1988
Porsche 928 Performance Portfolio 1977-1994
Porsche 944 Gold Portfolio1981-1991
Range Rover Gold Portfolio 1970-1992
Reliant Scimitar 1964-1986
Riley Gold Portfolio 1924-1939
Riley 1.5 & 2.5 Litre Gold Portfolio 1945-1955
Rolls Royce Silver Cloud & Bentley 'S' Series Gold Portfolio 1955-1965
Rolls Royce Silver Shadow Gold P. 1965-1980
Rolls Royce & Bentley Gold P. 1980-1989
Rover P4 1949-1959
Rover P4 1955-1964
Rover 3 & 3.5 Litre Gold Portfolio 1958-1973
Rover 2000 & 2200 1963-1977
Rover 3500 1968-1977
Rover 3500 & Vitesse 1976-1986
Saab Sonett Collection No.1 1966-1974
Saab Turbo 1976-1983
Studebaker Gold Portfolio 1947-1966
Studebaker Hawks & Larks 1956-1963
Avanti 1962-1990
Sunbeam Tiger & Alpine Gold P. 1959-1967
Toyota MR2 1984-1988
Toyota Land Cruiser 1956-1984
Triumph TR2 & TR3 Gold Portfolio 1952-1961
Triumph TR4, TR5, TR250 1961-1968
Triumph TR6 Gold Portfolio 1969-1976
Triumph TR7 & TR8 Gold Portfolio 1975-1982
Triumph Herald 1959-1971
Triumph Vitesse 1962-1971
Triumph Spitfire Gold Portfolio 1962-1980
Triumph 2000, 2.5, 2500 1963-1977
Triumph GT6 Gold Portfolio 1966-1974
Triumph Stag 1970-1980
TVR Gold Portfolio 1959-1986
TVR Performance Portfolio 1986-1994
VW Beetle Gold Portfolio1935-1967
VW Beetle Gold Portfolio1968-1991
VW Beetle Collection No.1 1970-1982
VW Karmann Ghia 1955-1982
VW Bus, Camper, Van 1954-1967
VW Bus, Camper, Van 1968-1979
VW Bus, Camper, Van 1979-1989
VW Scirocco 1974-1981
VW Golf GTI 1976-1986
Volvo PV444 & PV544 1945-1965
Volvo Amazon-120 Gold Portfolio 1956-1970
Volvo 1800 Gold Portfolio 1960-1973

BROOKLANDS ROAD & TRACK SERIES

Road & Track on Alfa Romeo 1949-1963
Road & Track on Alfa Romeo 1964-1970
Road & Track on Alfa Romeo 1971-1976
Road & Track on Alfa Romeo 1977-1989
Road & Track on Aston Martin 1962-1990
R & T on Auburn Cord and Duesenburg 1952-84
Road & Track on Audi & Auto Union 1952-1980
Road & Track on Audi & Auto Union 1980-1986
Road & Track on Austin Healey 1953-1970
Road & Track on BMW Cars 1966-1974
Road & Track on BMW Cars 1975-1978
Road & Track on BMW Cars 1979-1983
R & T on Cobra, Shelby & Ford GT40 1962-1992
Road & Track on Corvette 1953-1967
Road & Track on Corvette 1968-1982
Road & Track on Corvette 1982-1986
Road & Track on Corvette 1986-1990
Road & Track on Datsun Z 1970-1983
Road & Track on Ferrari 1975-1981
Road & Track on Ferrari 1981-1984
Road & Track on Ferrari 1984-1988
Road & Track on Fiat Sports Cars 1968-1987
Road & Track on Jaguar 1950-1960
Road & Track on Jaguar 1961-1968
Road & Track on Jaguar 1968-1974
Road & Track on Jaguar 1974-1982
Road & Track on Jaguar 1983-1989
Road & Track on Lamborghini 1964-1985
Road & Track on Lotus 1972-1981
Road & Track on Maserati 1952-1974
Road & Track on Maserati 1975-1983
R & T on Mazda RX7 & MX5 Miata 1986-1991
Road & Track on Mercedes 1952-1962
Road & Track on Mercedes 1963-1970
Road & Track on Mercedes 1971-1979
Road & Track on Mercedes 1980-1987
Road & Track on MG Sports Cars 1949-1961
Road & Track on MG Sports Cars 1962-1980
Road & Track on Mustang 1964-1977
R & T on Nissan 300-ZX & Turbo 1984-1989
Road & Track on Peugeot 1955-1986
Road & Track on Pontiac 1960-1983
Road & Track on Porsche 1951-1967
Road & Track on Porsche 1968-1971
Road & Track on Porsche 1972-1975
Road & Track on Porsche 1975-1978
Road & Track on Porsche 1979-1982
Road & Track on Porsche 1982-1985
Road & Track on Porsche 1985-1988
R & T on Rolls Royce & Bentley 1950-1965
R & T on Rolls Royce & Bentley 1966-1984
Road & Track on Saab 1972-1992
R & T on Toyota Sports & GT Cars 1966-1984
R & T on Triumph Sports Cars 1953-1967
R & T on Triumph Sports Cars 1967-1974
R & T on Triumph Sports Cars 1974-1982
Road & Track on Volkswagen 1951-1968
Road & Track on Volkswagen 1968-1978
Road & Track on Volkswagen 1978-1985
Road & Track on Volvo 1957-1974
Road & Track on Volvo 1977-1994
R&T - Henry Manney at Large & Abroad
R&T - Peter Egan's "Side Glances"

BROOKLANDS CAR AND DRIVER SERIES

Car and Driver on BMW 1955-1977
Car and Driver on BMW 1977-1985
C and D on Cobra, Shelby & Ford GT40 1963-84
Car and Driver on Corvette 1956-1967
Car and Driver on Corvette 1968-1977
Car and Driver on Corvette 1978-1982
Car and Driver on Corvette 1983-1988
C and D on Datsun Z 1600 & 2000 1966-1984
Car and Driver on Ferrari 1955-1962
Car and Driver on Ferrari 1963-1975
Car and Driver on Ferrari 1976-1983
Car and Driver on Mopar 1956-1967
Car and Driver on Mopar 1968-1975
Car and Driver on Mustang 1964-1972
Car and Driver on Pontiac 1961-1975
Car and Driver on Porsche 1955-1962
Car and Driver on Porsche 1963-1970
Car and Driver on Porsche 1970-1976
Car and Driver on Porsche 1977-1981
Car and Driver on Porsche 1982-1986
Car and Driver on Saab 1956-1985
Car and Driver on Volvo 1955-1986

BROOKLANDS PRACTICAL CLASSICS SERIES

PC on Austin A40 Restoration
PC on Land Rover Restoration
PC on Metalworking in Restoration
PC on Midget/Sprite Restoration
PC on Mini Cooper Restoration
PC on MGB Restoration
PC on Morris Minor Restoration
PC on Sunbeam Rapier Restoration
PC on Triumph Herald/Vitesse
PC on Spitfire Restoration
PC on Beetle Restoration
PC on 1930s Car Restoration

BROOKLANDS HOT ROD 'MUSCLECAR & HI-PO ENGINES' SERIES

Chevy 265 & 283
Chevy 302 & 327
Chevy 348 & 409
Chevy 350 & 400
Chevy 396 & 427
Chevy 454 thru 512
Chrysler Hemi
Chrysler 273, 318, 340 & 360
Chrysler 361, 383, 400, 413, 426, 440
Ford 289, 302, Boss 302 & 351W
Ford 351C & Boss 351
Ford Big Block

BROOKLANDS RESTORATION SERIES

Auto Restoration Tips & Techniques
Basic Bodywork Tips & Techniques
Basic Painting Tips & Techniques
Camaro Restoration Tips & Techniques
Chevrolet High Performance Tips & Techniques
Chevy Engine Swapping Tips & Techniques
Chevy-GMC Pickup Repair
Chrysler Engine Swapping Tips & Techniques
Custom Painting Tips & Techniques
Engine Swapping Tips & Techniques
Ford Pickup Repair
How to Build a Street Rod
Land Rover Restoration Tips & Techniques
MG 'T' Series Restoration Guide
Mustang Restoration Tips & Techniques
Performance Tuning - Chevrolets of the '60s
Performance Tuning - Pontiacs of the '60s

BROOKLANDS MILITARY VEHICLES SERIES

Allied Military Vehicles No.1 1942-1945
Allied Military Vehicles No.2 1941-1946
Complete WW2 Military Jeep Manual
Dodge Military Vehicles No.1 1940-1945
Hail To The Jeep
Land Rovers in Military Service
Mil. & Civ Amphibians 1940-1990
Off Road Jeeps: Civ. & Mil. 1944-1971
US Military Vehicles 1941-1945
US Army Military Vehicles WW2-TM9-2800
VW Kubelwagen Military Portfolio1940-1990
WW2 Jeep Military Portfolio 1941-1945

794

CONTENTS

5	Delta Luxury Hatchback	Lancia Motor Club		1994
7	Small Car Refinements - Lancia Delta	Autosport	Oct. 16	1979
8	Lancia's Delectable Delta	Motor	Sept. 8	1979
12	Driving the Delta	Autocar	Oct. 13	1979
14	Lancia Delta 1500 Road Test	Motor	July 26	1980
20	Can the Delta Earn its Wings? Road Test	Autosport	Nov. 22	1980
22	Delta - Delights Delayed? 12,000 Miles On	Motor	May 2	1981
27	Lancia Delta, Honda Quintet, Ford Escort Comparison Test	What Car?	July	1981
32	Lancia Delta Automatic Road Test	Autocar	July 10	1982
38	Lancia Delta 1.5LX Road & Track Impressions	Practical Motorist	Jan.	1983
40	Lancia Delta, Toyota Corolla, Citroen BX, Datsun Stanza	What Car?	Jan.	1984
	Renault 11GTX Comparison Test			
47	Prisma	Lancia Motor Club		1994
49	Bridging the Gap? On the Road	Motor	Jan 1	1983
50	Lancia Prisma Road Test	Autocar	July 30	1983
56	Lancia Prisma 1500 Road Impressions	Motor Sport	Nov.	1983
57	Lancia Prisma 1600 Road Test	Car South Africa	Nov.	1984
60	New Prismas Show Promise	Car South Africa	June	1986
62	Lancia Prisma 1600ie Road Test	Motor	Sept. 13	1986
66	Lancia Prisma LX - Design by Degree Test Update	Autocar	Oct. 8	1986
69	Delta Hot Hatches	Lancia Motor Club		1994
71	Fancier Lancia On the Road	Motor	Nov. 27	1982
72	Lancia Delta 1600GT	Cars & Car Conversions	Mar.	1984
74	Lancia's Second Wind: Delta HF Turbo	Performance Car	Mar.	1984
76	Lancia Delta HF Turbo: Power Packed Hatch Road Test	Autosport	Aug. 23	1984
78	Lancia Delta HF Turbo Road Test	Performance Car	Nov.	1984
84	Lancia Delta 1600 GT	Motor	Dec. 22	1984
87	Lancia Delta HF Turbo ie Test Update	Autocar	Aug. 27	1986
92	Lancia Delta HF Turbo vs Mitsubishi Colt Turbo	Fast Lane	Jan.	1987
	Comparison Test			
95	A Hard Day's Night - Lancia Delta GTie	NZ Car	July	1989
97	Lancia Delta HF Turbo Road Test	Fast Lane	Mar.	1989
98	Lancia Delta GTie, Honda Integra EX 16, Peugeot 309 SRi,	What Car?	May	1987
	VW Golf GTi Comparison Test			
106	Lancia Delta GT1600ie Buying Used	Autocar & Motor	Jan. 15	1992
108	Delta Four Wheel Drive	Lancia Motor Club		1994
110	Lancia's Delta Force	What Car?	Aug.	1986
112	Lancia Delta HF 4WD - Road Test	Autocar & Motor	Oct. 22	1986
118	Lancia Delta HF Integrale	Car South Africa	July	1988
121	Lancia Delta HF 4WD - Shaken and Stirred... Stage Test	Motor	Nov. 14	1987
126	Lancia Delta HF Integrale Road Test	Autosport	Mar. 31	1988
128	Lancia Delta HF Integrale Road Test	Motor Sport	Aug.	1988
130	Lancia Delta Integrale vs Ferrari Testarossa - David and Goliath	Autocar	July 27	1988
	Comparison Test			
136	Lancia Delta HF Integrale 16v - Staying One Step Ahead	Motor Sport	June	1989
137	Lancia Delta Integrale 16v Road Test	Autocar & Motor	July 5	1989
140	Lancia Delta Integrale Rally Car - Evolving Integrale	4x4 Driver	Aug.	1989
142	Delta Force Road Test	Sports Car International	Nov.	1989
146	Lancia vs Audi vs Mitubishi - 4 by One	Car & Car Conversions	Mar.	1990
	Comparison Test			
150	Lancia Beefs up the Integrale	Autosport	Oct. 3	1991
152	Lancia Delta HF Integrale Long Term Test	Autocar & Motor	Nov. 6	1991
156	Simply the Best? - Lancia Delta HF Integrale	Motor Sport	May	1992
	Long Term Test			
160	Rally Refugees - Lancia Delta Integrale	Whar Car?	June	1992
	vs Nissan Sunny GTi-R Comparison Test			
162	Lancia Delta HF 4WD Turbo & Integrale Road Test	Performance Car	April	1991
166	Lancia Delta Owners View	Performance Car	April	1991
167	Lancia Intergrale vs Escort RS Cosworth - Lords of the Forest	Fast Lane	Jan.	1994
	Kings of the Road? Comparison Test			

ACKNOWLEDGEMENTS

Our aim at Brooklands Books is to make available to motoring enthusiasts that printed material which would otherwise be hard for them to find. In doing this, we are fortunate to have an extensive archive to draw on; but we are always delighted to receive assistance in putting our collections together. On this occasion, we had considerable help from Brian Long of the Lancia Motor Club, who not only suggested articles for inclusion but also provided valuable overview material which introduces each of the sections of this book.

In addition, we must acknowledge our debt to those who granted us permission to reproduce their copyright material. Sincere thanks therefore go to the managements of *Autocar, Autocar and Motor, Autosport, Car and Car Conversions, Car South Africa, Fast Lane, 4x4 Driver, Motor, Motor Sport, NZ Car, Performance Car, Practical Motorist, Sports Car International* and *What Car?*

R M Clarke

When Lancia's Delta first appeared in 1979, it looked like an interesting variation on an existing theme. For here was a family-sized hatchback with equipment levels which marked it out as a pocket-sized luxury car. For further distinction, it had sharp styling by Giugiaro of Ital Design, and road testers rapidly discovered that it also had quite remarkable handling allied to the sort of performance expected from a Lancia.

Over the next few years, Lancia introduced both smaller-engined and larger-engined versions of the original Delta 1500, plus a three-box saloon derivative called the Prisma. The stage seemed set for the car to have a relatively unspectacular career as one of the more interesting family saloons of the 1980s. But in the mid-1980s, Lancia changed all that.

The catalyst was the domination of international rallying by the four-wheel drive Audi Quattro. Lancia resolved to take up the challenge, and developed the formidable Delta S4, a mid-engined four-wheel drive rally special based loosely on the Delta. It was an immediate success, but rule changes forced Lancia to develop a rally competitor which was more closely based on the production Delta. So in 1986 they came up with the first HF 4WD Delta, and this later became the Integrale. In the second half of the 1980s and the early years of the 1990s, the Deltas - competing in the familiar colours of Lancia's Martini sponsors - dominated the international rally scene.

As the rally cars had to be matched by production variants, the ordinary enthusiast benefits, too. Roadgoing versions of the Integrale (always with left-hand drive) were sold through the showrooms, and offered an exhilaration in everyday motoring which was simply not available in the products of other manufacturers.

Not surprisingly, road testers loved the Intergrale, and used every opportunity they could to get their hands on one and to write about it. Much of what they wrote - and of what they thought about lesser examples of the Delta - is contained within the covers of this book. Reading it will help to explain exactly what was so special about the Lancia hatchback which evolved into a world-beating rally car.

James Taylor

Delta Luxury Hatchback

The Delta was introduced at the Frankfurt motor show in September 1979. With a high standard of finish and equipment, it was aimed at the top end of the medium car market, in the tradition of the Aprilia, Appia and Fulvia saloons which had enjoyed success in this sector since the thirties. Unlike its predecessors, the Delta was of hatchback configuration rather than a conventional saloon, but this reflected the growing popularity of the hatchback among car buyers. However, the Delta had five doors rather than the three of most of its competitors.

Although based on the same floorpan as Ritmo/Strada and using the same single overhead camshaft engines, the Delta was a distinctively different car. The floorpan was modified to use a suspension layout derived from that which had proved highly successful in Lancia's Beta range. The engines, nominally 1500 and 1300cc capacity, were also modified to give more power and torque in Lancia form. The styling, very much in the modern idiom, was provided by Giorgio Giugiaro of Ital Design.

The result was a car which not only looked very different from the Ritmo but was also much superior in its performance, handling and equipment. The success of Lancia's concept was confirmed by motoring journalists throughout Europe who voted it "Car of the Year 1980" by a wide margin.

The Delta was launched in three versions; 1500, 1300 five-speed and 1300 four-speed. The 1500 had a 1498cc four-cylinder engine producing 85bhp (DIN) at 5,800rpm which was equipped with electronic ignition and drove the front wheels through a five-speed close-ratio gearbox. The car was provided with wool velour upholstery, was fully carpeted throughout and boasted such extra touches of comfort as twin courtesy lights for rear seat passengers and a light and vanity mirror in the glove box. The heating and ventilation system had been developed in conjunction with Saab, who sold the car in Sweden as the Saab 600. The 1300 five-speed had the same trim and equipment as the 1500, but utilised a 1301cc engine with a power output of 75bhp at 5,800rpm. The gearbox was not the same as that of the 1500; it was effectively a four-speed unit with an overdrive fifth gear for greater economy. The 1300 four speed model used this same gearbox without the overdrive fifth speed, and also had a less luxurious level of trim.

Right-hand drive Deltas reached the UK in 1980, but only the 1500 was imported. Upgraded versions of the 1300 and 1500, identified by the letters LX after the model number, appeared in 1982. Mechanically the same as the standard cars, the LX models had alloy wheels, a steel sun-roof and electric

Delta 1500 - The original luxury hatchback.

front windows added to the specification. Again, only the 1500 model was imported into the UK. A three-speed automatic gearbox was offered for the 1500 in 1982; this feature was initially available only on the 1500LX in the UK. The LX models were discontinued at the end of 1982, but this notation was revived in 1986 for the 1300 and in 1991 for the 1500.

The short life of the initial LX models was the result of a restyling exercise for the Delta range in November 1982 after which only the 1300 five-speed and 1500 automatic were produced. In 1983 the 1300 fivespeed was imported into the UK for the first time. The restyled version of the original 1300 was also slightly upgraded with a power output of 78bhp and a claimed maximum speed of 100mph rather than the 96mph of the earlier car.

In 1984 the 1500-engined Delta was discontinued altogether in the UK though it continued in production in left-hand drive form. A minor change to the 1300, which had no effect on its performance, was a reduction in engine capacity from 1301cc to 1299cc achieved by reducing the stroke from 55.5 to 55.4mm. In 1986 the Delta underwent a further "facelift" with minor external styling changes and a revised facia. Improvements were also made to the front suspension geometry and spring characteristics. At this time the 1300 was joined by the revived 1300LX with more luxurious equipment, while the 1300 engine reverted to 1301cc and continued at this capacity for both standard and LX models in the UK until they were discontinued in 1991. The revised cars were known as Nuova (New) Deltas in Italy, but this designation was not used in the UK. The 1300 models were not the only Deltas sold in the UK at this time as they had been joined by the higher-performance cars in 1983, described later.

Delta 1300 - Introduced in the UK in 1983

The 1300 was not the smallest capacity Delta; that distinction belongs to the 1100, produced specially for the Greek market. The smaller capacity was obtained by reducing the bore of the 1300 unit from 86.4 to 80mm, giving a swept volume of 1116cc and a power output of 64bhp. This version was produced from 1986 to 1990. Another variant of the Delta not seen in the UK was the diesel-powered 1900 TDS, which is described with its Prisma counterpart in the following section.

In 1991 the Delta range was again revised but by this time only the four-wheel drive model was available in the UK. In other markets the 1500 reverted to a manual gearbox, with no automatic option, while both the 1300 and 1500 were available in LX trim only. Production ended in 1993, more than thirteen years after the model was introduced, when an entirely new replacement was launched.

Models covered in this section: 1300, 1300LX, 1500, 1500LX, 1500 auto, 1100 and 1900 TDS

Production Data

Model	In production to	from	No. built
1300/1300LX	1979	1982	85,627
1500/1500LX	1979	1982	66,615
1300	1982	1986	62,627
1500	1982	1986	4,154
1100	1986	1990	3,532
1300/1300LX	1986	1991	86,428
1500	1986	1991	391
1900 TDS	1985	1991	20,163
1300LX	1991	1993	892 (at end 1992)
1500LX	1991	1993	9,888 " " "

new cars
edited by John Bolster

The body shape is the work of Giugiaro, who has created a style which seems unlikely to date.

Small car refinement

Specification and performance data
Car Tested: Lancia Delta 5-door saloon, 1500, 1300 5-speed, 1300 4-speed.
Engine: 1500 86.4 x 63.9mm (1498cc). Compression ratio 9.2:1. 85bhp DIN at 5800rpm. **1300:** 86.4 x 55.5mm (1301cc). Compression ratio 9.1:1. 75bhp DIN at 5800rpm. Belt-driven overhead camshaft and twin-choke carburettor (both sizes). Electronic ignition (**1500**).
Transmission: Single dry plate clutch driving gearbox primary shaft. Gearbox in unit with engine, transversely mounted and driving front wheels. Ratios: **1500:** 0.959, 1.163, 1.550, 2.235 and 3.583:1. **1300:** 0.863, 1.042, 1.454, 2.235 and 3.583:1. **1300 4-speed:** 1.042, 1.454, 2.235 and 3.583:1. Helical spur gear final drive, ratio 3.765:1.
Chassis: Combined steel body and chassis. MacPherson independent suspension of all four wheels, with anti-roll bars at both ends. Rack and pinion steering. Servo-assisted disc/drum brakes, with diagonally-split circuits and dual rear control valve. Bolt-on steel wheels, fitted 165/70 SR 13 tyres (**1300 4-speed:** 145 SR 13).
Dimensions: Wheelbase 8ft 1.4in. Track 4ft 7.1in. Overall length 12ft 9in. Width 5ft 3.8in. Weight: **1500** 2152lb, **1300** 2141lb, **1300 4-speed** 2108lb.
Performance: 1500: 100+mph. 0-62mph 12.5s. **1300:** 96mph, 0-62mph 15.0s. (makers figures).
Fuel Consumption: Official figures at 56mph, 75mph, and urban cycle respectively: **1500** 40.3mpg, 31mpg, 26.4mpg. **1300 5-speed** 44.1mpg, 32.8mpg, 27.7mpg. **1300 4-speed** 40.9mpg, 31mpg, 27.7mpg.

Above: Clever frontal design has retained the traditional grille. Below: The refined interior.

Since the firm of Lancia was integrated into the Fiat empire, its function has been to produce the up-market models of the range. The great success of the Beta and Gamma has proved the soundness of that policy. Nevertheless, in this age of energy conservation, there is a new demand for economical cars of moderate overall dimensions, but with the refinement and luxury associated with bigger machines. Lancia have not hesitated to grasp this exciting opportunity and the Delta is just such an up-market small car.

In case anybody thinks that this is merely a Ritmo/Strada with a college education, it must be made clear that it is a new Lancia design and not merely a Fiat derivative. It was logical to use the basic engine and transmission package of the Ritmo, with Lancia refinements, but the body shell is totally different. The rear suspension is strictly Lancia, with coil springs instead of the Fiat transverse leaf spring, and there are parallel links and a trailing arm each side, beneath the MacPherson strut; anti-roll bars are fitted front and rear.

The Fiat-based engine has new Lancia manifolds, with a Weber twin-choke carburettor, and air is fed to the inlet silencer at a controlled temperature. In the case of the 1500, a fully electronic ignition system, on magnetic impulse and inductive discharge principles, is featured.

There is a basic model with the 1300 engine (actually 1301cc to avoid small-car limits on the Autostrada) which has a four-speed gearbox, but this is unlikely to come to the UK, though I have included it in the data panel for completeness. The five-speed 1300 and 1500 models have differently-spaced ratios, and it might, at first glance, seem crazy that the smaller engine carries the higher gearing. The explanation is that the 1500 is geared for performance, reaching its maximum speed on fifth gear, whereas the 1300 attains its top velocity on fourth, and fifth is strictly an economy or cruising gear.

In addition to the availability of Fiat technology, Lancia also have a close technical rapport with Saab. This has resulted in a heating and ventilation system on Scandinavian lines, which can deliver cool breathing air while the feet are warmed. To make the interior conditions independent of vehicle speed, air is drawn from low-pressure areas and ram effect is not employed.

The body shape is the work of Giugiaro, incorporating four main doors and a tail gate. The bumpers are of reinforced glassfibre and finished in the body colour, with rubber strips to absorb minor shocks. To achieve an individual shape, Giugiaro has reverted to the pronounced tumble-home of the side panels employed by the British coachbuilders of the 1920s, notably Barker. However, this is a style which is unlikely to date, unlike the late-lamented wedge. The upholstery material is seamless velour, and the instruments, in all but the cheapest version, include an electronic rev-counter.

Road impressions

Though I tried the 1300 models on short trips, my main test-drive took place in the mountains, with a return via the Autostrada, over a useful selection of Italian roads. For this exercise, I used the 1500, which is the high-performance Delta.

Perhaps one thinks of sunshine in Italy, but on this occasion heavy rain and fog were predominant. I was able briefly to assure myself that this is a 100mph car, but perhaps the test was more valuable in trying out the roadholding on slippery surfaces. The little machine has a surprisingly spacious interior and the controls are well arranged, with an adjustable angle for the steering column.

The gearchange has fifth out to the right and forward, but although this is certainly the preferred pattern nowadays, the lever has a long travel and is somewhat springy in action. Fourth speed is an excellent gear for winding roads and overtaking, with a maximum of just over 90mph at the permitted 6200rph. Fifth is not much used in the cut and thrust of Italian driving, but it allows the peak 5800rpm to be attained, under perhaps slightly favourable conditions, at 103mph. It will be seen, therefore, that the 1500 is ideally geared for fast driving, unlike the five-speed 1300, which is over-geared for economy.

The 1500 is an effective vehicle on wet roads and its handling gives every confidence. The weather conditions did not allow me to set the brakes on fire, but I formed the opinion that this is a well-balanced car that can be enjoyed even in fog and torrential rain. It is mechanically quiet, with very little road and wind noise, while the ride is comfortable. Perhaps there was a suspicion of scuttle-shake over cobbles, but there is a remarkable absence of rumbling and booming and refinement is the Delta's middle name. ■

Lancia's delectable DELTA

The competition within the small family car class hots up. Lancia's new Delta uses the Fiat Ritmo/Strada engine and transmission but is otherwise Lancia. Rex Greenslade describes this spacious "elite" newcomer

Cutaway drawing by Graham Cooke

The Delta revealed. Notable features are the MacPherson struts front and rear and the long transverse arms locating their bottoms at the rear. The Delta will be offered only as a five door, but with three engine/transmission combinations

IT'S MORE than seven years since Lancia was absorbed into the Fiat Group. The launch of the new Lancia Delta is final proof that the fears that Lancia would lose its identity in the merger were unfounded. True, of the three Lancias introduced since 1972 — Beta, Gamma and Delta — only the Gamma has a wholly Lancia engine, but both of the other models are very different from the Fiat models that use the same basic engine.

The role of Lancia within the Fiat group is to produce the "elite" cars of each particular class, so that although the Delta has the Fiat Ritmo/Strada engine (albeit in substantially more powerful form) the Delta is designed to appeal to those customers who want a small "prestige" car. Lancia estimate the small car class as being four million cars a year in Europe and Lancia will be aiming unashamedly at the top slice.

Like the Ritmo, the Delta has front-wheel drive, the engine is mounted transversely, the suspension is by MacPherson struts, steering is by rack and pinion and most versions have a five-speed gearbox. But there the similarities end, for the Delta has a completely different body and structure, and the struts are located and sprung differently. It is all clothed in an attractive five-door body with the now familiar Lancia

two-box styling. Three models of the Delta will be offered, these being the 1500 (fitted with a close ratio five-speed box), the 1300 five-speed (but with fifth as a cruising gear) and the 1300 four-speed.

Styling

Yet another design from the pen of Giugiaro of Ital Design, the Delta styling aims to produce a compact yet roomy car with good aerodynamics. Just how slippery the shape is we can't judge as no C_d figures have been released, but there is no doubting its roominess and good packaging, not least because the wheelbase of 97.4 in is larger than average for the class, and is incorporated in an overall length that is itself less than the class norm.

The signature of Giugiaro is evident in the relationship between the front windscreen pillar, bonnet line and front door hinge line; overtones of the VW Scirocco and Alfasud Sprint, also Giugiaro designs, are unmistakeable. The heavy rear pillar is less successful but the two-box styling does allow the use of a rear tailgate that opens right down to bumper level. There is a one-piece, impact-resistant bumper/spoiler at the front, the joining of which to the front wing is emphasised by a bodyside crease that runs back to the top of the rear bumper.

Structure

In typical modern style, the Delta has a very strong central passenger "cell" protected by a crumple zone at each end. The wing panels are bolted in place for ease of assembly and replacement and thus play a relatively minor part in the car's crash strength. There are stiffening members within the doors to resist intrusion in a

Above: The influence of Giugiaro's styling is unmistakable around the bottom of the windscreen. Below: The rear door opens right down to bumper level. The front and rear bumpers are plastic and can withstand low-speed knocks

side impact and the 10-gallon fuel tank is sited under the back seat, partially protected in a rear end impact by the spare wheel in the boot floor.

The main body structure consists of a one-piece floorpan of pressed steel stiffened by three deep cross-members welded into place from above to form footings for the seats, the seat belt anchorages and the rear suspension mountings. The other most important structural pressing is the front bulkhead which, at 0.050 in, is

GENERAL SPECIFICATION

ENGINE	1500	1300 5-speed	1300 4-Speed
Cylinders		4 in-line	
Capacity, cc	1498	1301	1301
Bore/stroke, mm	86.4/63.9	86.4/55.5	86.4/55.5
Cooling		Water	
Block		Cast iron	
Head		Aluminium alloy	
Valves		Sohc	
Cam drive		Toothed belt	
Compression	9.2:1	9.1:1	9.1:1
Carburetter	Weber 34 DAT 7	Weber 32 DAT 7	Weber 32 DAT 7
Bearings		5	
Bhp/rpm	85/5800	75/5800	75/5800
Lb ft/rpm	90.4/3500	77.4/3500	77.4/3500
TRANSMISSION			
Type	5-speed	5-speed	4-speed
Clutch dia	7.48 in	7.15 in	7.15 in
Actuation	Mechanical	Mechanical	Mechanical
Internal ratios			
Top	0.959:1	0.863:1	1.042:1
4th	1.163:1	1.042:1	—
3rd	1.550:1	1.454:1	1.454:1
2nd	2.235:1	2.235:1	2.235:1
1st	3.583:1	3.583:1	3.583:1
Reverse	3.714:1	3.714:1	3.714:1
Final drive		3.765:1	
BODY/CHASSIS			
Construction		All steel integral	
SUSPENSION			
Front		Independent by MacPherson strut, lower arms and anti-roll bar. Coil springs	
Rear		Independent by MacPherson strut, transverse links and longitudinal reaction rods. Coil springs	
STEERING			
Type		Rack and pinion	
BRAKES			
Front		8.94 in dia disc, floating caliper	
Rear		7.28 in dia drums	
Park		On rear	
Servo		Vacuum servo	
Circuit		Diagonally split	
Rear valve		Yes	
Adjustment		Automatic	
WHEELS/TYRES			
Type		Pressed steel, 5B x 13	
Tyres	165/70 SR 13	165/70 SR 13	145 SR 13
ELECTRICAL			
Battery		12v 45Ah	
Generator		45A	
WEIGHTS			
Kerb (DIN), lb	2152	2141	2108
Payload, lb		1049	
DIMENSIONS			
Wheelbase, in		97.4	
Overall length, in		153	
Overall width, in		63.8	
Front track, in		55.1	
Rear track, in		55.1	
Unladen height, in		54.3	
PERFORMANCE (manufacturer's figures)			
Max speed, mph	100+	96+	96+
Mph/1000 rpm in top	17.8	19.8	16.3
Stdg 400 metres, sec	18.2	19.2	19.2

unusually thick, say Lancia, in this class of car. This is designed to insulate the passenger compartment from engine noise and vibration (and is helped considerably in these respects by a supplementary removeable bulkhead within the engine bay) and also to absorb and distribute running and impact loads from the nose structure.

Corrosion protection

In view of the severity of Swedish winters and the widespread use of salt there, it is not surprising that it is in the areas of heating and rust-proofing that the recently announced technical liaison between Lancia and Saab has borne fruit. Even so, Lancia emphasise that their rust protection is not a palliative or after-thought for production cars, rather that the whole design was studied carefully to minimise the chances of rust finding a hold. All joints are treated with an adhesive and a plastic sealant for instance, and the most exposed parts of the body are either made from galvanised sheet steel or steel pre-treated by the "Zincrometal" process. This Zincrometal steel — about 40 lb of which is used in the Delta — has its surface enriched with zinc powder and chromium and is said to have anti-corrosion properties superior to any type of protective paint.

In production, electrophoretic primer is used, anti-corrosion oil is sprayed into all closed box sections and the entire underbody is protected by a PVC coating. Polyethylene guards are fitted inside the front wheelarches. All of the external brightwork (except the stainless steel front grille) is made of anodised aluminium.

Interior

Lancia claim that the Delta is one of the roomiest cars in the class, and a comparison of their quoted figures for maximum front seat legroom and minimum rear seat kneeroom with those we've measured on rival cars appears to back this up.

(Inches)	Delta	Ritmo	Golf	Escort	Alfasud
Max front legroom	42.1	39	39	37	40
Min rear kneeroom	25.6	24	20	26	24

In addition, Lancia also claim that interior width is on a par with cars from the next class up.

Folding the back seat (which is a two-stage operation, the cushion being pushed up and forward, the backrest falling into place behind) increases the luggage capacity from 9.2 to 35.3 cu ft, Lancia say, and a split rear seat backrest is an option. The rear suspension struts do intrude into the luggage area but the space behind them in each rear quarter is cleverly used for the storing of the water reservoir for the rear wash/wipe system (standard except on 1300 4-speed models) and the toolkit. The boot is thus square-shaped and free of intrusions or bumps.

Extensive development was concentrated on reducing noise within the passenger compartment, both by finite element analysis of the structure and the use of laser holography to determine the vibration spectrum of the engine/transmission package. Lancia claim particular success in minimising low frequency rumbling within the car notably by using special rubber bushes for the engine mounts and by mounting the engine torque reaction member to the front cross-member rather than to the bulkhead itself. Five-speed models have an additional layer of sound-deadening under the bonnet and the exhaust manifold is linked to the silencer system via a stainless steel flexible section.

The main upholstery material will be seamless velour which is attached to the seat cushion and backrest by a draw-string. The result is a unique pleated effect. The same material is used for the door trim panels and for the headlining, in the latter case the method of attachment being entirely new and the subject of several patents. The front seats have integral, adjustable head restraints and a handwheel adjustment for the seat back angle.

Covering the facia is a soft-foam matt-finish material, applied by a new (for Europe) process called "slush moulding". The dashboard itself is split into two functional sections: instruments and minor controls. The instruments are sited in front of the driver while the minor switchgear and controls are placed in the centre. The lights, indicators and wash-wipe switchgear is incorporated into three column-mounted fingertip stalks.

In 5-speed models the instrumentation is comprehensive, including a rev counter, a speedometer with trip meter, an oil pressure gauge, a water temperature gauge, a fuel gauge, a voltmeter and 12 warning lights. The 1300 4-speed model doesn't have a rev counter, and the normal roof-mounted digital clock is replaced by an ordinary analogue clock.

Lancia also co-operated with Saab in the design and development of the heating and ventilation system. Testing was carried out in temperatures down to −40 deg C in Sweden and the system was designed to fulfill the following requirements: rapid warm-up and high output both for heating and de-icing; optimum temperature stratification, ie warm feet/cool face; a temperature "buffer" down each side to protect occupants from cooling effect of windows; and minimum change of flow with speed. An integral air conditioning system can be fitted if deemed necessary in the future. At present the system is controlled by three levers plus a four-position switch for the fan. Apart from the windscreen slots, air outlets are provided in the two footwells, and into the front doors for side-window demisting. There are four face-level vents.

Mechanical layout

The engines used are both developed from those of the Fiat Ritmo/Strada and share the same general design having four cylinders, five main bearings, an alloy cylinder head and a toothed-belt driven single overhead camshaft. In the Delta, however, the power and torque outputs are enhanced considerably by the adoption of a twin-choke carburetter and new inlet and exhaust manifolds. The 1500 engine also has breakerless electronic ignition. As a result of these changes, the Delta engines produce 75 bhp (1300) and 85 bhp (1500), both at 5800 rpm; equivalent Ritmo/Strada outputs are 65 and 75 bhp.

Following current Lancia practice, there is rack and pinion steering and the suspension is by MacPherson struts front and rear with the lower ends of the struts exceptionally well located. In the front the lower location is by a very wide wishbone, at the rear by twin transverse links and a single trailing link at each side, identical to the Beta layout, in fact. In all four struts, the coil spring surrounds the damper but is offset to minimise side force on the damper piston. There are anti-roll bars front and rear. Steel 13 x 5 in wheels are standard, shod with 165/70 SR 13 tubeless tyres on 5-speed models; the 1300 4-speed has 145 SR 13 tyres.

The braking system is split diagonally, each circuit serving one front disc and the opposite rear drum. There are dual rear suspension-height-sensitive proportioning valves to limit line pressure to the rear wheels. To match this brake layout, there is virtually no steering "offset" — the extended kingpin axis meets the ground almost exactly in the centre of the tyre contact patch.

The facia is divided into two discreet areas. One contains the comprehensive instrumentation, the other the minor controls

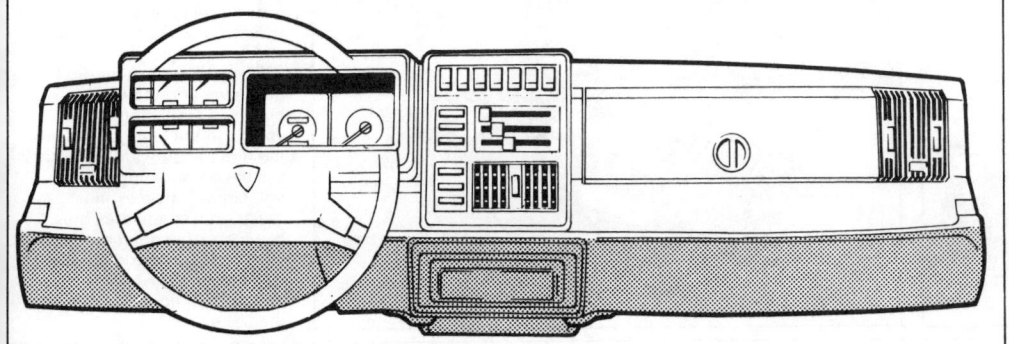

The bumpers front and rear are moulded in glass fibre and designed to survive parking impacts of up to 2½mph

Driving the Delta

Big car standards from Lancia's new small car — but you can't buy one until next year

By Ray Hutton

IN THIS AGE of rationalization and standardization it is encouraging to find that the big wheels of the European motor industry can still produce cars of distinctive character. Fiat Auto was set up to bring the Group's car manufacturing under the same umbrella and could easily have resulted in badge engineered or simply reskinned versions of Fiats to be sold as Lancias. But like Citroen, Peugeot and Talbot in the PSA Group, the influence of the corporation on the individual companies is subtle. Lancia, always fiercely independent in engineering, have to draw from a Fiat components "pool" but continue with their own design team. They wanted a small car, to slot into their range below the Beta and play the role of the Aprilia, Appia and Fulvia of the past. It could have been a Ritmo-Strada by a different name but it isn't.

When you look round the Lancia Delta there are a number of familiar bits. Under the bonnet there is a familiar engine, though with something different in the way of carburettors from the Ritmo original. It is even made in a Fiat factory — Lingotto. But there the resemblance ends.

Lancia describe their role within Fiat as "producing the elite cars" of the Group. They are tooling up for 40 per cent increase in output, and 200,000 sales next year. The high-specification, luxury small car is they say "the car of the moment." It is a social thing; an awareness perhaps that it is unnecessarily wasteful to have a car bigger than you need. If the same luxury, the same equipment, the same aura, can come from something smaller, people will convert to it. A small car needn't be cheap, noisy or unrefined. That is the premise on which the Delta was based. They admit that its lines and its design are conservative (particularly by comparison with the cubist-modern Ritmo) but believe that too is right for its market. It is perhaps what the Triumph Dolomite of the 1980s should be.

The Delta was described in detail in *Autocar* 8 September. 153in. long, or about the size of an Alfasud, it has a body by Guigaro's Ital-Design. Clean and tidy in modern five-door hatchback style, it has affinities with several others of its type (not surprising in view of Ital's work for other clients) and is distinguished by a version of the traditional Lancia grille, neatly integrated with the front lamp units. Given the desire to keep the overall size down, the uniformity of its shape with similar cars is understandable; it unfortunately shares with some others a high-set rear window which seriously limits rear vision when parking.

In summary, it has transverse-mounted, four cylinder engines of 1,301 and 1,498 c.c. with twin-choke carburettors and maximum power of 75 and 85 bhp respectively, 10 more than two versions of the Ritmo. There are three different gear-sets: a basic four-speeder that goes with the smaller engine in the base model, and a five-speed gearbox with overdrive fifth for the 1300 and almost direct fifth (plus different fourth and third ratios) for the 1500. Suspension is by MacPherson struts front and rear, but unlike the Ritmo (which has a transverse leaf spring) the Delta has long transverse arms at the rear and coil springs. There are anti-roll bars front and rear. Brakes are servo-assisted disc front, drum rear, with a diagonal split system.

Particular importance has been placed on ride comfort and running quietness. This is reflected in suspension bushes and mountings, the "independent" mounting of the gear-lever and the extra bulkhead in the engine compartment (compared with the Ritmo). Outwardly the cars still have something of the traditional Lancia exhaust "rasp" but inside they are commendably quiet at touring speeds, as we found out last week when British journalists had the first opportunity to drive the Delta.

Below: Rear wash-wipe systems are standard on all models

Right: Giugiaro's Ital Design five-door body styling gives the Delta a length of just under 13ft. The optional alloy wheels on the car in the background look rather like steel ones with alloy centre caps

Welcoming

As our Italian correspondent said at the time of our description, first impressions of the Delta are good. The seats are comfortable, the surroundings inside nicely toning and pleasingly designed. The trim is nicely finished and the facia and controls neatly integrated; there is an air of thoroughness, of quality. The driving position is not of the worst Italian kind. The steering column can be adjusted up and down through 2½in. and unlike in some of the Betas, this driver did not find himself driving the Delta knees-splayed.

On the move, the steering immediately impresses. Lancias generally have nice steering and the Delta is no exception. Sweet through 3.8 turns from lock to lock, it is remarkably light by the standards of its class — and much lighter than the Ritmo. Other controls are similarly light and easy in operation — some thought the brakes rather too light and easy in operation — but, like the Beta, the quality of the gear change seemed to vary from car to car. At its best, the Delta gearchange is no better than average. Some we tried baulked on the third-to-second downchange and had an awkward movement into and out of fifth. Right-hand-drive cars may feel different again.

We tried a 1300 five-speed first. Its performance is adequate without being quick. It is around 1 cwt heavier than the Ritmo and would seem to be slightly slower than our long term Strada 75 with the same power output. On the autostrada, running through the tunnels and galleries of the mountains behind the Italian Riviera, the 1300 needed fourth rather than fifth gear most of the time to keep up with the higher speed traffic. Like the Ritmo, whose ratios it shares, this version has a high overdrive top gear (0.863 to 1) and achieves its maximum speed of something over 90 mph in fourth gear which is almost direct. The four-speed 1300 Delta tried briefly later has the same four ratios without the overdrive and for the driver who isn't in a hurry would not be expected to produce the same kind of fuel consumption figures as the five-speed. The Italian Official Fuel Consumption figures at 56 mph are 44.1 mpg for the five-speed and 40.9 for the four-speed.

The 1500 is immediately more lively, not only because of its extra power but also because it has a "sports-type" five-speed with top at 0.959 to 1 and proportionally lower

fourth and third. The effect, however, is to improve performance (100 mph and 0.60 mph in 12.5 sec are claimed) and noise levels, though all the cars have a fair amount of transmission whine in the lower gears.

Twisting mountain roads confirmed the precision of the steering, the effectiveness of the brakes and the small amount of body roll. This is a car of particularly good road manners from a group (of small front-drive saloons) which are generally safe and predictable. Throttle-off mid-corner evokes little adverse reaction and the ride is impressive. Regular undulations, like cobbles, can set some vibrations through the facia but otherwise the efforts at insulating the suspension from the cabin seem to have paid off.

Inside, the Delta is comfortable for four. The seat cushions and backs are unusual in not having seams at their edges but being stitched in square panels and pulled over the foam structure and tightened with a draw-string. Rear leg-room is good considering the size of car and at six foot I could sit behind the driver's seat as set for me — though headroom is marginal. The heating and ventilating system is the first technical outcome of Lancia's links with Saab (who will call the Delta their own in Sweden), is adaptable and has good output, though last week's weather didn't provide much of a test for this. The facia is neatly done and incorporates a padded knee roll which forms an oddments shelf across the full width (but as with all facia-top trays, contents reflect in the screen). The instrument warning lights and switchgear all have their specific place and Lancias are helped here by all versions having the same high level of equipment and so they do not have to leave the "blanks" which are such an unsightly feature of the cheaper models of some other people's ranges. The cheapest four-speed 1300 has a clock instead of the rev counter matching the speedometer and does not have the LED-display digital clock (with stopwatch facility) that sits above the rear-view mirror in the other models.

The switchgear is arranged so that essential items are on column-mounted stalks and the less often used controls are on a push button panel at the facia centre. The main light switch is a square section column stalk which could do with some more obvious indication that it is designed to be twisted.

The attention to interior detail with such things as grab handles that recess into the cant-rail, and a service reminder panel above the mirror is as one would expect from an up-market Lancia. The easily folded-down rear seat and resulting flat luggage area is a useful utility borrowed from less elaborate cars. A split rear seat can be specified as an option, as can headlamp wiper/washers, electric front windows, sunroof, light alloy wheels and even air conditioning on the 1500.

The Delta goes on sale in Italy on 5 November and as we went to press the domestic prices had not been revealed. For a car of its size and type it will not be cheap. We were told the price would be in the 8-8.5 million lire range (£4,500-4,800) or about equivalent to the Audi 80 and

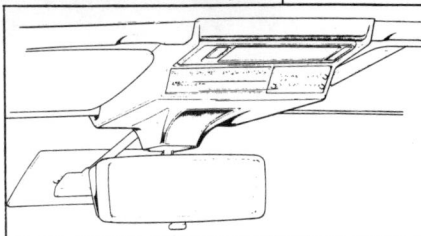

Alfa Romeo Giuiletta. That suggests that the Beta 1300 will eventually be displaced by the newcomer at the bottom of the Lancia range. On past form, there will be high performance and an ES-type "fully loaded" model in time. But no Deltas will be coming to Britain until well into next year — perhaps not until the summer. By then the luxury small car may be even more in vogue than it is now.

The transverse engine carries the name Lancia rather than Fiat on it. The distributor is driven from the end of the camshaft. Note the dual bulkheads, to reduce internal noise levels

Far left: Seats have moulded backs, with cloth drawn over them without conventional seam edges

Left: There are three separate instrument panels, with speedometer and rev counter in the largest. Row of press button switches are above the heating and ventilation controls in the centre

Console above rear view mirror carries the interior lamp, digital clock, with stopwatch facility, and wipe-clean "note pad" to jot down tyre pressures. Italian cars also have integral licence holder

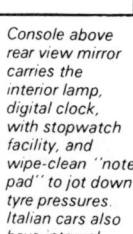

Star Road Test
LANCIA DELTA 1500

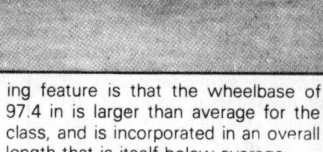

Lancia's Delta is 1980's Car of The Year. With no outstanding qualities, though many good ones, it is a hard car to fault

THE INTEGRATION of Lancia into the giant Fiat Group in 1972 has inevitably blurred the family lines and engineering traditions behind the famous *shield and flag* insignia. Is the new Delta a worthy successor to the Appia and Fulvia of the 'fifties and 'sixties, or merely a Fiat Strada re-packaged for a classier clientèle? Perhaps it is really too cleverly crafted an amalgam of parts and qualities to be so casually pigeon-holed. That it arrives when Lancia most needs it — at the start of a potentially crippling recession — is a convincing demonstration of the effective planning and efficient design-development-production process Fiat's financial muscle has ensured.

To have a compact prestige car at this time is probably as important to Lancia as the Car of The Year accolade the Delta has already received Because of the general depression, Europe's car market is rapidly being compressed towards the 1.5-litre class: Lancia is looking to the Delta to exploit this shift by luring car buyers forced to cut their motoring costs yet reluctant to sacrifice their standard of motoring.

The "moving buyer" must be a prime target for all car makers suffering by their larger, costlier models languishing unsold in the showrooms. In Lancia UK's case, with sales of the excellent Gamma severely depressed and the Beta's reputation blighted by rust problems, nothing less than a direct hit with the Delta will count for much.

The five-door-only Delta is available with three engine/transmission combinations: the 1500 (fitted with a close-ratio five-speed gearbox), the 1300 five-speed (but with fifth as a cruising gear) and the 1300 four-speed. Both the engines are basically Fiat Strada units modified by Lancia to produce an extra 10 bhp in each case, 75 bhp (1300 cc) and 85 bhp (1500 cc).

In its mechanical layout, the Delta is only partially Strada-based: its transverse engine and front-wheel drive package is derived directly from that of the Fiat model and the Delta also shares the Strada's rack and pinion steering, four and five-speed gearboxes and disc/drum braking system. But there the close similarities end, for the all-independent suspension's MacPherson struts are located and sprung differently and owe more to Lancia's engineers than to Fiat's.

Unlike other Lancias, the Delta's square-cut two-box shape is the work of Ital Design (Bertone was responsible for the Beta, Pininfarina the Gamma). The design aims were to produce a compact yet roomy car with good aerodynamics, and one interesting feature is that the wheelbase of 97.4 in is larger than average for the class, and is incorporated in an overall length that is itself below average.

Initial availability of the Delta in the UK will be restricted to the fully equipped, five-speed 1500 which boasts a split rear seat backrest and alloy wheels as part of its specification. At £4,995 it is undeniably expensive for a 1½-litre car, though in this class capacity is a poor guide to capability and, as already explained, the *raison d'etre* of the Delta is to provide big car standards of refinement and civility in a compact, economical package. Purpose-built for this end, it holds the upper hand over fwd rivals like VW's Golf 1.3 GLS (£4,572), Opel's Kadett 1.3 Berlina (£5,087), Peugeot's 305 SR (£4,838) and Talbot's Horizon 1.3 GLS (£4,808) all of which are up-market versions of models with more modest aspirations. But that's not to underestimate the opposition: most of our selected rivals are civilised beyond their originally intended roles, though the rear-driven Alfa Romeo Giulietta 1.6 (£5,100), like the Delta, is aimed directly at the prestige car market.

PERFORMANCE

★★★★ The Delta's transversely-mounted 1,498 cc engine is developed from that used in the Fiat Strada 75CL and shares the same general design, having four cylinders, five main bearings and a belt-driven overhead camshaft. In the Delta, however, the power and torque outputs are increased by the adoption of a twin-choke carburetter, new inlet and exhaust manifolds, and a breakerless electronic ignition system. With a compression ratio of 9.2:1 it develops 85 bhp (DIN) at 5,800 rpm and 90.4 lb ft of torque at 3,500, respectively 10 bhp and 3 lb ft of torque more than in Fiat tune.

The automatic choke ensures prompt starts from cold, though it takes too long to cut-out. Warm-up is generally hesitation free. Unlike Fiat's version of this engine, which needs to be revved hard for brisk progress, the

Delta relies on a better spread of torque and closer gear ratios to achieve its superior performance. Characteristically, its commendable smoothness and refinement give way to mild harshness as the rev counter's 6,500 rpm red line is approached.

The Delta 1500 is geared for performance rather than economy, its 98.9 mph top speed corresponding to 5,550 rpm in fifth, just 250 rpm below peak power. This almost exactly matches Lancia's claimed 100 mph and makes the Delta faster than the Opel Kadett (98.6 mph), the Peugeot 305 (95.1), the Talbot Horizon (96.2) or the VW Golf (91.2) Only the significantly more powerful twin-cam Alfa Romeo Giulietta is faster at 104.8 mph.

With a 0-60 mph acceleration time of 12.2 sec, the Delta is again marginally quicker than the Opel (12.6 sec), Peugeot (13.2), Talbot (13.0) and VW (13.2), though not as quick as the Alfa (10.5 sec).

On low speed flexibility, the Delta fares equally well against the rivals we've chosen, its 30-50 mph time (in fourth) almost matching the five-speed Alfa's 8.2 sec, for instance. In fifth (geared at 17.8 mph/1000 rpm), the Delta has no clear-cut advantage over its mostly lower-geared and smaller-engined four-speed rivals, though its 30-50 and 50-70 mph times of 11.3 and 13.1 sec respectively look respectable enough against the same increment times for the Opel (11.5/13.4 sec), the Peugeot (11.2/13.6) and the VW (10.1/13.0) — though the Talbot is significantly quicker (9.9/12.6).

Subjectively, the engine characteristics and gearing are well matched, relaxed high speed cruising not having been sacrificed for good top gear flexibility.

ECONOMY

 We achieved 26.6 mpg overall. Although, as always, the result of very hard driving, this is a little below par for the class.

The touring consumption figure of 28.2 mpg is poor. This figure is computed from steady speed figures and represents the consumption that an owner could expect to achieve by driving gently. This touring figure is bettered by most rivals, some by a substantial margin. With the Delta's larger-than-average 10.0 gallon tank, this gives a potential range of 280 miles. In practice, the poorly damped fuel gauge doesn't inspire a great deal of confidence, setting 230 miles as the effective limit.

TRANSMISSION

The Delta's gearchange demands firm, positive movements as the gate is narrow and the action can be rubbery at times. Some testers complained that it was too easy to go from second to fifth gear, but with familiarity the gearchange proved to be pleasantly fast.

If you press the engine to its 6,500 maximum, the change-up points in the lower gears are 32, 52 and 75 mph. The engine won't quite pull maximum revs in 4th — 6,500 rpm would correspond to 99.4 mph.

As with the Strada, whine was noticeable in the lower gears, especially on the overrun. The clutch action is smooth and well-cushioned.

HANDLING

The suspension is by MacPherson struts front and rear with the lower ends of the struts exceptionally well located. In the front the lower location is by a very wide wishbone, and at the rear by twin transverse links and a single trailing link at each side — identical to the Beta layout, in fact. In all four struts, the coil spring surrounds the damper. There are anti-roll bars front and rear. Alloy 13 × 5 in wheels are standard, shod, on the test car, with Michelin XZX 165/70 SR 13 tyres. The result is a taut yet resilient suspension system which endows the Delta with superbly surefooted cornering in all surfaces, wet or dry.

With 3.8 turns lock to lock the steering seems slightly low geared around town, though at higher speeds the Delta's chassis is so responsive and possesses so much feel that you are seldom aware of having to wind on lock.

Like the very best handling of contemporary compact fwd cars — the Alfasud and Opel Kadett/Vauxhall Astra — the Delta's road manners give little indication of which wheels are being driven. When accelerating hard out of a tight corner in a low gear, however, the steering does tug, though there is insufficient torque to make it *fight* in your hands. For its great stability and poise, the Delta also approaches the best in the class, understeering only

MOTOR ROAD TEST No. 29/80

 excellent	 good	average	poor	bad

Make: Lancia   **Model:** Delta 1500
Maker: Lancia & C SpA, Via Vincenzo Lancia 27, 1014 Turin, Italy
Concessionaire: Lancia (England) Ltd, Great West Road, Brentford, Middlesex TW8 9DJ.
Price: £4,009.00 plus £334.08 Car Tax and £651.46 VAT equals £4,994.54.

mildly up to and on its very high limits of adhesion, the tail moving progressively out after that. The Delta is only less than forgiving if you brake in mid-corner: then, the transition to oversteer is abrupt enough to catch out the unwary. Corrective lock must be applied quickly and accurately.

BRAKES

★★
★★

The braking system is split diagonally, each circuit serving one front disc and the opposite rear drum. There are dual weight-sensitive proportioning valves to limit line pressure to the rear wheels.

In normal use the brakes performed well, proving adequately powerful and progressive. Our objective tests at MIRA confirmed these assessments, a 76 lb pedal pressure achieving a 0.97 g stop from 30 mph and only 70 lb being necessary to give 0.83 retardation from 70 mph, though in both cases the rear wheels were already locking. While slightly better than average, these maximum g stops put a question mark over the effectiveness of the Delta's anti-lock arrangement at the rear.

The handbrake, which gave a slightly-below-average 0.31 g retardation from 30 mph, successfully held the car pointing up or down on the 1-in-3 test hill.

The ease with which the Delta sailed through our fade test, which is severe enough to induce strong fade in some cars was most impressive The pedal pressures slowly increased at various stages of the test, an initial leap from 43 to 64 lb more or less stabilising after the sixth stop before climbing again slightly to 68 lb at the end. Apart from a faint smell of burning pads, the brakes registered no other symptoms of stress. A soaking in the watersplash caused some drop in efficiency, but the brakes quickly recovered.

ACCOMMODATION

★★
★★

Lancia claim that the Delta is one of the roomiest cars in its class, and we wouldn't argue with that. In a car less than 13 ft long, interior space utilisation needs to be good and the fact that rear headroom is marginal for six-footers shouldn't be allowed to detract from Lancia's generally excellent packaging. In particular, shoulder room is very generous, as is legroom at the front. With the front seats set to suit our tallest tester (6ft 3in), the rear seats were better suited to passengers of average height or less.

The four doors open wide for easy access, and the high-lifting tailgate (which starts at bumper level) reveals a fully-trimmed boot. Luggage capacity can be dramatically increased by folding down one or both of the individual rear backrests (which is a two-stage operation, the cushion being pushed up and the backrests falling into place). The rear suspension struts intrude into the load platform, but the space behind them in each rear quarter is usefully employed for the storing of the water reservoir for the rear wash/wipe system and the tool kit.

The provision for the stowage of oddments inside the car is excellent, with a sensibly-proportioned non-locking glove box, a full width shelf-cum-tray forming the lower moulding of the facia, and deep door pockets.

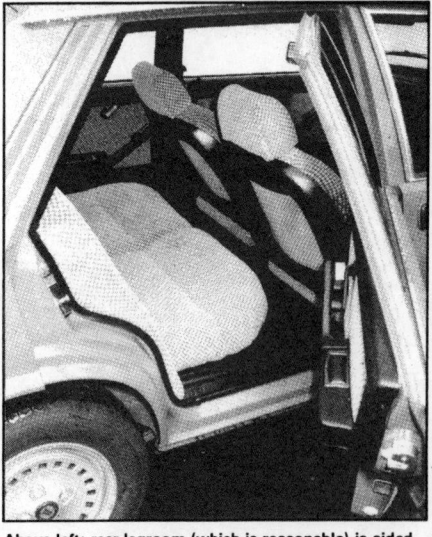

Above left: rear legroom (which is reasonable) is aided by the recessed backrests for the front seats. The latter, above right, have sensibly proportioned cushions and good shaping which makes them very comfortable.

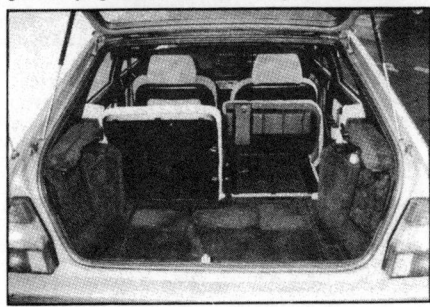

Left: luggage versatility is aided by rear seat backrests which fold down individually

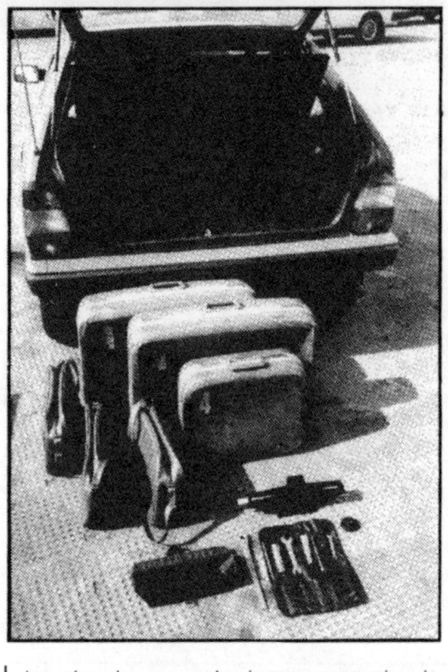

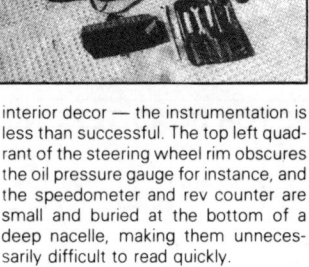

Right: with the rear seat backrests erect, the boot swallowed only 7.1 cu ft of test luggage, boot width suffering from the intrusion of the rear suspension housings

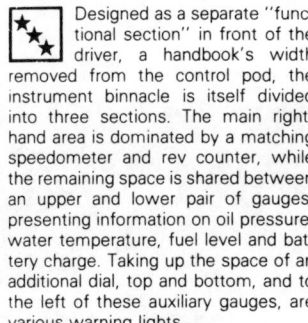

RIDE COMFORT

★★
★★

Although firm, the Delta's ride is never uncomfortable. It copes remarkably well with all types of surface, soaking up short sharp irregularities as easily and competently as it smooths out long-amplitude undulations, all with a commendable lack of hub-bub from the suspension. It is slightly fidgety at low speed around town and deep potholes are heard as well as felt, but apart from that it seldom draws attention to itself.

AT THE WHEEL

★★
★★

All but the very tallest of our testers were able to get satisfactorily comfortable behind the wheel for the relationship between the major controls (unusually for an Italian car) is good, and the steering is adjustable for height.

The seamless velour-covered front seats (the material is attached to the seat cushion and backrest by a drawstring, creating an unusual pleated effect) are well shaped and comfortable, though some of our testers would have preferred more side support. Adjustable head restraints are standard and a handwheel finely adjusts the backrest angle.

Three steering column stalks are provided, one on the left controlling the two-speed wipers and electric screen washers, and the two on the right operating the winkers and headlights. The stalks themselves are well placed and are light yet positive in action.

The rest of the switchgear comprises a neat but hard-to-identify horizontal row of rectangular buttons running along the top edge of the centrally-mounted pod which houses the heating and ventilation controls.

VISIBILITY

★★
★★

In most respects, the Delta is easy to see out of and place on the road, its low waist and generous glass area also contributing to the interior's light and airy atmosphere. One potential blind spot, however, is created by the heavy rear three-quarter panels, though diligent use of the standard door mirrors can overcome this.

Although the windscreen wiper pattern is good, the blades of our test car lifted from the screen at about 80 mph. The headlights were more impressive being bright and well-defined on main beam, though short on range on dip.

INSTRUMENTS

★★
★★

Designed as a separate "functional section" in front of the driver, a handbook's width removed from the control pod, the instrument binnacle is itself divided into three sections. The main right-hand area is dominated by a matching speedometer and rev counter, while the remaining space is shared between an upper and lower pair of gauges, presenting information on oil pressure, water temperature, fuel level and battery charge. Taking up the space of an additional dial, top and bottom, and to the left of these auxiliary gauges, are various warning lights.

Although boldly styled — in striking contrast to the otherwise restrained interior decor — the instrumentation is less than successful. The top left quadrant of the steering wheel rim obscures the oil pressure gauge for instance, and the speedometer and rev counter are small and buried at the bottom of a deep nacelle, making them unnecessarily difficult to read quickly.

HEATING

★★
★★

Lancia linked up with Saab to design and develop the heating and ventilation system for the Delta, clearly not wishing to end up with a system as poor as the Fiat Strada's. Lancia's brief to Saab included the following requirements: rapid warm-up and high output both for heating and de-icing; optimum temperature stratification ie, warm feet and cool face; a temperature "buffer" down each side to protect occupants from the cooling effect of the windows; and minimum change of flow with speed.

As far as heat output is concerned the system certainly matches its claims. Warm air can be obtained quickly from a cold start and throughput is good even at low speed without fan assistance. When the engine is warm, output is prodigious — though it can be finely regulated and effectively directed to the windscreen and side windows.

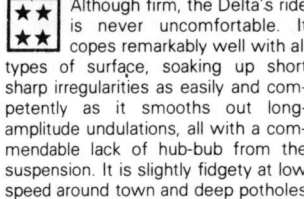
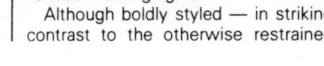

MOTOR ROAD TEST NO 29/80 ● LANCIA DELTA 1500

VENTILATION

★★★

Fresh air enters the car through four face-level vents which are individually adjustable for direction though the system of thumbwheels and slides that does this seems unnecessarily complicated. The system does not induce sufficient volumes of air under ram pressure on hot days, flow from the outer vents seeming suspiciously feeble, but it can be finely adjusted and is well diffused. The flow can be boosted by the four-speed fan but, even then, flow from the outer vents remains woefully weak, and that from the centre vents poor on the driver's side, though much better for the front passenger.

NOISE

★★
★★

The elimination of noise within the passenger compartment of the Delta has been the target of extensive development. Lancia's technicians have used laser holography and 'finite element analysis' in attempting to minimise engine/transmission vibrations and claim particular success in cutting down low-frequency rumbling within the car.

On the whole, Lancia's efforts have paid off, for the Delta is one of the more refined cars in its class, free from irritating boom periods and intrusive mechanical harshness, while suffering little from wind noise, road roar and bump thump at speed. Our dB measurements bear this out. But they also point to the fact that, when extended in the lower gears, the engine becomes much noisier though the *quality* of the engine note is never unpleasant.

FINISH

Inside, the Delta is both tasteful and plush, with good co-ordination of materials and mouldings creating a well-integrated appearance. The attractive check velour covering the seats is also used for the door trim panels, and for the head-lining, where the method of attachment is entirely new and the subject of several patents. The standard of finish is what you would expect of a £5,000 car.

The standard of finish is impressive outside as well. The doors shut with a satisfying clunk, the paintwork is bright and crisp and the panels fit snugly together.

FITTINGS

It almost goes without saying that any small car setting out to attract up-market buyers trading down in size must be well equipped: the Delta doesn't disappoint. Noteworthy items of standard equipment include: adjustable steering column, alloy wheels, a cigar lighter, electronic ignition, remote adjustment door mirrors, a laminated windscreen, a quartz electronic digital clock, a rear wash/wipe system and tinted glass.

IN SERVICE

In addition to Lancia's usual 12-month/unlimited mileage warranty on mechanical and other components, every Delta customer until September 30, will be entitled to free routine ser-

PERFORMANCE

CONDITIONS
Weather	Fine, wind 13-18 mph
Temperature	64°F
Barometer	29.4 in Hg
Surface	Dry tarmacadam

MAXIMUM SPEEDS
	mph	kph
Banked Circuit	98.9	159.1
Best ¼ mile	102.7	165.2
Terminal Speeds:		
at ¼ mile	73	117.4
at kilometre	89	143
at mile	95	153
Speed in gears (at 6500 rpm):		
1st	32	51
2nd	52	84
3rd	75	121

ACCELERATION FROM REST
mph	sec	kph	sec
0-30	3.5	0-40	2.7
0-40	5.6	0-60	5.0
0-50	8.8	0-80	8.3
0-60	12.2	0-100	12.9
0-70	16.7	0-120	19.2
0-80	23.5	0-140	32.6
0-90	37.0		
Stand'g ¼	18.4	Stand'g km	34.8

ACCELERATION IN TOP
mph	sec	kph	sec
20-40	11.9	40-60	8.0
30-50	11.3	60-80	6.1
40-60	12.1	80-100	8.3
50-70	13.1	100-120	8.4
60-80	16.0		

ACCELERATION IN 4th
mph	sec	kph	sec
20-40	8.5	40-60	5.5
30-50	8.4	60-80	5.0
40-60	9.0	80-100	6.1
50-70	9.8	100-120	6.8
60-80	12.3	120-140	11.0
70-90	21.3		

FUEL CONSUMPTION
Touring*	28.2 mpg
	10.0 litres/100 km
Overall	26.6 mpg
	10.6 litres/100 km
Govt tests	28.4 mpg (urban)
	40.3 mpg (56 mph)
Fuel grade	30.2 mpg (75 mph)
	97 octane
	4 star rating
Tank capacity	10.0 galls
	45.5 litres
Max range	282 miles
	454 km
Test distance	528 miles
	849 km

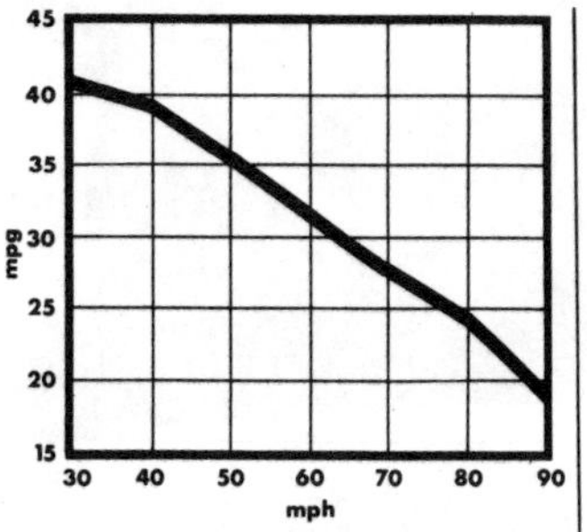

*Consumption midway between 30 mph and maximum less 5 per cent for acceleration.

BRAKES
Pedal pressure, stopping distance and average deceleration from 30 mph (48 kph).

lb	kg	ft	m	g
20	9.1	103	31	0.29
40	18.2	52	16	0.58
60	27.2	36	11	0.83
76	34.5	31	94	0.97
Handbrake		97	27	0.31

Maximum from 70 mph (113 kph)
| 70 | 31.8 | 197 | 60 | 0.83 |

FADE
Twenty 20 0.6g stops at 45 sec intervals from speed midway between 40 mph (64 kph) and maximum (69 mph, 111 kph) at gross vehicle weight.

	lb	kg
Pedal force at start	43	19
Pedal force at 10th stop	50	23
Pedal force at 20th stop	68	31

STEERING
Torque at wheel rim when parking and when cornering on 216 ft diameter circle.

	lb ft
Parking	17.0
Cornering at 0.1g	3.0
0.3g	5.5
0.6g	8.5

Turning circle between kerbs
	ft	m
left	35.4	10.7
right	34.3	10.5
lock to lock	3.8 turns	
50ft diam. circle	1.1 turns	

CLUTCH
	in	cm
Total pedal travel	5.0	12.7
Maximum pedal load	28lb	12.7 kg

NOISE
	dBA	Motor rating*
30 mph	64	10
50 mph	71	17
70 mph	74	21
Max revs in 2nd	84	41

*A rating where 1 = 30 dBA and 100 = 96 dBA, and where double the number means double the loudness

SPEEDOMETER (mph)
Speedo	30	40	50	60	70	80	90
True mph	28	37	46	56	66	76	85

Distance recorder: 0.1 per cent fast

WEIGHT
	cwt	kg
Unladen weight*	20.1	1021
Weight as tested	23.8	1209

*with fuel for approx 50 miles

Performance tests carried out by Motor's staff at the Motor Industry Research Association proving ground, Lindley.

Test Data: World Copyright reserved; no unauthorised reproduction in whole or part.

GENERAL SPECIFICATION

ENGINE
Cylinders	4 in-line
Capacity	1498 cc (91.41 cu in)
Bore/stroke	86.4/63.9 mm (3.40/2.52 in)
Cooling	Water
Block	Cast iron
Head	Light alloy
Valves	Sohc
Cam drive	Belt
Compression	9.2:1
Carburetter	Weber 34 DAT 8 twin-choke downdraught
Bearings	5 main
Max power	85 bhp (DIN) at 5800 rpm
Max torque	90.4 lb ft (DIN) at 3500 rpm

TRANSMISSION
Type	5-speed manual,
Clutch dia	7.5 in (20.3 mm)
Actuation	Cable

Internal ratios and mph/1000 rpm
Top	0.96:1/17.8
4th	1.16:1/15.3
3rd	1.55:1/11.5
2nd	2.23:1/8.0
1st	3.58:1/5.0
Rev	3.72:1
Final drive	3.76:1

BODY/CHASSIS
Construction	Unitary, all-steel
Protection	Adhesive and plastic sealant; galvanising and Zin-crometal process on exposed body panels.

SUSPENSION
Front	Ind by MacPherson struts, lower arms and anti-roll bar. Coil springs.
Rear	Ind by MacPherson struts, transverse links and longitudinal reaction rods. Coil springs

STEERING
Type	Rack and pinion
Assistance	None

BRAKES
Front	Disc, 8.9 in dia
Rear	Drum, 7.3 in dia
Park	On rear drums
Servo	Yes
Circuit	Diagonally split
Rear valve	Yes
Adjustment	Automatic

WHEELS/TYRES
Type	Light alloy, 5J×13
Tyres	165/70 SR 13
Pressures	26/26 psi F/R (normal) 29/29 psi F/R (full load)

ELECTRICAL
Battery	12V, 45 Ah
Earth	Negative
Generator	Alternator 45A
Fuses	14
Headlights	
type	Halogen
dip	110 W total
main	120 W total

GUARANTEE
Duration ..12 months, unlimited mileage

MAINTENANCE
Schedule: ....every 6000 miles
Free service: ....at 600 miles

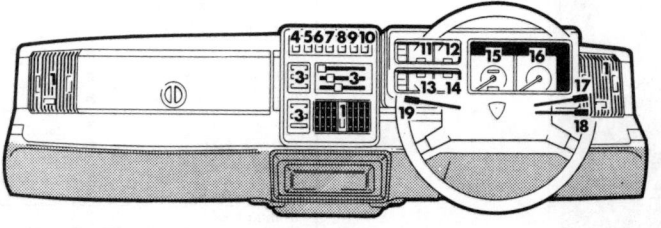

1 Fresh air vents
3 Heating and ventilation controls and instrument panel rheostat
4 Fog light switch
5 Reversing light switch
6 Rear wash/wipe switch
7 Heated rear window switch
8 Spare switch
9 Hazard warning light switch
10 Switch to check operation of main warning lights
11 Oil pressure gauge
12 Water temperature gauge
13 Fuel gauge
14 Voltmeter
15 Speedometer
16 Rev counter
17 Headlight/flash stalk
18 Winker stalk
19 Wash/wipe stalk

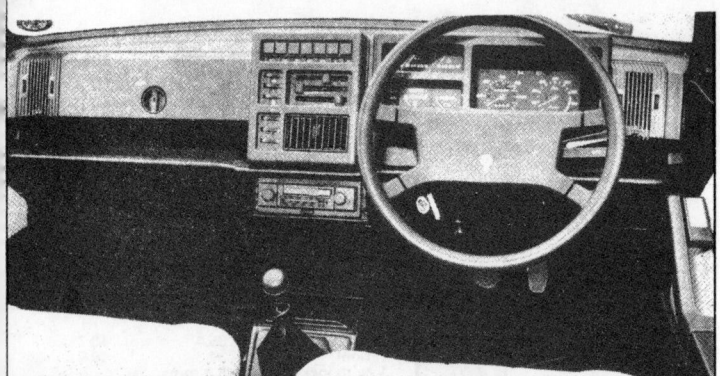

Above: the facia is in two distinct sections. In practice the controls (especially horizontal thumbwheels) are fiddly to use. Instruments, left, are clear and comprehensive. The 'control section, below, houses the heating and ventilation controls and the minor switchgear

vicing for the first two years/20,000 miles under the "Delta Deal". Servicing is required every 12,000 miles, though the oil has to be changed every 6,000 miles.

Then there is Lancia's 6-year Corrosion Prevention Warranty which involves treatment with the Cadulac process and re-treatment at 23 and 42 months. All of this goes on top of Lancia's own already extensive anti-corrosion treatment, developed with the expert help of Saab, whereby all joints are treated with an adhesive and a plastic sealant, for instance, and the most exposed parts of the body are either made from galvanised sheet steel or steel treated by the "Zincrometal" process. This time round, Lancia — badly hurt by the Beta episode — is obviously leaving nothing to chance.

Although the engine bay is not particularly spacious, most normal service items can be easily reached.

CONCLUSIONS

The Delta is a good car. One of its greatest strengths is that it does nothing badly and most things very well though its surprising thirst *is* mildly disappointing, even considering its above average performance.

Sceptics who might be tempted to label the Delta an up-market Fiat Strada should think again for, while certain aspects of its performance do betray humble origins (engine characteristics/gearchange) its character belongs more to the traditional Lancia stable with fine handling, a good ride, impressive refinement, plush trim and excellent finish. It is both a civilised and dignified small car and one that it would be no hardship to 'trade-down' to. The question of its Car of The Year status is open to debate. We should have a clearer idea of its relative merits when it comes face to face with the Opel Kadett (the car it pushed into second place) in a forthcoming Group Test.

Left: the engine bay is crowded, but most major service items can be easily reached

Comparisons

PERFORMANCE	Lancia	Alfa	Opel	Peugeot	Talbot	VW
Max speed, mph	98.9	104.8	98.6	95.1	96.2	91.2
Max in 4th	—	97	—	—	—	—
3rd	75	74	76	82	74	76
2nd	52	51	50	56	49	50
1st	32	30	30	32	29	30
0-60 mph, secs	12.2	10.5	12.6	13.2	13.0	13.2
30-50 mph in 4th, secs	8.4	8.2	11.5	11.2	9.9	10.1
50-70 mph in top, secs	13.1	12.4	13.4	13.6	12.6	13.0
Weight, cwt	20.1	21.9	17.7	19.0	18.5	15.7
Turning circle, ft*	34.8	33.3	32.0	30.2	30.8	30.2
50ft circle, turns	1.1	1.25	1.3	1.2	1.35	1.2
Boot capacity, cu.ft.	7.1	9.5	10.3	10.3	7.4	6.6

*mean of left and right

COSTS & SERVICE	Lancia	Alfa	Opel	Peugeot	Talbot	VW
Price, inc VAT & tax, £	4995	5100	5087	4838	4808	4572
Insurance group	N/A	7	4	4	4	4
Overall mpg	26.6	24.8	29.8	25.3	31.6	29.3
Touring mpg	28.2	30.7	36.9	28.9	35.9	34.9
Fuel grade (stars)	4	4	4	4	4	2
Tank capacity, gals	10	11.0	9.5	9.5	10.3	9.9
Service interval, miles	6,000	6,000	6,000	5,000	5,000	5,000
No of dealers	127	137	237	272	600	350
Set brake pads (front) £*	N/A	9.20	14.64	14.09	14.38	15.85
Complete clutch £*	70.45	95.45	44.61§	97.72	45.61	54.69§
Complete exhaust £*	44.89	101.78	64.51	63.86	51.29	68.07
Front wing panel £*	45.25	44.28	39.50	36.88	40.25	38.52
Oil filter, £*	3.02	4.37	3.10	3.15	3.91	3.27
Starter motor, £*	52.32§	48.30§	62.33§	41.88§	43.70§	75.66§
Windscreen, £*	61.73**	58.65**	70.15**	52.10**	56.35**	45.94**

**Laminated §Exchange †Front silencer and tailpipe only *Inc VAT but not labour charges

EQUIPMENT	Lancia	Alfa	Opel	Peugeot	Talbot	VW
Adjustable steering	●	●				
Air Conditioning						
Alloy Wheels	●		●			
Central door locking						
Cigar lighter	●	●	●	●	●	●
Clock	●	●	●	●	●	●
Cloth trim	●	●	●	●	●	●
Dipping mirror	●	●	●	●	●	●
Driver seat height adjust						
Driver seat tilt adjust						
Electric window lifters						
Fresh air vents	●	●	●	●	●	●
Headlamp washers			●			
Head restraints	●	●	●	●	●	●
Heated rear window	●	●	●	●	●	●
Intermit/flick wipe	●	●	●	●		●
Laminated screen	●	●			●	
Locker	●	●	●	●	●	●
Passenger door mirror	●					
Petrol filler lock	●					●
Power steering						
Radio					●	
Rear central armrest						
Rear courtesy light	●	●				
Rear fog light	●					
Rear wash/wipe	●				●	●
Remote mirror adjust	●			●	●	
Rev counter	●	●	●		●	●
Reverse lights	●	●	●	●	●	●
Seat belts — rear						
Seat recline	●	●	●	●	●	●
Sliding roof						
Tape player					●	
Tinted glass	●	●	●			●
Vanity mirror	●	●	●		●	●

The Rivals

Other rivals include the Alfasud 1.5 (£4,300), the Citroën GSA Pallas (£4,575), the Honda Accord 1.6 Executive (£4,790), the Renault 14 TS (£4,530) and the Ford Escort 1.6 Ghia (£4,726).

LANCIA DELTA — £4,995

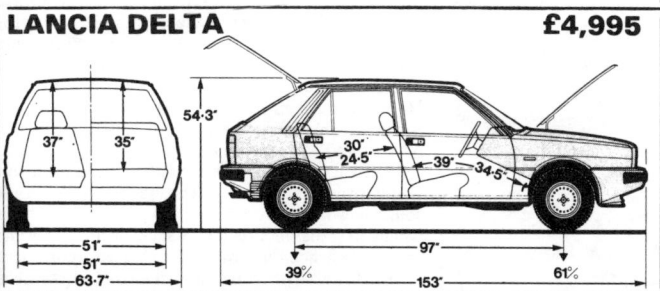
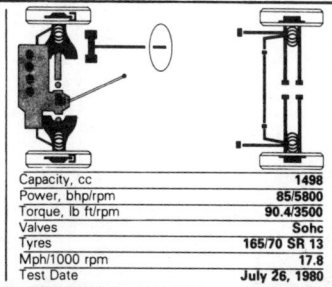

Capacity, cc	1498
Power, bhp/rpm	85/5800
Torque, lb ft/rpm	90.4/3500
Valves	Sohc
Tyres	165/70 SR 13
Mph/1000 rpm	17.8
Test Date	July 26, 1980

The 1980 Car of The Year, Lancia's Delta is partially Fiat Strada based but wears a distinctive five door body styled by Giugiaro and is chasing a different market. A civilised and dignified compact hatchback, its brisk performance is matched by taut handling and a supple ride. Good packaging and high levels of trim and equipment are further plus points though economy is only fair and boot size only average.

ALFA ROMEO GIULIETTA 1.6 — £5,100

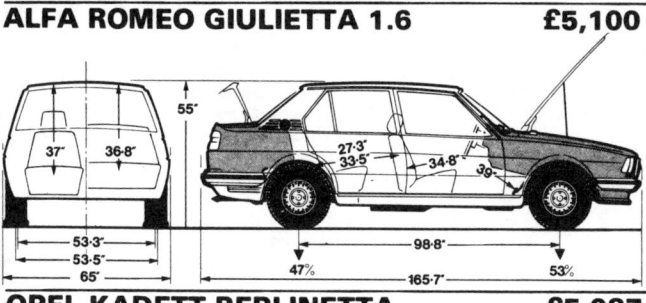
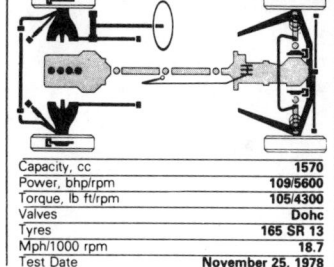

Capacity, cc	1570
Power, bhp/rpm	109/5600
Torque, lb ft/rpm	105/4300
Valves	Dohc
Tyres	165 SR 13
Mph/1000 rpm	18.7
Test Date	November 25, 1978

Mid-range Alfa filling the gap between the 'Sud and the Alfetta manages very good all-round performance from its 1600 cc with reasonable economy. It has a spacious interior (though smallish boot), a comfortable ride and is a relaxed high speed cruiser. Open road handling is both safe and satisfying, while the brakes are excellent. It has some faults — low geared steering at town speeds, poor ventilation — but overall it is a car we rate highly.

OPEL KADETT BERLINETTA — £5,087

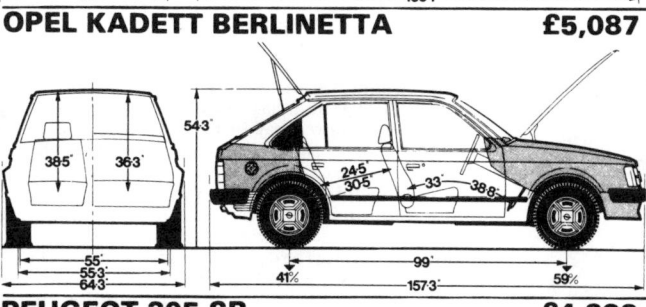
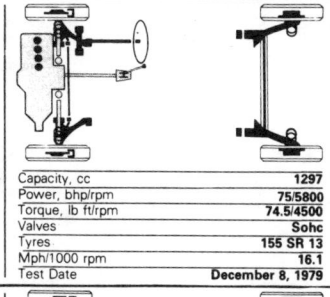

Capacity, cc	1297
Power, bhp/rpm	75/5800
Torque, lb ft/rpm	74.5/4500
Valves	Sohc
Tyres	155 SR 13
Mph/1000 rpm	16.1
Test Date	December 8, 1979

In this form with the more powerful 1.3S engine, Opel's front-wheel drive, second generation Kadett combines excellent performance with good economy (we've quoted figures for the mechanically identical Vauxhall Astra) in a package with very good accommodation, taut, precise handling, quick and positive gearchange. Ride comfort is nothing special, and the GL we tested needed better refinement, though Berlina should be better.

PEUGEOT 305 SR — £4,838

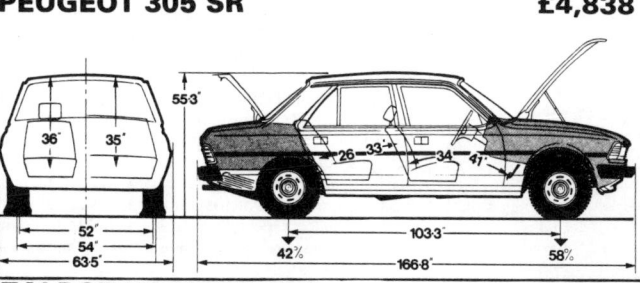
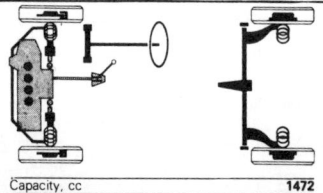

Capacity, cc	1472
Power, bhp/rpm	74/6000
Torque, lb ft/rpm	86/3000
Valves	Sohc
Tyres	145 SR 14
Mph/1000 rpm	17.8
Test Date	May 13, 1978

Like many of its up-market small car rivals in this class, this expensive but well appointed 305 has only fair performance for the price, with mediocre economy. Excellent gearchange, powerful brakes and comfortable, roomy interior are its best features: ride also excellent on some surfaces, but harsh and noisy on others. Noise suppression only fair, handling safe but stodgy. Messy instruments, good heating and excellent ventilation.

TALBOT HORIZON GLS — £4,808

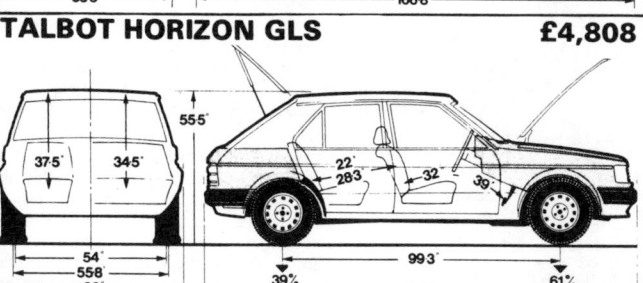
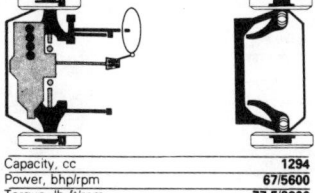

Capacity, cc	1294
Power, bhp/rpm	67/5600
Torque, lb ft/rpm	77.5/2800
Valves	Pushrod ohv
Tyres	145 SR 13
Mph/1000 rpm	16.5
Test Date	October 14, 1978

Talbot's contender in this market is plushy trimmed and extremely well equipped in GLS form, and 1300 engine gives impressive blend of performance and economy; refined cruising marred by boom periods at some speeds. Smooth but rather noisy ride, and sure-footed handling that is spoiled by very low-geared (though light) steering. Not as roomy as might be expected, and driving position is cramped.

VW GOLF 1.3 GLS — £4,572

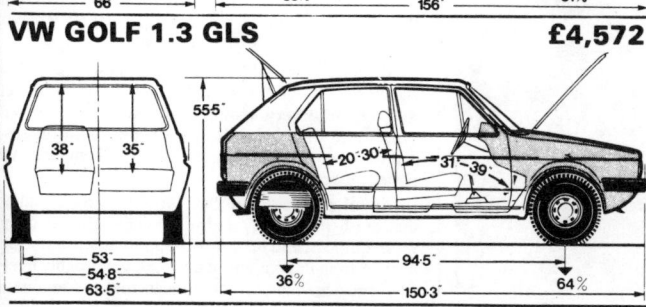
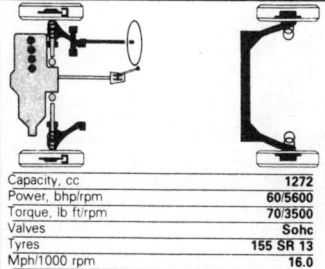

Capacity, cc	1272
Power, bhp/rpm	60/5600
Torque, lb ft/rpm	70/3500
Valves	Sohc
Tyres	155 SR 13
Mph/1000 rpm	16.0
Test Date	December 22, 1979

The latest GLS (and LS) Golf now has a 1300 engine, which is noisier than the previous 1500, but performance is only slightly down and competitive in its class (the £5,444 GTi is much quicker, of course): fuel economy fair. Not as roomy as some rivals but still a comfortable car with a supple ride and good heating and ventilation; and a driver's car with eager engine, balanced handling, delightful gearchange, good control layout. Not as well equipped as some similarly priced rivals.

Road Test

Adjudged the 1980 Car of the Year, Lancia's Delta brings refinement to everyday motoring.

Can the Delta earn its wings?

JOHN BOLSTER puts the Lancia Delta through its paces and finds out why it was judged 1980 Car of the Year. A trip to Paris for the *Salon* revealed the car to be a lively performer with excellent road holding but with some detail faults.

As it is a couple of inches shorter than its cousin, the Fiat Strada, interior space is not over generous.

Having already driven the Lancia Delta, both in Italy and Scotland, I looked forward to a full road test. This was to be quite a searching appraisal, for in addition to the test routine the Delta was to take me and my wife to Paris for the Salon. Our average speed broke no records because, after saving seconds on every corner and gearchange, we were apt to remember a favourite restaurant and a delay of three hours would ensue. However, we competed in the Paris Grand Prix, which is just as competitive as it always was, though the French have slowed down on the *Autoroutes*.

The Lancia Delta is a new kind of car, designed to suit a modern demand. The man who used to run a V8 needs something more economical now, but though he is willing to settle for a relatively small engine, he wants the distinguished appearance, comfort, and refinement to which he has become accustomed. Giugiaro has excelled himself with a body shape which is just a bit snooty and tends to make other 1500cc cars look slightly common. It's a four-door hatchback, with rear seats that split and fold, if one needs more luggage space than the reasonably roomy boot provides.

The Delta has a slightly longer wheelbase than the Fiat Ritmo/Strada, to which it is distantly related, but it is actually a couple of inches shorter in overall length, resulting in a rear seat which provides adequate but not particularly generous space for the passengers. The engine is basically the Fiat unit, with its belt-driven overhead camshaft, but it has a higher compression ratio, larger manifolds with a different carburettor, and electronic ignition. All this is good for an extra 10bhp, and the five-speed gearbox also has special ratios for the Lancia application.

The interior is attractively trimmed and the seats comfortable, while the steering column is adjustable for rake. A good driving position is easy to achieve, though a small driver may sit rather low. There is a good all-round view but the rear quarters were a little blind for close parking in Paris streets. At this point, I must be critical of certain details. The speedometer dial is so deeply recessed that the front passenger cannot read the mileage recorder when navigating, which is maddening. The digital clock is roof-mounted, but too far back so that the driver cannot read it safely in traffic. Finally, the excellent and easily controlled heater is not allied with a proper ventilation system, so cool breathing air and warm feet are not obtainable simultaneously. Yet, Saab are supposed to have helped over this matter.

Although the Delta is heavier than the Ritmo, it can out-perform it because the Lancia engineers have breathed on the power unit to good effect. On a long straight, it is possible to push the speedometer needle past 100mph but my stopwatch said that 98mph is the honest maximum, which is good for a substantial saloon of only 1500cc, especially as fifth gear holds the engine well below peak revs. A cruising speed of 90mph seems a natural pace.

The gearchange is very pleasant but it seems to work best if not hurried unduly. Fourth is a splendid gear that can be used continuously away from the *Autoroute*. Driven reasonably, the car is exceptionally quiet, but the engine suddenly becomes noisy when pressed close to its limit.

Excellent roadholding is one of the best features, on wet or dry roads. The cornering power is notably high but the angle of roll is very moderate. A neutral response turns to understeer at the limit, but the car always seems to have some unexpected grip in reserve. It is a most forgiving machine, even tolerating braking in the wrong places and clumsy throttle control, and though the steering is not high-geared, this is never noticeable except when manoeuvring.

Although the ride at first felt quite firm, the suspension suited the French roads particularly well and we were very comfortable. Road noise is well suppressed and wind noise is moderate. Of the brakes I can only say that one forgets all about them, which is praise indeed.

It is unfortunately true that there is no way of bending the mathematics of motoring. The Delta, being well-equipped and luxuriously appointed, is naturally heavier than its competitors, yet it has a notably lively performance. This can only mean that the engine is

Giugiaro's pen has created a design that makes other hatchbacks look common.

unusually powerful for its size, and power means petrol. A 30mpg consumption is possible, but the carefree driver may drop down towards 25mpg, which is not brilliant for a 1500cc car.

The latest crop of transverse-engined, front-wheel drive cars are all outstanding roadholders, but the Lancia Delta is up among the best of them. It is so stable that it can be driven hard without disturbing the passengers, so that one tends to travel fast all the time. Perhaps it would be more economical if the roadholding were less exceptional!

In conclusion, I must tell of a bad fright that the Delta gave me. In the midst of Paris traffic, I was horrified to see the fuel gauge registering zero, *merde alors!* I died a thousand deaths, for I could imagine the bedlam of hundreds of klaxons as I brought the city to a standstill. Then the needle crept up the scale again and gave me no more trouble, but that was a moment I would rather forget. ■

Above left: the engine is basically a higher compression version of the Strada unit. Above right: dashboard layout is neat but heating and ventilation are poor. Below: simple yet elegant.

LANCIA DELTA
£5,200
Specifications

Cylinders capacity	4 1498cc
Bore x stroke	86.4 x 63.9mm
Valve gear	Belt-driven single overhead camshaft
Compression ratio	9.2:1
Fuel system	Weber 34 DAT 8 carburettor
Power system	85bhp DIN at 6200rpm
Torque/rpm	90lbf ft 350rpm
Gear ratios	0.959, 1.163, 1.55, 2.23, and 3.583:1
Final drive	Helical spur gears, ratio 3.765:1
Steering	Rack and pinion
Brakes	Servo-assisted disc/drum
Tyres	165/70 SR 13 tyres
Suspension (F)	Independent, MacPherson with anti-roll bar
(R)	Independent, MacPherson with anti-roll bar

Dimensions

Wheelbase	97.4ins
Track	55.1ins
Length	153ins
Width	63.8ins
Weight	2175lbs
Boot	7.1cuft

Performance

Max in 5th	98mph
Max in 4th	92mph
Max in 2nd	50mph
Max in 1st	31mph
0-30mph	3.4s
0-50mph	7.8s
0-60mph	11.6s
0-80mph	22.6s
Standing ¼ mile in 5th	18.0s
40-60mph in 5th	11.1s

Fuel

Testing	24-30mpg
Govt	25.5/40.3/31.0mpg

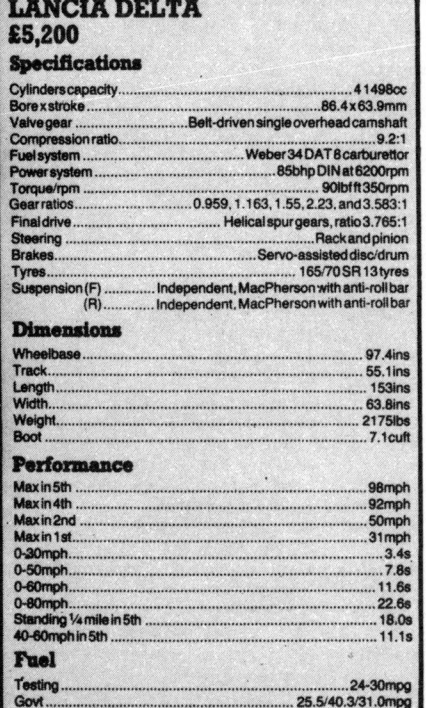

12,000 Miles On

DELTA—DELIGHTS DELAYED?

Niggling faults during its first 12,000 miles have detracted from our Lancia's appeal — but the promise is there. John Thorpe reports on seven months of ownership

IT WAS, I surmised, not so much a Delta, more a creek — and I was well and truly stranded without a paddle. Or, to be more accurate, without a throttle. I glared first at the odometer — it recorded just 62.9 miles from collection — and then at the snapped throttle cable which had brought my brand-spanking new Car of the Year 1980 to a standstill. A bird twittered in the evening sunlight, far away in the Sussex woods. It could have been a cuckoo . . .

I hadn't worked for *Practical Motorist* for 14 years without learning a trick or two, though, and eventually a crafty do-it-yourself dodge — ah well, *bodge*, if you insist — provided just enough control to allow the Delta to be driven the remaining 13 miles home, there to be abandoned while the ponderously-slow Lancia parts organisation searched its computers for a new cable. That aspect, though, we'll leave till later. The point is, that my long-term test Delta had got off to a pretty disastrous start — one which could well have soured our relationship for good. That it didn't is, of course, only partly a tribute to my own patient nature. The Delta itself also takes some of the credit, and though in the end I shed few tears over relinquishing its keys I didn't send out to order the champagne, either. And that, at least, is a considerable advance over my feelings as I stood by the roadside that mellow summer evening!

Lancia is, of course, a prestigious name — true Lancia enthusiasts might perhaps put that into the past tense — which is now borne by cars that, at basis, are really more luxurious Fiats. The Delta could justifiably be regarded as an up-market Strada, since it has basically the same engine — though with the benefit of a twin-choke Weber carburetter — and is built on the same running chassis, albeit one on which the rear suspension system has been redesigned.

In my admittedly simplistic world, I tend to think that if you are using somebody else's engine and somebody else's chassis then you are building somebody else's car and that's the end of it. However, the Delta does have a completely different Giugiaro-designed bodywork; an entirely different standard of finish and trim; and — according to colleagues who have driven both cars — a completely different "feel". I *still* think it's a rather sportier Strada . . .

It was launched in Britain in the summer of 1980, having "grown" a righthand drive conversion which includes, among other things, an oddly complex bell crank cross-link from the brake pedal to the nearside-mounted servo and master cylinder — one of which Heath Robinson himself would have been proud. Otherwise, little appears to have been altered, even where sheer ergonomics should have demanded a rethink, as with the ancillary switchgear on the central facia.

ON DELIVERY

★★ "Guess what colour your Italian car is!" challenged David Windsor, who had collected it for me in Central London. It had been supplied by Waterloo Carriage Co. Ltd., 42 The Cut, London SE1, and they had certainly gone to town with the polish. "Rossini" shone, a beautiful deep Italian racing red.

I decided to delay my check for delivery faults until the end of my trip home — I wanted to get to grips with this potent-looking hatchback. Perhaps it was just as well that I did, or I might have had second thoughts about driving it!

The first delivery fault, of course, didn't even wait to be discovered. It just jumped up and bit me, that

wretched throttle cable snapping just when I was fully committed to overtaking another car. Suddenly, there was a complete loss of power, plus some panic braking on my part as I sought the safety of the nearside verge.

Cables don't snap without a reason, but it was not until the replacement made a tardy appearance over a week later (it had taken the combined efforts of three Lancia agents to obtain one) that the full significance of the fault became apparent. The throttle pedal is mounted on a bracket, which carries a right-angled abutment against which the pedal rests. The cable is attached to a peg on the lefthand side of the pedal, and normally passes about half an inch above the abutment. On assembly, the bracket on our car had been so distorted that the abutment was angled sharply upwards, and — incredibly — somebody on the Turin assembly line had *forced* the cable into place, so that it sawed backwards and forwards on the sharp metal edge of the pedal stop. Leastways, it did until the strands were cut through — and I can testify that the process takes just 62.9 miles to complete!

This discovery, of course, meant that the new cable which I had been sent to fit myself would now have to be installed by a Lancia agent, who could reshape the bracket. This work was carried out by Keen and Betts, Ltd., of Shoreham by Sea, West Sussex. They collected the car; did the work competently — modifying the new cable, which didn't actually fit! — and then spoiled a good impression by a curmudgeonly display of petulance towards my wife when they delivered it to my home.

My own post-delivery inspection was done at this point. Apart from that little matter of the cable and bracket, I ended up listing 20 faults. Admittedly, some were minor, but items such as the soundproofing mat on the bonnet lid already frayed, and dropping insulating material into the engine compartment; both electrical conduits into the hatch detached from their housings, opening up two sizeable leakage paths into the car; the headlining not inserted into the trim around the hatch aperture; and half a dozen spare electrical connections dangling loosely under the facia all suggested sloppy assembly and slack inspection, both by the factory and at the dealer.

Delving deeper, I didn't feel very impressed, either, by finding securing buttons missing from the soundproofing; odd nuts and bolts scattered around the engine compartment and the boot; or by the fact that the underbonnet had not been dewaxed, and was both sticky and filthy. But at least it shouldn't rust!

There were more serious faults, too. The exhaust system was rattling — after both Keen and Betts and Waterloo Carriage had tried, and failed, to rectify it, this fault was cured by an Audi dealer (Jack Barclay European, of Battersea) who simply replaced the bolts that the Lancia agents hadn't noticed were missing!

The brakes were spongy in operation, and the internal rear-view mirror vibrated so much that it was virtually useless at any speed over 30 mph. Twelve thousand miles later, both faults remain . . .

LIFE WITH THE DELTA

★★★ Despite its unpromising beginning, I was determined to give the Delta a fair chance. It might yet, I thought, prove itself to be a triumph of design over manufacture. And Italian cars do have a reputation of being real driver's cars.

Certainly, I had already made myself comfortable. Unlike many of its compatriots, the Delta can accommodate a North European without making him feel that he is crouched, rather than seated, with the pedals too close and the wheel too high. Internally, the Delta is highly civilised — leastways, in the two front seats.

The driver is positively cosseted. There is ample fore-and-aft seat adjustment; a wheel gives infinitely-variable control for the rake of the seatback; and the steering column is adjustable for height at the flick of a lever. Both door-mounted external mirrors have internal adjustment for angle, laterally and longitudinally. A few experiments soon produced a comfortable and efficient driving position, with almost everything just where I wanted it.

I liked the seating and internal trim, too. For years, I used to bore my colleagues by telling them how superior cloth. upholstery is either to plastic (obviously) or leather (which always makes me sick!) Now that cloth is widely accepted, it was pleasant to find that the Delta makes very full use of it. Smart checked velour for the seats, the door trims and the headlining not only looks distinctive, but also provides a mini-environment that is cool in the summer and snug in the winter. It's enhanced by a seat design that tends to envelop you like a well-used and favourite armchair, giving good lumbar support — for which my long-suffering back was duly thankful. I'd rate the Delta as an excellent car for long journeys.

The front passenger seat is equally comfortable, as I found during the course of a memorable witchride through the Surrey lanes with one Greenslade, R., in the hot seat. While I clung manfully to the grab handle, it was convincingly demonstrated that though the Delta would make an excellent ground trainer for a kamikaze squadron, *I* would not make a notably good recruit . . ! Even so, despite RG's spirited Tricentrol-style cornering, I was able to brace myself into a comfortable position and enjoy at least parts of the ride without sliding from side to side or bouncing around.

It's a pity that the head restraints spoil the effect. They are adjustable for height, but when they are raised the internal friction mechanism is far too weak in operation, so a head snapped back on to the restraint merely pushes it rearwards. Unfortunately, the top of the seat is then perfectly positioned to deliver a rabbit punch, right on the back of the neck! Left lowered, they are badly positioned, so you can't really win.

Rear-seat comfort is marred, for anybody over 5ft 8in in height, by lack of head clearance under the hatch. Either one sits, head bowed, or one slumps in the seat. Shorter passengers, of course, have no such problem and find the rear of the Delta as pleasant as the front, though a mite short of foot clearance under the front seats.

Part of the credit is, of course, down to the ride provided by the MacPherson strut all-independent suspension. Lancia have obviously worked hard on this aspect, and have produced a smallish car that offers a big-car ride. It's an area in which I tend to get fussy — harsh suspension systems are a pet aversion with me — and early in the car's life I felt that the Delta's springing gave an uneasy, pitching motion on bumpy lefthanders. However, it settled down within 3,000 miles, and the ride became delightfully smooth, with very little body roll and virtually no transmission of road shocks. On my favourite country-lane "yumps", where most of the test cars I drive will lumber into the air at around 50 mph, the Delta sticks to the road until around the 60 mph mark — *and* the springing doesn't bottom when it lands. Very impressive.

As with the ride, so with the handling. The Delta will go round corners faster than I would normally want to take them, and it does so without drama save under one particular set of circumstances. It can be caught out by a combination of bumps, gradient, and a sharp lefthander, when its normally neutral handling can turn to an oversteer that has the tail swinging out with very little warning. Rex agreed, but felt that on righthanders he had detected a tendency to understeer instead.

Normally, only sheer abuse will make the car misbehave; but for the record, I don't recommend braking sharply half-way through a lefthander being taken briskly . . .

As a generalisation, I'd say that the Delta is one of the best cars I have ever driven when the road and weather conditions are bad. The only time I've felt any anxiety at all was when I saw a van exiting from a righthander, uncomfortably close, sideways, on ice. Cadence braking, and the Delta's inherent stability, had me safely at a standstill on that ice-rink surface well before the van took to the ditch, 20 yards ahead.

MOTOR LONG TERM TEST ● LANCIA DELTA

Total mileage: 12,100　　**Overall mpg:** 28.7
When bought: August 1980　　**Touring mpg:** 29.4
Price when bought: £4,995　　**Days off road:** 21
Value now: £3,800　　**Extra visits to dealer:** 2

Make: Lancia. **Model:** Delta 1500
Maker: Lancia & C., s.p.a, Via Vincenzo Lancia 27, Turin, Italy.
Price: £4009.00 plus £334.08 Car Tax plus £651.46 VAT equals £4,994.54.

 excellent good average poor  bad

When torrential rain swamps the motorway, my Delta has been the car that hasn't actually *needed* to slow down. The grip offered by its CEAT tyres is totally reassuring. In short, I am going to miss the drivability which it has offered, and which has so largely offset the annoyances caused by poor detail finishing.

Unfortunately, one important dynamic aspect has to be excepted: the brakes. True, they have never failed to stop the car, but the sponginess of the pedal has been a constant source of irritation. I suspect that the root of the problem is the multiplicity of pivots and linkages in the "remote" operating mechanism.

Completely joyous bendswinging tends to be spoiled when one is not convinced that there will be instant control response, and a second failing makes this a Delta problem — the gearchange, normally good, can for no apparent reason suddenly become obstructive when dropping down from fourth to third. Tastes differ, but I personally don't find it amusing to be setting a car up for a corner, only to find that I am in neutral, and that the brake pedal has a good inch of lost travel and that the gears just won't re-engage. It's happened! Which is a pity, because the Delta is a car that can sometimes make a direct and very basic appeal to one's sporting instincts. I feel it is better-mannered than the equivalent Alfasud — the 1.5 — and more comfortable too. It has an engine which refuses to run out of steam, even though it is hampered by a not-quite-ideal choice of second and third gear ratios. Though the unit itself is responsive, acceleration is beginning to tail off in second at a point that is a mite early for a change into third, so overtaking in the "overlap" area between the two needs to be carefully judged. It takes longer than you think!

Otherwise, the Lancia's Fiat-based engine is impressive. From reliable first-time start to final switch-off, it is smooth and responsive, and mine has never so much as skipped a beat. Heavy sound-proofing makes it mechanically remote from the driver — you can hear some exhaust noise, but virtually nothing from below the bonnet — and at a steady 70mph motorway cruise the motor is turning over at a relaxed 4,100 rpm.

Designed in conjunction with Saab, the heating system (after I had reset the control cable to the water valve) proved capable of producing a blast of hot air that would not have disgraced the winds of the desert. However, the distribution system has bemused and baffled everybody who has driven the car. Hot air can be had from windscreen vents, centre vents, under-facia vents, facia end vents, and side window vents — but, short of cutting the heated flow off completely, nobody has been able to obtain cold air. The quick answer is to open a window — the Delta is one of those excellent cars in which a fully-open window provides ventilation without buffetting — which is *very* pleasant when cruising through the forest roads of North Sussex on a summer's day: but what happens when it rains?

A good load carrier, the Delta has only one disadvantage when the hatchback theory is put into practice. The rear seat is split, and can be folded completely or half folded as circumstances require, and the hatch is well supported on telescopic struts. But, its designers must have watched too many Wild West films — a moment of forgetfulness, a failure to duck, and the extremely solid catch projecting from the bottom of the hatch will carry out an efficient scalping job. I know — from *very* painful experience.

One needs to watch out for the razor-sharp edges of the luggage compartment side-panels. These have removable "doors" concealing cubby holes for the tool kit and the jack, but you reach in only at the risk of slashing your fingers or your wrists. And while the scissor-type jack works well, lifting the whole side of the car on a single central jacking point, the ratchet-type handle has raw, potentially dangerous, right-angled edges.

I was annoyed to find that, despite its "luxury" price tag, the Delta had no radio as standard. That was soon rectified — I fitted a Radiomobile Model 420 stereo cassette tape player, which has an inbuilt medium-wave and long-wave radio. And (that *Practical Motorist* training again!) I decided to fit it myself, since the Delta wiring diagram indicated that all necessary wiring was in place.

Hmm. For a start, the sockets on the wiring and those on the set didn't match. A handful of Scotchlok connectors soon cured that. Then, I searched for — but couldn't discover — the aerial lead which allegedly ran from the recommended mounting point at the rear to the under-facia. The rear end I could find: the forward end, however, defied all my efforts to locate it — until I accidentally knocked off the cover over the footbrake crosslinkage, and discovered my missing aerial lead coiled, snakelike, behind it.

Linking it up to the set added one more Delta demerit. Predictably, the aerial lead didn't work! At which point I gave it best, and operated on a tapes-only basis until Jack Barclay European saved the day for me by running a new lead right through the car.

RELIABILITY

★★★☆ Not without heart searching, I have concluded that the Delta is just about average in terms of reliability, though I must admit to being swayed mainly by the fact that nothing *major* has gone wrong. The list of minor failures, however, is a formidable one: and a private buyer might well think that my conclusion is over-generous. Read and judge.

The delivery faults already noted have, for the main part, remained uncorrected. So have most of the faults that have developed since, though virtually everything had been reported to the supplying dealer by the time the car had had its 6,200 mile service.

While I was waiting for my tape player, my main source of entertainment had been the rear screen wiper/washer. Originally, pressing its control button resulted in a single sweep of the wiper, accompanied by a jet of water. Not very impressive. As the washer was operating throughout the sweep, the screen was left wet every time the wiper parked and was re-wetted every time the wiper blade moved — but at least this self-defeating piece of technology *worked*. By 3,000 miles, however, it had

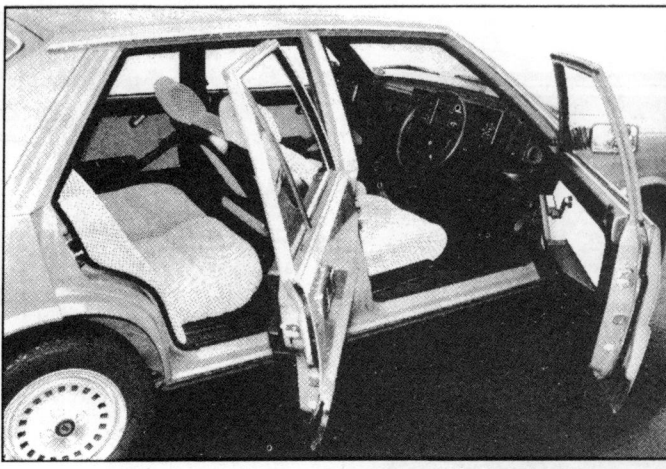

developed a random action. Press the button, and it might provide one wipe, four wipes, twenty wipes, or constant wiping — the last-named requiring the ignition to be switched off before the wiper would park. All good fun on a crowded motorway; and I assumed that this was some form of gambling device built specially for Mafia staff cars.

Eventually, the inevitable happened. David Windsor, in all innocence, pressed the button and found himself with a screen wiper that had gone critical and run amok! Before he could pull in and stop, the motor had sheared the splines from the hub of the wiper and the arm was trailing lamely behind the car on the end of its washer pipe. That was in December. The following April we were *still* waiting for Waterloo Carriage to come up with the replacement parts.

Wind noise had always been noticeable, and our request that the driver's door should be rehung to cure it was ignored. The wind rammed its message home when the car was at a standstill, with the driver's door wide open. A sudden gust actually snapped the door strap pivot pin in two. It was reported in December, at the main service — and, again, it had still not been corrected by April.

By now, the front wipers were emitting an annoying knocking noise (it eventually cured itself), and the driver's-side window winder handle twice flew off, to land in my lap.

By 4,000 miles the speedometer was making death-rattle noises, which again eventually cured themselves. The bulb illuminating the heater slide controls blew. It was due to be renewed at the main service, but was still out of action when the car was collected from Waterloo Carriage ("Sorry. The electrician's on holiday!" they said).

Also left undone was the requested replacement of an exterior mirror glass which had been shattered when a lunatic squeezed himself through the gap between the Delta and a parked car — well, *almost*. Our mirrors met, and he disappeared into the night without stopping, while I surveyed a shattered driver's-side mirror. After four months, no spare had materialised. Still, we were still waiting for the new under-bonnet soundproofing mat, under guarantee, complained about the previous August!

One further delivery fault came to light in February, when a cut sidewall demanded that the spare wheel be

Comfortable seating (above), but rear seat headroom is restricted by the sharply-raked rear hatch. The facia is neat, but the dials have no markings and the ancillary switches are located for lefthand drive

brought into use. The result was eyeball-shaking vibration at any speed from 60 mph onwards. Yes, you've guessed — the factory hadn't bothered to balance the spare wheel!

By now, too, the driver's head restraint was waving limply on its stalks, and the stainless steel gutter trim above the driver's door had lifted at a joint and was loose.

After 11,000 miles came up on the clock, I started congratulating myself that nothing appeared to have gone wrong for weeks, and that the Delta's teething troubles were over. Came, too, the day when — after a long dry spell — overnight rain suggested that it was time to pick up my rolled umbrella from the back seat, where it had lain unused. So how come an unused umbrella was soaking wet? A tell-tale line of water droplets along the upper edge of the hatch told its own mute story. The seal was faulty, and it wasn't just my umbrella that had suffered as a result. The Delta now had a rear seat that could have made a fair substitute for a water bed!

We referred the problem to Waterloo Carriage, who decided that a new set of hatch hinges would be required. A modified pattern had been introduced after some of the early Deltas (like ours!) had deposited their hatches on the public highway. None was in stock: so a new appointment was made and the problem passed to News Ed. Howard Walker, who by then had "inherited" the car. He takes up the tale in "Second Opinion."

Then came the dark, rainy afternoon when — with 11,500 miles recorded — I barely limped back to the office from Southend with the battery condition meter recording a steadily-flattening battery, though the generator warning light remained unlit. I made it — just. Even so, the car had to be pushed out of the lift and on to our rooftop car park, since there was insufficient life to turn the engine. Waterloo Carriage diagnosed a stretched fan belt as the cause, fitted a new one, and charged us for it. Some guarantee!

In one respect, though, Lancia really do seem to have got everything right. The finish stood up well to a period of ownership which had included an entire winter during which the car stood in the open day and night. At the end of it, the red coachwork still had depth and sheen and all the brightwork remained sparkling. A few stone chips on the air dam had opened up the white undercoating, but did not expose the metal, and there has been no sign of rust anywhere. The alloy wheels show no signs of "white corrosion" either.

Internally, the trim is still smart — the driver's seat perhaps a little rumpled, as you would expect — and it would take very little work to put the car into showroom condition.

SERVICING

★★ All regular servicing — apart from the "emergency" treatment by Keen and Betts — was carried out by Waterloo Carriage. Theirs — as the preceding section may suggest — was not an impressive performance, though the purely routine items appeared to have been competently done.

We had a first service at 1,000 miles; a second service at 6,200 miles; and the third service at 12,000 miles. In between, nothing was done — and at the services nothing was cured. The only improvement made to the spongey braking action (our major "mechanical" complaint) was effected by Jack Barclay European (who are not even Lancia agents), who reset the rear brakes to alleviate the condition.

COSTS

★★ Notionally, the Delta was supposed to benefit from free servicing, our costs being restricted to materials only. The Delta demands eight pints of lubricant, so allowing for an oil filter change at each of our three services, and for a set of plugs and an air filter at 12,000 miles, total servicing costs should have worked out at £39.13.

Theory is one thing, practice another — and the third service proved to be a law unto itself, since the front brake pads also had to be renewed. That added £16.81 to the total for materials, and 4½ hours' labour charges added a further £59.80, bringing our total servicing costs up to £115.74. Reasonable — but more than we had anticipated.

Tyre wear on the Delta is light. After 12,000 miles, the front treads registered 6mm and the rear treads just under 8mm, compared with 9mm when new. Changing the covers front to rear at 24,000 miles should enable the average Delta owner to obtain 36,000 miles use before one tyre needs to be changed and the spare brought into use, and this would have to be followed by a completely new set at 60,000 miles, with a part-worn tyre reverting to duty as a spare. Five tyres, then, in 60,000 miles gives, conveniently, the equivalent of one cover in 12,000. Snag! If your particular Delta chances to have Goodyears as original equipment, or Ceats like ours, your replacement will have to be Dunlop or Michelin. Neither Goodyear nor Pirelli covers are imported in these sizes, according to our tyre supplier.

One of the criticisms levelled at the Delta when it first appeared was that its fuel consumption was unduly high. Our original road-test car managed only 26.6mpg overall; so it is pleasant to be able to report that in this respect, at least, our long-term car has shown a welcome improvement. Until heavier-footed colleagues laid their clogs upon it, about halfway through the test, my car was returning just over 29mpg, and over the entire 12,000 miles it averaged 28.7mpg. That's still pretty heavy for a 1500cc car — but at least it betters two of our chosen five rivals.

READER FEEDBACK

Response was less than we had hoped, only 23 Delta owners replying to our questionnaire. Of their cars, only one had been bought secondhand and a majority — 18, in all — were between 4 months and 6 months old. Seventeen cars had covered more than 2,000 miles but less than 5,000 miles, and only two had done more than 11,000 miles when the replies were submitted. The average mileage covered was 6,000. Twenty of the cars were privately owned; 11 purchasers had paid the full price, and 11 had obtained a discount. Servicing was carried out by a manufacturer's agent in all but three cases. Only 10 owners rated their agents as Good, compared with 11 who said they were either Fair or Bad; and of the 14 owners whose Deltas had required spares, 12 said flat-footedly that the spares service was bad.

Owners' ratings*

Acceleration	6.9
Cruising ability	7.5
Steering	7.3
Roadholding	8.7
Handling	8.4
Braking	5.9
Gearchange	6.4
Clutch action	6.5
Gear ratios	7.3
Ride comfort	7.5
Seat comfort	7.3
Driving position	7.7
Heating	7.3
Ventilation	6.0
Noise at 70 mph	6.7
Instruments	6.9
Minor controls	5.9
Fuel consumption	4.7
Tyre wear	6.5
Visibility	5.9
Lights	8.1
Boot space	4.6
Reliability	8.1
Paintwork	8.0
Rusting	9.0
Styling	8.2

COSTS

		£
PETROL	418 gallons at £1.53 per gallon	639.54
OIL	between services:	—
SERVICING	at 600 miles	11.14
	at 6,000 miles	11.14
	at 12,000 miles	93.46
TYRES	25 per cent worn: £24 per tyre (discounted)	24.00
ROAD FUND LICENCE	for 12 months	60.00
TOTAL	for 12,000 miles	839.28
BASIC COST PER MILE		7.0p

PERFORMANCE

CONDITIONS
Weather: Wind 8-18 mph
Temperature: 46°F
Barometer: 29.8 in Hg
Surface: Dry tarmacadam

MAXIMUM SPEEDS

	Staff car mph	RT car mph
Banked Circuit	100.2*	98.9
Best ¼ mile	126.9	102.7
Terminal Speeds:		
at ¼ mile	73	73
at kilometre	89	89
at mile	95	95
Speed in gears (at 6,500 rpm):		
1st	32	32
2nd	52	52
3rd	75	75
4th	94	94

*See Text

ACCELERATION FROM REST

	Staff car	RT car
mph	sec	sec
0-30	3.4	3.5
0-40	5.4	5.6
0-50	8.5	8.8
0-60	11.4	12.2
0-70	15.4	16.7
0-80	22.2	23.5
0-90	34.0	37.0
Standing ¼	18.0	18.4
Standing km	33.9	34.8

ACCELERATION IN TOP

	Staff car	RT car
mph	sec	sec
20-40	10.5	11.9
30-50	10.6	11.3
40-60	10.8	12.1
50-70	11.9	13.1
60-80	15.5	16.0

FUEL CONSUMPTION

	Staff car	RT car
Touring*	29.4 mpg	28.2

Overall	28.7 mpg	26.6
Govt tests	28.4 mpg (urban)	
	40.3 mpg (56 mph)	
	30.2 mpg (75 mph)	
Fuel grade	97 octane	
	4 star rating	
Tank capacity	10.0 galls	
	45.5 litres	
Max range	294 miles	
	473 km	
Test distance	12,000 miles	
	19,308 km	

*Consumption midway between 30 mph and maximum, less 5 per cent for acceleration.

NOISE

	dBA	Motor rating*
30 mph	64	10
50 mph	71	17
70 mph	74	21
Max revs in 2nd	84	41

*A rating where 1=30 dBA and 100=96 dBA, and where double the number means double the loudness.

SPEEDOMETER (mph)
Speedo: 30 40 50 60 70 80
True mph: 29 39 48 56 66 76

Distance recorder: 2% over-read

WEIGHT

	cwt	kg
Unladen weight*	20.1	1021
Weight as tested	23.8	1209

*with fuel for approx 50 miles

Performance tests carried out by Motor's staff at the Motor Industry Research Association proving ground, Lindley.

Test Data: World copyright reserved; no unauthorised reproduction in whole or in part.

THE RIVALS

Included below is comparative information on five front-wheel-drive rivals to the Delta — the Alfasud 1.5 Ti Veloce, Austin Allegro 1.7 HLS, Citroën GSA Pallas 1.3, Ford Escort 1.6 GL and Opel Kadett Berlina 1.3S.

COMPARISONS	Lancia	Alfa	Austin	Citroën	Ford	Opel
Max speed, mph	100.2	106.9	100.0	95.0	103.0	98.1
Max in 4th	94	92	—	—	—	—
3rd	75	71	73	74	86	76
2nd	52	50	50	49	58	50
1st	32	29	31	29	35	30
0-60 mph, secs	11.4	9.9	11.0	14.6	10.7	11.7
30-50 mph in 4th, secs	7.8	7.1	8.1	11.4	8.9	10.5
50-70 mph in top, secs	11.9	10.3	11.4	17.6	10.0	12.6
Weight, cwt	20.1	18.0	16.7	19.1	17.0	17.7
Turning circle, ft	34.7	—	—	—	31.8	32.0
50ft circle, turns	1.1	—	—	—	1.2	1.3
Boot cu. ft.	7.1	9.4	9.2	11.9	10.3	10.3

IN SERVICE	Lancia	Alfa	Austin	Citroën	Ford	Opel
Price, inc VAT & tax, £	4,995	5,100	5,195	4,880	4,721	5,051
Insurance group	5	6	4	4	5	4
Overall mpg	28.7	26.1	28.6	31.3	32.2	35.0
Touring mpg	29.4	34.6	31.5	36.5	33.6	35.2
Fuel grade (stars)	4	4	4	4	4	4
Tank capacity, gals	10.0	11.0	10.5	9.5	9.0	9.5
Service interval, miles	6,000	6,000	6,000	5,000	6,000	6,000
No of dealers	134	130	1,900	260	1,241	230
Recom service time**	4.0	2.5	2.6	5.1	2.5	5.1

**In hours, for 12,000 miles

* Owners were asked to rate from "Excellent" to "Bad". The scores are based on giving "Excellent" 10, "Good" 7, "Average" 4, "Poor" 2 and "Bad" 0.

Faults
Owners having at least one with:
Engine	11
Transmission	6
Steering, suspension	2
Brakes	9
Electrical	15
Body, paint, trim	17
Fittings, trim	14
Instruments	10

Unfortunately, our sample proved to be really too small to be representative of individual faults. However, those faults mentioned by 4 or more owners were:— difficulty in engaging gears (by 5 owners — though 2 said the problem wore off); headlight bulb needed replacing (4 owners); scratched/dented/damaged/wing/bonnet/door on delivery (4 owners); miscellaneous rattles either inside or out (8 owners); and speedo cable noise/replacement (8 owners).

Time off road
None	9
Up to 1 day	6
2/3 days	5
4 or more days	3
Average (days)	2.3

Servicing by manufacturer's agent
Good	10
Fair	8
Bad	3

Warranty work
None	3
At least some	20

Satisfied with warranty work?
Yes	13
No	7

Would you buy another Delta?
Yes	13
No/not sure	10

If no/not sure, why?
Poor service/parts/PDI/aftersales	5
Poor performance	3
Other reasons	1

MANUFACTURER'S COMMENT

IT IS gratifying that despite an unfortunate initial problem the Delta's many excellent qualities became apparent during the test period.

The Lancia Delta was conceived as a comfortable — even luxurious — yet thoroughly practical five-door hatchback with performance, ride and road holding that is expected of the Lancia marque. The success of the design is reflected not only in its overwhelming choice as Car of the Year by the 52 international judges (despite Howard Walker!) but by the fact that the more adverse road conditions became, the better the Delta performed. From the description, Rex Greenslade's spirited drive was a formidable test for any car — an ordeal that the Delta seemed happy to take in its stride.

It is unfortunate that the car's more positive features have been to a degree masked by the dealer's performance. It appears that the service extended to Motor during the test period is below the standard that is expected from a Lancia dealer.

Finally, we must take John Thorpe gently to task on one or two points. The running chassis for the Delta is entirely dissimilar to the Fiat Strada's — which explains the different feel; and, of course, the air dam is produced in corrosion-free moulded resin, not metal.

Replacement tyres *are* available from both Pirelli and Goodyear, in addition to Dunlop and Michelin.

SECOND OPINION

SOMEWHERE UNDER that stylish Italian bodywork, there's a good car trying desperately to get out. Hopefully it will emerge during my tenure, for as from the 12,000 mile mark, the Delta Lady is under my care.

Reading John Thorpe's conclusions, though, you could well be forgiven for consoling me — "lacking in colour... curiously anonymous... personality not one of its virtues". Well, he *does* think Citroën Dyanes are the best thing since Dolly Parton! (Actually, I think French cars in general were the best thing until Kiri Te Kanawa; but I suppose the principle's the same! — JT).

I liked the road test Delta and I like this one. You can drive it hard and fast and it will satisfy the sporting blood in any driver, darting from one bend to the next, responding with precision to the slightest movement of the wheel and sweeping through bends like a thoroughbred. Its torquey, free-revving engine responds well to the throttle pedal, enhancing the car with lively, gutsy performance matched by a light and generally precise gearchange.

But, equally important, the car is just as adept in its role of a relaxed and refined motorway cruiser. In fifth, the Delta eats up the motorway miles and even at speeds well over the legal limit causes little stress or strain on the driver or passengers.

If it lacks inherent personality it is simply because it has the ability to tune itself to the demands of the person behind the wheel. To my mind, that adds up to a sound car.

During the next few weeks though, I'll be trying my best to let the "good car" emerge, for at the moment it is being restrained by poor and inattentive servicing from a dealer who seems only interested in carrying out the bare minimum on the job sheet.

As John Thorpe points out, there is still no satisfaction in the saga of the broken rear wiper, the shattered door mirror, the leaking rear hatch, the damaged underbonnet sound-deadening and the wind noise from the driver's door. Dammit, these faults have been going on for months!

At 12,000 miles the brake pads were replaced yet, less than 200 miles on, every application of the brakes produces an ear-piercing shriek that would match that of the most vocal of peacocks.

A securing pin for the driver's door tie-strap may not have been in stock, but surely any mechanic worth his salt would have realised that even a makeshift pin would have been better than letting the car go out with nothing to prevent the door from slamming against the wing when the door was opened in a strong wind.

When the car was returned a week after the 12,000-mile service, to have its weak hatchback hinges replaced, I asked for a slight dent in the leading edge of the driver's door to be attended to, in the hope that it would help cure the wind noise. I asked for the stodgy brake pedal to be checked. Neither job was carried out.

Whether Waterloo Carriage Company thinks that Sutton is just around the corner and we can "pop" the car in at the drop of a hat, I'm not quite sure. But it takes two members of staff (one to ferry the Delta deliverer back to the office) half a morning to "pop" the car in, and the same amount of time to get it back.

I like the car a lot and it suits my needs down to the ground. Perhaps with better attention from another dealer, my initial favourable impressions will become lasting.

Howard Walker

Hatchback flexibility. A useful luggage compartment can be extended by lowering the split rear seat, and the parcel shelf simply lifts out. Below: heater controls are awkward to operate

CONCLUSIONS

It would be pleasant to report that, at length, this paradoxical car produced its full potential and became a classic in the making. Unfortunately, endings like that are for fairy tales rather than road tests. After seven months of Delta ownership I could still not really recommend it, though I have more than a passing idea that a good secondhand Delta — whose teething troubles are over — would be a pretty good buy in later years. Even so, I still do not understand how it ever came to be a Car of the Year, unless 1980 was a singularly bad year for cars . . .

But — paradox again — the Delta is by no means a bad car. Dynamically, it is a good one. It corners well, rides well, responds well — though I still think it is sluggish for its size, with a disappointing lack of low-speed pulling power. Even so, its point-to-point performance is more than just acceptable, and I have never regarded the car as anything but safe and comfortable.

And yet — and yet: in all those 30-odd weeks, never once did I feel the least compulsion to jump into the Delta and go for a drive just for the fun of it. More than one test car, during that time, tempted me out for just that few miles more at the wheel (often when I *should* have been doing other things, like house repairs). When it was the Delta in the driveway, the woodworking got done! Yet some of those cars, by any objective assessment, did not share all of the Delta's qualities.

What they did have — and what the Delta lacks — is colour. Under bad road conditions it comes nearest to endearing itself to its driver, but most of the time it remains curiously anonymous. Its makers have given it so many virtues to offset its more obvious defects, but personality is not one of them. Perhaps it's the one quality that cannot be reproduced by computers or robots: one that results only from often fallible, but sometimes inspired, human brains and hands . . .

Group test: Honda Quintet, Ford Escort, Lancia Delta

Take five

Honda's new luxury hatchback will be in very limited supply. Will the wait be worthwhile or are there better and easier to come by rivals?

It is perhaps a sign of the times that even Germany, that pillar of strength of the European car community, is beginning to feel the chill wind of Japanese competition; for Britain, already accustomed to a high level of import penetration, it is no more than cold comfort to see stronger economies shaken by the same force that gripped us years ago.

Most European countries have by now arrived at some sort of agreement with Japan limiting the volume of exports; in Britain the level has been held at a more or less constant 10-12 per cent of the total car market. But while the political sensitivity of Japanese car imports has remained at the same high pitch, the nature of the invasion is beginning to change. Models such as the Mazda 323, the Colt Mirage and of course most of the Honda range have achieved a level of sophistication equal to that of the best European cars and are now selling on their own considerable merits rather than merely as superficially attractive gap-fillers.

The result of these restrictions has been a gross imbalance between demand and supply: Honda estimate that they could sell their whole Civic allocation four times over, for instance. But Honda's allocation has been pruned still more sharply, down from 23,000 cars last year to a mere 14,000 for the whole of 1981. So with waiting lists for Honda cars running into years rather than months the commercial logic of introducing yet another new model is questionable, to say the least.

But the new model is here, and at £4900 the five-door Quintet hatchback certainly appears to make economic sense alongside its opponents. Though the body is all new, the Quintet makes use of the Civic/Prelude 'building brick' floorpan and is powered by the familiar 1602cc overhead cam four used in the Accord and the Prelude coupé. In fact, the Quintet is Honda's fourth car in this sector.

Honda see the Quintet not as a strong seller – hardly surprising since their hands are tied on the sales front – but push it as a car to "elevate the company's image in society". Cynics, meanwhile, see a discreet move up-market behind this grandiose facade, to an area of higher profits on a given car volume.

Waiting lists are bound to build up rapidly as only 4000 Quintets are to be brought into the UK this year. They key question is, therefore, is the Quintet worth waiting for? Is it worthwhile, in the words of the company's current advertising campaign, pulling strings to get one?

Here we put the matter to the test and compare the Quintet with some capable and award-winning opposition – the Lancia Delta and the Ford Escort, Europe's past two Car of the Year winners and prime runners in the luxury small hatchbacks market that Honda are out to contest.

THE CARS

January's price cut brings the Lancia Delta back below the £5000 barrier to within a whisker of the Quintet's £4990; the 1500cc Delta and the 1600 Honda are the only versions of their ranges available in the UK. By comparison, a veritable plethora of Escort permutations is offered, running from the basic 1100 three-door at £3723 to the sporting XR3 at £5692. Closest in appeal to the Delta and Quintet's concept of sophisticated medium-hatchback motoring is the Escort 1.6 Ghia – the top model in the range but priced, following a recent increase, a rather daunting £586 above the Quintet, at £5576.

Despite the differences in their origins, all three are very similar in engineering. All are, of course, transversely engined front-wheel-drive designs; all have rack and pinion steering, all-independent suspension by MacPherson struts all round, and servo-assisted disc/drum braking. Even in engine design the similarities continue, though the Ford's advanced CVH unit has the advantage of maintainence-free hydraulic tappets. Ford are the only manufacturer not to go along with the trend to five-speed gearboxes, however.

The Honda sits on the shortest wheelbase but substantial front and rear overhang makes it the longest overall and gives it a rather bulky look: arguably the most attractive is the compact but elegant Lancia – longest in wheelbase but shortest in overall length. The Escort comes midway between these extremes but is the lightest overall even in its plush Ghia form, though the imposing list of optional extras fitted to the test car must have added noticeably to its weight.

Included in the Honda's price is an impressive catalogue of standard equipment including an electrically operated sunroof, radio and five-speed transmission. Hondamatic semi-automatic transmission costs an extra £295.

The Escort is only available as a four-speed manual, the Delta as a five-speed; the Escort's Ghia trim pack runs to a radio-cassette and manual sunroof, plus a liberal dosage of chrome trim on the exterior.

PERFORMANCE

While none is intended as a full-blooded sports saloon, all three set out to provide the high standards of performance now associated with well-appointed mid-range cars with an up-market image to maintain. And in this respect the Escort must be judged the most successful, despite having the least powerful engine; though the margin between it and the Lancia is so small as to make little effective difference on the open road.

The Honda, too, would be judged lively in less rapid company. Its overhead camshaft 1600cc engine gives some 80 bhp, running smoothly and cleanly – apart from some hesitation when the carburettor's second choke

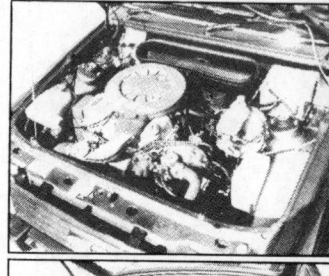

FORD ESCORT 1.6 GHIA

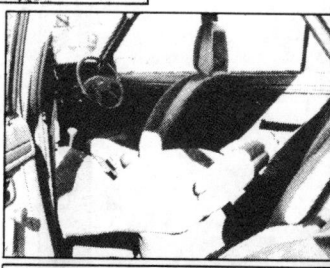

Escort's CVH engine gives good performance but is less refined than some rivals; luggage space is big but sill is high and seat is not split. Interior (below) is the roomiest and driving position is good though wheel is close to taller drivers' knees

comes in – under most conditions. It is far from being quiet, however, and 4000 rpm heralds the start of an annoyingly boomy phase of its rev range.

The Quintet reaches a respectable maximum speed of around 97 mph but is not as quick off the mark as either the Ford or Lancia. It reaches 60mph in 12.3 sec, well within the compass of third gear, but is noisy in the process; little is gained by revving the engine beyond its 5300 rpm power peak.

The engine pulls cleanly, if not particularly strongly from low revs; the gearchange is better than the Delta's but not as good as previous Hondas', with the shift from fourth to fifth and back being occasionally tricky. The clutch, however, is unbelievably light – just as if its cable had snapped.

At speed the Quintet is not particularly quiet, despite its high gearing; the engine roar makes it necessary to turn the radio up. At lower speeds it is more relaxed.

The Lancia Delta immediately feels sweeter and more refined than its rivals, the 85 bhp development of Fiat's Strada engine being worlds apart in terms of liveliness and all-round sparkle. The gearchange is rather stiff but its five close-set ratios allow the Delta driver to make best use of the engine's zestful, puncy character – not that low-down pulling power is in any way lacking. Only at very high speeds does the Delta's engine show any harshness, but again, it is only truly quiet when cruising at a steady speed.

The Escort feels – and is, by a slender margin – the most rapid performer. Its advanced power unit does not perhaps have the refinement expected of a freshly developed design, but its power characteristics are excellent. There is plenty of low-down urge, making frequent use of the good four-speed gearbox unnecessary, and there is sufficient mid-range power to achieve a very quick 0-60mph time of 11.3 sec (fractionally slower than the slightly lighter estate tested in March) with little effort. It is not a smooth engine, however, and its harshness increases noticeably beyond a rather modest 5000 rpm threshold.

HANDLING AND RIDE

Two European winners here, but the Japanese Honda is certainly no loser. Its all-independent suspension is softer than the Ford's or Lancia's, and gives a surprisingly comfortable ride on good or average road surfaces; the Quintet feels to be a very smooth-riding, comfortable car – until poorer roads are encountered. Only then do the shortcomings of the damping begin to be felt, in the form of jarring and a general restlessness.

The Quintet's handling, however, is much better than standard Japanese fare. The rack and pinion steering is much firmer and more direct than before, and successfully avoids the excessive lightness and vagueness once the hallmark of Japanese systems. Steering response is more immediate, too, and the Quintet is positive and enjoyable to handle, despite some body roll. Understeer is well tempered – again, unusual for a Japanese car – but the other side of this coin is a slightly unnerving tail-end lightness at high speed.

There is little to choose between

HONDA QUINTET

the high standards set by both the Delta and the Escort. The Italian car has the edge in ride and general refinement, the Anglo-German car that in steering and cornering.

Effective damping rather than hard springing makes the Delta's ride seem initially firm but it proves to be a very pleasant compromise between comfort over all types of road and an enjoyable precision and stability of handling to please the enthusiastic driver. Our only quibble is with the steering: at 3.7 turns between locks it is reasonably high geared, but suffers for some reason from excessive lightness around the centre position and transmits an unwelcome amount of front-drive tugging when the car is powered through tight bends.

Much criticism has been levelled at the Escort's ride, tending to mask the favourable comments on its handling. Our Ghia model appeared to ride better than other versions we have tested, and on its wider tyres displayed terrific levels of roadholding – almost up to the standard of the sporting XR3. Low-speed heaviness is a consequence of the steering's high gearing, but this is more than outweighed by its excellent precision and the responsive handling that it generates. Though the damping was on occasions caught out and could produce a surprise tail-end hop or bounce, the Escort is an enjoyable car to drive fast. Less enjoyable for passengers, however, who comment unfavourably on the hardness of the ride – though again the Ghia version appears to be less of an offender.

ACCOMMODATION

With a relatively long wheelbase in a compact body, the Lancia Delta has all the ingredients of a thoroughly space-efficient small car. But though the tape measure shows it to be the equal of the slightly bulkier Escort in cabin length, subjective assessment does not bear this out.

Both the Escort and the Delta will make the Honda buyer feel short-changed, however. Inside the Quintet's long body is a disappointingly cramped passenger compartment, little more spacious than that of the five-door Civic with which it shares the basic floorpan pressing and certainly no improvement over the Accord models with which it overlaps in Honda's curiously structured range.

The front seats are thin and not particularly comfortable; there is a reasonable amount of rearward seat travel but the very limited tilt adjustment cannot hide the serious lack of headroom, caused by the low roofline and the electric

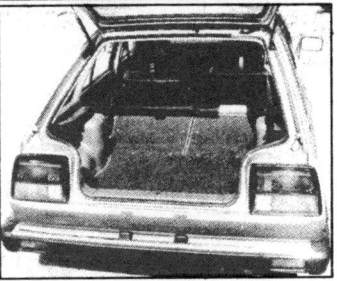

Quintet engine is smooth and very economical but becomes boomy at high revs. Lamp clusters obstruct good boot. Quintet's main drawback is poor interior space, though handling and ride are improved. Seat adjusters are complex but electric sunroof is standard

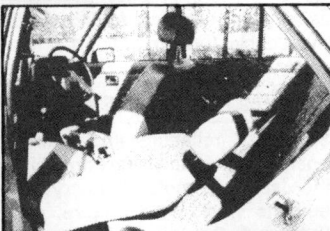

sunroof's intrusion. Drivers of six foot and over can write the Quintet off their lists unless they are willing to risk a cricked neck or adopt a Formula One reclining driving position. In the rear it's the same story: thin seats, poor headroom and cramped knees – especially with a tall driver at the wheel.

The Lancia's plush interior is close-fitting rather than tight; the front seats are soft and well padded, and have plenty of front-rear adjustment; the driving position is better than in many Italian cars, though the steering wheel is unnecessarily large in diameter. Front headroom is better than the Honda's but still not quite enough for a tall person; in the rear it is almost as restricted as the Japanese car. Rear knee room is better, too, but the only car of the three to pass the 'six footer behind a six footer' test is the Escort. Rear seat space on the Ford is noticeably less restricted, though at some cost in the front seats' rearward adjustment, and there is sufficient headroom both front and rear. The front seats themselves are firm but support well, and prove comfortable – apart from the thick headrest pads.

However, the smallish steering wheel is uncomfortably close to the driver's knees, making manoeuvring tricky for the taller driver.

Both the Honda and the Lancia have practical luggage loading arrangements, with split folding rear seats and a low sill to the hatchback tailgate. The Honda's is less effective in practice, however, as only the seat backs fold forward, the Lancia having a double fold action for more space.

LIVING WITH THE CARS

Simplicity, reliability and ease of operation are the virtues on which the Japanese motor industry has built its world-wide empire, and the Honda Quintet is of course no deserter of the tradition.

A manual choke ensures easy cold starting and a stable warm-up tickover, and the engine displays few vices – apart from a peculiar hesitation on pick-up just after a period of over-run. Steering improvements make for easier and more precise driving in crowded traffic conditions and the clutch and gearbox are delightfully light and smooth in use. However, a handicap in town and on angled junctions are the four side roof pillars aft of the driver which, though thin when viewed from the wide, are quite deep and add up to a substantial blind spot when looking diagonally backwards.

Wind noise on the move is less than on the Delta, for instance, but we found that the otherwise practical radio aerial, retracting conveniently into the windscreen pillar, generated a lot of whistle when in the up position.

Thanks to the high fifth gear the engine's irritating boom period does not occur until well beyond the motorway speed limit, but the Quintet is still not quite as quiet as some of its rivals at speed.

Gadget-minded owners will find paradise in the Quintet. Apart from the convenient electrically operated sunroof (surprisingly free from buffeting), the digital clock and the interior door mirror adjuster, there is also a remote bootlid opener, a headlamp levelling knob (oddly given pride of place in front of the gearlever) and a useful car 'plan' in the instrument panel which indicates if any of the five doors are not properly closed. The heating and ventilation system is versatile and very effective. However, many of the minor controls are scattered and illogical in their positioning, and the sophisticated-looking seat adjustment controls cannot disguise the basically poor design of the seating and the lack of space inside the car.

The Delta has much more flair and one is immediately conscious of being in something different and altogether more classy. A lot of functions are crammed into the compact instrument layout which is on the whole clear, though the seven switches on the crowded central bank are too close together and too poorly marked to be easy to hit correctly first time. The heating and ventilation are the poorest of the three cars. The Delta is very well provided with shelves, ledges and lockers, though these run quite close to the passengers' knees. On the move the Delta feels the most vivacious and nimble of the three, and is generally the most fun to drive.

The Escort's well-finished but rather characterless interior is probably the best planned of the three: instruments are clear and reasonably plentiful, heating is effective (ventilation being further aided by the standard sunroof) and a useful feature is the variable screen wiper delay knob. However, the minor switchgear is concentrated along the bottom edge of the facia, where it is poorly labelled, poorly lit and frequently obscured by the steering wheel. A further annoyance is the lack of audible warning for the indicators.

LANCIA DELTA

The Escort feels lively on the road, but is not as spirited as the Lancia – despite what the stopwatches say. Its steering is very much heavier – especially around town – and it has the feeling of a bigger, heavier car. Adding to the big-car feel are up-market extras such as central door locking and electric windows. A much-needed tailgate wash/wipe system, standard on the other two, is a surprising ommission from the luxury Ghia Escort's specification, and costs a full £79 extra.

As practical, working vehicles none comes up to the high standards set by the Vauxhall Astra/Opel Kadett: though both the Quintet and Delta have usefully low loading sills and practical split folding rear seats, the width of the hatch is limited in the Honda by the tail light units and in the Lancia by the suspension towers. The Escort has more space, but a high sill and a one-piece seat back.

Lively Lancia is the most enjoyable to drive; Fiat-derived engine is smooth and sweet. Boot space is restricted but seats have practical split action. Interior is plush and attractive, with thoughtful detailing. Switches are too close together, however

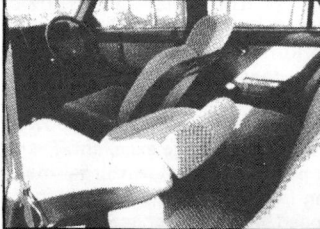

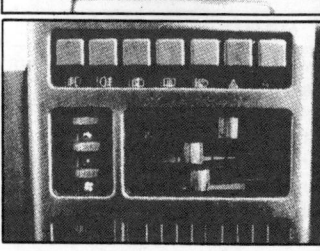

Group test: Honda Quintet, Ford Escort, Lancia Delta

COSTS

Despite the uncertainty of its exact country of origin due to Ford's multinational manufacturing policy, the Escort still has most of the advantages of a domestically produced car and as such must win the costs race hands down.

It is by no means the most economical on fuel – that honour belongs to the five-speed, overdriven Honda – nor is it particularly cheap to buy in the first instance. But it will be easier and cheaper to service and repair (1.7 hrs labour time every 12,000 miles as opposed to the Lancia's 4.0 every 12,000 and 0.5 for each 6000 lubrication service and the Honda's rather frequent 7500 miles) and it is bound to be slower to depreciate – even in expensive Ghia guise. Insurance wil be less costly, too, but we are still surprised that the new Escorts are grouped so highly.

The Honda requires more and costlier servicing than the Delta, but counters with excellent 32mpg fuel economy – on two star – as opposed to the Italian car's hard-driven 28 and the Escort's disappointing 26, both on four-star. The Honda's potential for economy could be greater, as we were able to record over 33 on a motorway run into a headwind.

An important point in the Honda's favour is its price: at £5195 it is good value by any standards and offers many more features than any of its near rivals.

VERDICT

If hearts were to rule minds, there would be no question about our choice: the seductive charms of the sophisticated Delta would have us hooked. But concern for the economics rather than the aesthetics of driving would lead into the safe, predictable and well managed Ford camp; yet even a cursory glance at the respective specifications and prices would have the Honda head the list.

As it is, none of the three is totally satisfactory. The Honda, in particular, came as a major disappointment. It *is* well priced, well equipped and well built, but it is just too cramped for four normal-sized adults. More importantly, it lacks the extra flair and sophistication that has for so many years kept Hondas that key step ahead of their Japanese rivals – and that much closer to the European standard. We think that the four year old Accord is still the better car: it has more room inside, and the Quintet's only advantages are its sunroof and its better steering –

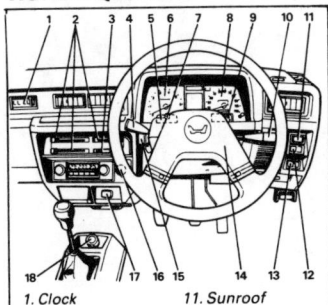

HONDA QUINTET

1. Clock
2. Heater controls
3. Heater fan
4. Wipers
5. Temperature
6. Rev counter
7. Hazard warning
8. Speedometer
9. Fuel gauge
10. Indicators/dip/flash/lights
11. Sunroof
12. Heated rear window
13. Rear fog light
14. Horn
15. Rear wash/wipe
16. Cigarette lighter
17. Choke
18. Headlamp levelling

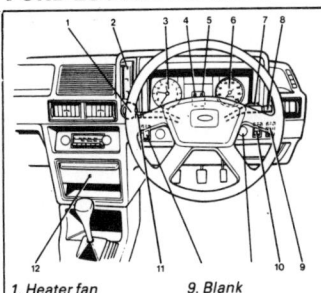

FORD ESCORT 1.6 GHIA

1. Heater fan
2. Heater controls
3. Rev counter
4. Fuel gauge
5. Temperature
6. Speedometer
7. Lights
8. Wipers
9. Blank
10. Heated rear window
11. Indicators/dip/flash/horn
12. Cigarette lighter
13. Rear fog lights
14. Wiper delay

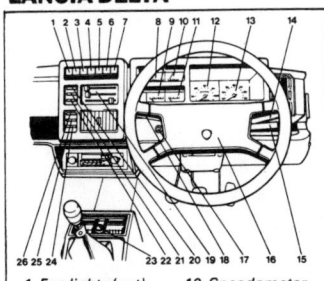

LANCIA DELTA

1. Fog lights (opt)
2. Rear fog lights
3. Rear wash/wipe
4. Heated rear window
5. Headlamp wash (option)
6. Hazard warning
7. Warning lights test
8. Oil pressure
9. Fuel gauge
10. Temperature
11. Voltmeter
12. Speedometer
13. Rev counter
14. Lights/dip/flash
15. Indicators
16. Horn
17. Trip reset
18. Wipers
19. Heater controls
20. Panel lights
21. Blank
22. Heater fan
23. Cigar lighter
24, 25, 26. Vent controls

plus the extra pair of doors.

The Ford is the most sensible choice – it has the most room inside and will be the cheapest to run – but it is pricey in Ghia form. And despite its strategically planned luxury trappings it doesn't convey that air of genuine chic and style that make the Italian Delta so very attractive.

CAR	Ford Escort 1.6 Ghia 5-door	Honda Quintet	Lancia Delta
PRICE	£5576	£5195	£4995
Other models	18 saloons, 7 estates	none	none
Price span	£3723-5692	—	—
PERFORMANCE			
Max speed (mph)	101	97	98
Max in 4th (mph)	—	92	92
Max in 3rd (mph)	83	74	71
Max in 2nd (mph)	55	48	52
Max in 1st (mph)	35	30	35
0-30 (sec)	4.0	4.0	3.6
0-40 (sec)	5.8	6.1	5.7
0-50 (sec)	8.5	8.8	8.0
0-60 (sec)	11.3	12.3	11.6
0-70 (sec)	15.1	17.0	15.9
0-80 (sec)	19.8	24.2	22.9
0-400 metres (sec)	18.4	18.7	18.5
Terminal speed (mph)	78	74	74
30-50 in 3rd/4th/5th (sec)	6.8/10.6/—	6.1/8.9/10.9	6.1/7.5/12.4
40-60 in 3rd/4th/5th (sec)	6.0/9.7/—	6.0/8.8/12.0	5.5/8.5/11.0
50-70 in 3rd/4th/5th (sec)	6.1/9.0/—	8.7/9.6/13.2	5.8/8.9/12.9
SPECIFICATIONS			
Cylinders/capacity (cc)	4/1596	4/1602	4/1498
Bore x stroke (mm)	80 x 79	77 x 86	86 x 64
Valve gear	ohc	ohc	ohc
Compression ratio	9.5 to 1	8.4 to 1	9.2 to 1
Induction	1 single choke	1 twin choke	1 twin choke
Power/rpm (bhp)	79/5800	80/5300	85/5800
Torque/rpm (lbs/ft)	92/3000	93/3500	90.4/3500
Steering	rack/pin	rack/pin	rack/pin
Turns lock to lock	3.7	3.3	3.7
Turning circle (ft)	34		34
Brakes	S/Di/Dr	S/Di/Dr	S/Di/Dr
Suspension front	I/McP	I/McP	I/McP
rear	I/McP	I/McP	I/McP/Ta/TL
COSTS			
Test mpg	21.5-27.7	28.7-33.2	28.3-28.9
Govt mpg City/56/75	30.4/47.1/36.7	30.4/47.9/34.4	28.4/40.3/30.2
Tank galls (grade)	9.0(4)	11.0(2)	10.0(4)
Major service miles (hours)	12,000(1.7)	7500(N/A)	12,000(4.0)
Parts costs (fitting hours)			
Front wing	£37.10(-)	£41.98(N/A)	£46.61(5.6)
Front bumper	£31.51(0.3)	£35.65(N/A)	£96.37(3.5)
Headlamp unit	£28.00(0.3)	£29.24(N/A)	£38.59(0.25)
Rear light lens	£18.49(0.3)	£21.35(N/A)	£4.40(0.15)
Front brake pads	£19.23(0.7)	£12.08(N/A)	£14.62(0.6)
Shock absorber	£27.31(0.7)	£45.61(N/A)	£31.80(1.1)
Windscreen	£33.70(0.9)	£58.99(N/A)	£73.66(1.25)
Exhaust system	£53.30(0.6)	£111.32(N/A)	£83.47(2.15)
Clutch unit	£43.41(2.2)	£93.59(N/A)	£46.98(4.0)
Alternator	£54.55(0.5)	£161.94(N/A)	£63.57(0.6)
Insurance group	5	5*	5
Warranty	12/UL+ Extra Cover option	12/UL	12/UL+ 6-year anti-corrosion
EQUIPMENT			
Alloy wheels	no	yes	no
Automatic choke	yes	no	yes
Automatic transmission	no	£310	no
Five-speed gearbox	no	yes	yes
Electronic ignition	yes	yes	yes
Central locking system	£163	no	no
Load-adjustable headlights	no	yes	no
Electric windows	£142	no	no
Laminated screen	yes	yes	yes
Tinted glass	£37	yes	yes
Petrol cap lock	£5	yes	yes
Adjustable steering column	no	no	yes
Radio	yes	yes	no
Split rear seats	no	yes	yes
Seat tilt adjustment	no	yes	no
Sunroof	yes	electric	no
Rear wash-wipe	£79	yes	yes
Headlamp wash-wipe	£75	no	no
DIMENSIONS			
Front headroom (ins)	35.5	33-34	35
Front legroom (ins)	35-40	34-41	36-42
Steering-wheel-seat (ins)	12-19	11-19	14-21
Rear headroom (ins)	34.5	32	32
Rear kneeroom (ins)	27-33	25-32	26-31
Length (ins)	159.8	161.8	153
Wheelbase (ins)	94.4	92.9	97.4
Height (ins)	52.6	53.4	54.3
Boot Load height (ins)	30	26	25
Overall width (ins)	62.5	63.6	63.8
Int. width (ins)	51	51	53
Weight (cwt)	17.5	17.6	19.4
Payload (lbs)	1003	1199	1050
Boot capacity (cu. ft)	20/48.7	N/A	9/35

KEY. Valve gear: *ohc*, overhead camshaft; **Steering:** *rack/pin*, rack and pinion; **Brakes:** *Di*, discs; *Dr*, drums; *S*, servo assistance; **Suspension:** *I*, independent; *McP*, MacPherson struts; *Ta*, trailing arm location; *TL*, transverse link; * estimated.

AutoTEST Lancia Delta automatic
Relaxed and refined

Lancia Delta Automatic
Lancia Delta announced in October 1979, and won 1980 Car Of The Year award. Giugiaro-styled five-door hatchback shares 1,498 c.c. engine used in the Strada 85 but in slightly higher compression form, with 90 lb. ft. torque at 3,500 rpm instead of 87 lb. ft. at 3,800 rpm. Peak power is 85 bhp (DIN) on both, but produced at 6,200 rpm on the Lancia instead of 5,800 rpm on the Fiat. Suspension is by MacPherson struts front and back, and transmission is either five speed manual or AP Beta based three speed automatic.

PRODUCED BY:
Lancia and Co, Fabbrica Automobili S.p.A
via Vincenzo Lancia 27
Turin 10141
Italy

SOLD IN THE UK BY:
Fiat Auto (UK) Ltd.
Great West Road,
Brentford,
Middlesex TW8 9DJ

AFTER THE delightful little Fulvia went out of production in 1976, Lancia did not produce another small car until they had been absorbed into the Fiat empire. The Delta came in late 1979 (and won the 1980 Car of the Year award). Its aspirations were very different; not a car of overtly sporty character, but one aiming to fulfil any need for a small market car while retaining some of the dynamic character of previous designs. The Automatic tested here was introduced in May this year.

The engine is common to the Fiat Strada Super 85; a belt-driven OHC unit measuring 86 x 64mm (1,498 c.c.) and delivering 85 bhp (DIN) at 6,200 rpm and 90 lb.ft.torque at 3,500 rpm, (against 85 bhp and 87 lb ft. at 5,500 and

3,700 rpm respectively in the Strada), mated to either five-speed manual or three-speed automatic transmission – designed and developed by AP but made mainly in Italy – with torque converter; a closely similar unit to that fitted in the Beta series.

Assuming no torque converter slip, overall gearing is virtually the same as in the manual car; 17.6 instead of 17.9 mph per 1,000 rpm on the 165/70-SR13 Michelin XZX tyres fitted to the test car. In other respects the Delta is unchanged. It is all Lancia; built on a 97.4in. wheelbase (1in. longer than the Strada), but is 2in. shorter overall. MacPherson struts are used for front and rear suspension; with bottom end location by wishbones at the front and twin transverse links plus forward facing tie bars at the rear. There are anti-roll bars front and rear.

The square edged, high tailed five door coachwork is the work of Guigiaro's Ital Design. Nothing has changed inside or out since its introduction. In our first test we commented that the Delta was a commendably refined car, a quality that probably owes something to the fact that the Delta is no lightweight. The road test car tipped MIRA's scales at precisely 20 cwt, distributed 61.3/38.7 front to rear.

Performance
Good for an auto

Flooring the throttle brings the automatic choke into operation. Start up proved immediate at all times, but we did occasionally find a tendency to stall on selecting Drive, and a hesitancy on driving away unless the engine was allowed to run for a moment or two before moving off. In other respects warm driveability was good.

Clearly those who opt for the automatic will not be expecting standards of acceleration available in the manual version, nevertheless the Delta automatic by no means disgraces itself against the manual opposition. It is one of the faster automatics in the class, this in spite of the test car having covered just 1,000 miles when the performance figures were taken, admittedly in near perfect weather conditions. Left in Drive, full throttle upchanges occured at 40 and 70 mph (5,500 and 5,800 rpm), a little way below peak power. In this mode the Delta went to 30, 60 and 90 in 5.3, 14.7 and 49.6 sec. Bearing in mind the free revving nature of the engine (and a 6,500 rpm red line) it came as no surprise to find that by using that limit and changing gear manually, an improvement in acceleration resulted from 50 mph onwards, a saving that grew linearly from 0.3 sec at 50 mph to 2.6 sec by the time the car had reached 90 mph. But even using this technique, the automatic proved some 3.2 seconds slower to 90 mph than the manual test car, having also taken 1.8 and 2.5 sec longer to reach 30 and 60 mph, than the five-speed Delta.

With this in mind it came as a pleasant surprise to find the Delta Automatic having a mean top speed of 96 mph, just 1 mph slower than its manual counterpart. However, on the open road (particularly when overtaking) and in the cut and thrust of urban road driving, the Delta Automatic is noticeably less accelerative.

Full and part throttle upchanges are smooth enough. The gearbox responds with commendable speed when looking for a downchange, however when full throttle is used, changedown quality is rather jerky.

On the test car kickdown was available from Top to 2 at 62 mph and 2 to 1 at 35 mph. There is no kickdown facility available from Intermediate to Low; a feature designed in by AP to provide low driving torque, and therefore minimal wheelspin when starting on ice and snow. If required such a change has to be executed manually. The T-handle selector is correctly gated, with free movement only possible between Drive and 2, thus preventing the inadvertent selection of Neutral or Low. The selector is precise if rather sticky in action.

Economy
No worse

Possibly because the Delta Automatic was used for more than the usual proportion of the test period on long runs we were able to exactly match the fuel consumption obtained with the manual. The overall of 26.6 mpg is about what one might expect from a small automatic and takes in interim brim figures ranging between 24 and 28 mpg. and we imagine only gentle drivers would be likely to obtain better than 30 mpg. The fuel tank holds 10 gallons giving the Delta a rather meagre safe range of 220-250 miles between fill ups. On the plus side the filler comfortably accepts full pump flow and brims to the top without delay. No change in the oil level occured during the test period.

Noise
Mainly engine

With most cars tending to have lighter bodywork these days (and often low profile tyres) one has come to expect higher levels of road induced roar, particularly on the coarse concrete motorway surfaces. The Delta is an exception to this trend. Road induced tyre roar is very well subdued, and there are no body boom periods, a quality that makes one rather more aware of the noticeable hum emanating from under the bonnet once the Delta is being pushed along at anything over the legal limit or when the car is being accelerated hard. Judging from the solid manner in which the dooors shut one can attribute the moderate levels of wind noise to effective sealing, though at anything above 80 mph even this does not prevent the distant rustle growing into an obvious noise source.

That said, the impression still remains of an above averagely refined and very solid small car with moderate levels of bump thump and no shakes and rattles over potholes and broken surfaces.

Road behaviour
A Lancia

It is in its handling and roadholding that the Delta displays some of the qualities one has come to expect of the marque. Nevertheless a few quirks do mar an otherwise dynamically very rewarding little car. With 3.8 turns of the wheel covering an unusually assymetric turning circle – 38 ft to the right

Rear shot emphasises shallow rear window. Hatch window wipe is standard (as are side turn indicators), while front and rear lower bumper cum apron sections are moulded in deformable material

Badging and grille follow a traditional Lancia theme. Bumpers have moulded rubber inserts, and there is a small lip spoiler incorporated in the front apron. Both door mirrors are internally adjustable

Delta is notable for neatness of rear light assemblies. Heaviness of rear apron is broken by six holes

There is no mistaking the Delta's sharp lined styling and multi holed cast aluminium wheels

compared with 33 ft to the left — the steering is low geared, and one finds that right angle bends require a surprising amount of wheel twirling to negotiate. At low speeds, when the loads through the steering mechanism are at their highest there is a noticeable degree of lost motion about the straight ahead position. Interestingly, this apparent lack of response becomes less evident as the loads lighten at higher speeds. The steering self centres from large lock angles quite smoothly but we did notice a rather odd variation in effort both when applying and revolving relatively small lock inputs particularly on gentle acceleration and deceleration when driving torque should not have been having any significant torque effect.

For most purposes though, the steering is nicely weighted and as implied it becomes noticeably better behaved as the speeds increase. Cornering roll is extremely well controlled and the Delta can be hustled through bends with the minimum of fuss. An enthusiastic driver will feel no unease at driving the car consistently at and around its limit of adhesion, since it is ultimately so well balanced. The wider than average for the class 165/70 aspect ratio Michelin XZX tyres give the car comparatively high levels of grip. At the limit of adhesion the Delta can be classed as an unwilling and very moderate understeerer. Backing off in mid-slide causes no erratic behaviour whatsoever, merely an immediate restoration of neutral balance and a tightening of line. One ex-

Hinging forward bonnet reveals crowded engine bay. Accessibility is nevertheless good with battery and fuses tucked next to nearside inner wing, a high mounted distributor, see through reservoirs, and a plainly visible wiper motor on offside scuttle panel

pects good straight line stability from front wheel drive cars, but the Delta is expecially good in this respect.

Such a high degree of handling control and vicelessness does however seem to have been paid for in some aspects of the Delta's ride. Worked hard over the larger undulations and variously cambered B road surfaces the suspension copes well, without coarse inputs, and with an overiding feeling of tautness and control. Yet there are times when quite ordinary looking but rippled urban inner city roads produce an unexpected bumpiness in the car. This "hard tyred" feel (tyre pressures were checked) is also occasionally present on some of the rougher concrete motorway surfaces when any sharp surface imperfection such as a surface joint produces an unusually jerky input inside the car, and can also generate a rubbery tremor in the shell.

As befits a car of the Delta's pedigree the brakes work well. On first acquaintance they seem a little over sensitive at the lower end of the scale with a 20lb pedal load sufficient to produce quite strong 0.45g retardation. Thereafter the response curve flattens out and it takes a hefty 80lb load to achieve the best crash stop deceleration, of 0.95 with the front brakes on the point of lock up.

Fade resistance is excellent — the Delta managing to run through *Autocar's* speed related fade test (10 consecutive 0.5 stops from 71 mph in this case), with only a small rise in pedal efforts halfway. As in most of today's braking systems the Delta has a pressure limiting valve to prevent premature rear locking when braking under certain load conditions. It also seems to provide the Delta with remarkably good brake balance, when panic stops from high speeds are made. Under such conditions it was very difficult to lock front or rear brakes, the car stopping straight and true with all four tyres scrabbling for grip — but not locked.

The handbrake works conventionally on the rear wheels. It managed a useful 0.3 retardation on the flat, however when facing down the 1 in 3 test slope, we encountered an all too common problem in heavily forward weight biased cars of insufficient traction between the lightly laden rear wheels and the road surface, resulting in the car sliding slowly downwards with rear wheels locked hard on. The Delta would handbrake park facing up the same slope and on being called upon to restart the Delta slowly and surely climbed to the top without wheelspin or fuss. Being an automatic, the car could of course be "mechanically" parked facing either way.

Behind the wheel
Sensible planning

Of the currently available Italian small cars, the Delta is probably least afflicted by an "Italianate" driving position. It would be nice to see the wheel adjustable for reach as well as rake (and a little more rearward seat adjustment) but drivers of most heights found they could obtain an acceptable seat-wheel-pedal relationship, though one complained of a lack of lumbar and lateral support — and of minimal headroom with the backrest set in the upright position.

The wheel itself is a rather nice four-spoke one, with horn press occupying the entire centre section. The facia is of neat "modular" design. Speedometer and revcounter are housed together on the right, leaving the split section on the left containing four separate instruments dealing with water temperature, oil pressure, fuel contents, and battery volts, plus a strip of warning lights on the far left. A minor criticism is that a tall driver selecting a low wheel angle cannot see the top two ancillary dials.

Minor controls are well laid out; the smaller right hand stalk

Modular instrument binnacle boasts neatly housed speedometer and revcounter (on right), oil pressure, water temperature, fuel contents gauges and battery voltmeter plus warning lights. Centre panel is topped by seven push-button switches (auxiliary lamps, heated rear window, etc) with heater controls, and two directionally controllable facia vents beneath. Stalk switches look after wipe/wash on left, main light switch and signalling on the right. Pressing any part of steering wheel centre crash pad operates horn. The "turn to zero" type trip reset button can be seen mounted on facia to left of this pad. T-handle selector incorporates "lift up" detente release

Shallow glovebox is complemented by generous facia trays and door bins. Piping for mounted side window demist vents can be seen

Rear seat passengers have comfortable seats with just tolerable knee-room, but headroom rates well below average, and door openings are comparatively narrow

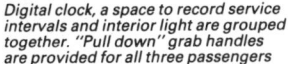

Digital clock, a space to record service intervals and interior light are grouped together. "Pull down" grab handles are provided for all three passengers

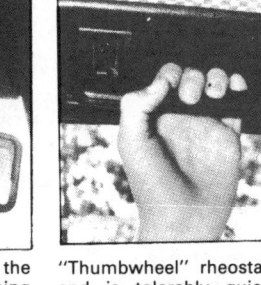
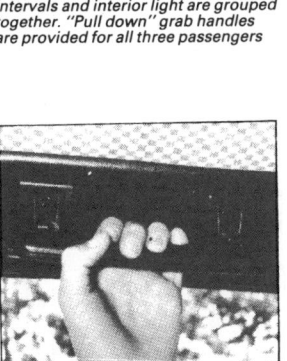

looks after signalling, and the larger one main lights switching and dip/main beam operation, which leaves the left stalk operating screen wipe/wash. Ancillary light and function switches (push button type) are tidily arranged in a row in the centre of the facia and immediately above the heating and ventilation controls — all within easy reach.

Three sliding and four turn wheel controls plus a multitude of vents are provided. When the system is set to run cold, a good flow of fresh air can be had but this is from the centre vents only, and fan assistance is normally required since there is very little ram flow available below the legal limit. The fan operates via a "Thumbwheel" rheostat control and is tolerably quiet except when running fast.

In spite of its apparent sophistication, the system lacks the crisp temperature and directional control we have come to expect in the better air blenders. The test car also exhibited rather poor flow to the footwell area — even with the facia vents switched off. Both the centre and outer facia vents are directly temperature sensitive, leaving the system with no independent source of fresh air whatsoever — a bad point.

Forward visibility is good, and with wipers that have been converted to a righthand drive pattern there is no problem in wet weather. In contrast the view out to the rear is frankly poor, thanks to the thick rear quarter pillaring and very shallow rear hatch window, a situation that is only slightly mitigated by two excellent internally adjustable door mirrors. One also feels the car simply lacks glass area by current standards, leading to some commenting on a rather claustrophobic feel about the interior, particularly when the car was filled with passengers.

Living with the Delta

The space provided for travelling paraphernalia on the other hand is remarkably generous. An admittedly small glovebox is complemented by large door bins and deep moulded facia trays running either side of the centre console. The Delta also has several useful features; reading lights for the rear passengers a digital clock mounted between the sun visors adjacent to a strip displaying recommended tyre pressures and a space for pencilling mileage for the next service. Each passenger has a pull down grab handle. The test car was fitted with a Voxon SW/MW/LW stereo radio. Reception and reproduction rated as better than average.

As a load carrier the Delta has a reasonably low rear sill and the flexibility of a split rear seat. A hinging parcel shelf covers items behind the rear seat when used as a four seater, and the entire rear load area is carpeted.

The spare wheel is contained conventionally under the rear floor leaving the jack and a six piece tool kit to be found separately behind the inner rear quarter panel trim panels.

The bonnet pull is on the right — another example of Lancia's consideration in executing the right hand drive conversion. The forward hinging bonnet props upright to expose a fairly crowded, if well laid out engine bay, with good access to dipstick, reservoirs, battery, fuses and distributor. Service intervals are at every 6,000 miles.

The Delta range

The Fiat/Lancia combine do not import the 1300 into Britain, which leaves us with only manual and automatic versions of the Delta 1500. They cost £5,429 and £5,867 respectively.

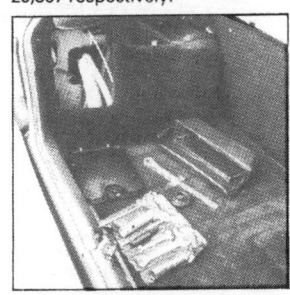

Rear floor is fully carpeted and the Delta has a split rear seat facility. Rear suspension intrudes noticeably into load area. Cubbyhole in inner rear wheel arch contains tools

HOW THE LANCIA DELTA AUTOMATIC PERFORMS

Figures taken at 892 miles by our own staff at the Motor Industry Research Association proving ground at Nuneaton. All Autocar test results are subject to world copyright and may not be reproduced in whole or part without the Editor's written permission.

TEST CONDITIONS:
Wind: 2 mph
Temperature: 21 deg C (70 deg F)
Barometer: 29.8in. Hg (1,012 mbar)
Humidity: 65 per cent
Surface: dry asphalt and concrete
Test distance: 666 miles

MAXIMUM SPEEDS

Gear	mph	kph	rpm
Top (mean)	96	153	5,450
(best)	97	154	5,500
2nd	82	132	6,500
1st	49	79	6,500

ACCELERATION

FROM REST

True mph	Time (sec)	Speedo mph
30	5.3	31
40	7.5	42
50	10.3	52
60	14.0	64
70	19.4	75
80	27.1	85
90	47.0	95

Standing ¼-mile: 20.2 sec, 71 mph
Standing km: 37.2 sec, 84 mph

IN EACH GEAR

mph	Top	2nd	1st
0-20	—	5.0	3.5
10-30	—	6.3	3.9
20-40	—	7.2	4.5
30-50	—	7.4	—
40-60	—	6.9	—
50-70	—	9.8	—
60-80	—	—	—
70-90	30.0	—	—

FUEL CONSUMPTION

Overall mpg: 26.6 (10.7 litres/100km)
5.8 mpl

Constant speed

mph	mpg	mpl	mph	mpg	mpl
30	45.7	10.0	70	27.4	6.0
40	43.3	9.5	80	23.5	5.2
50	39.3	8.6	90	20.0	4.4
60	32.3	7.1			

Autocar formula: Hard 23.9mpg
Driving conditions Average 29.3mpg
Gentle 34.6mpg

Grade of fuel: Premium, 4-star (97 RM)
Fuel tank: 10.0 Imp. galls (45 litres)
Mileage recorder reads: 3.8 per cent long

Official fuel consumption figures
(ECE laboratory test conditions; not necessarily related to Autocar figures)
Urban cycle: 24.9 mpg
Steady 56 mph: 36.1 mpg
Steady 75 mph: 27.2 mpg

OIL CONSUMPTION
(SAE 20/50) negligible

BRAKING

Fade (from 71 mph in neutral)
Pedal load for 0.5g stops in lb

	start/end		start/end
1	24-40	6	40-60
2	24-44	7	40-56
3	24-36	8	40-34
4	24-38	9	40-48
5	20-40	10	40-34

Response (from 30 mph in neutral)

Load	g	Distance
20 lb	0.45	67 ft
40 lb	0.55	55 ft
60 lb	0.75	40 ft
80 lb	0.95	32 ft
Handbrake	0.30	100 ft

CLUTCH
Max. gradient: 1 in 3

WEIGHT
Kerb, 20.0 cwt/2,240 lb/1,017 kg
(Distribution F/R, 61.3/38.7)
Test, 23.4 cwt/2,625 lb/1,192 kg
Max. payload, 1,049 lb/475 kg

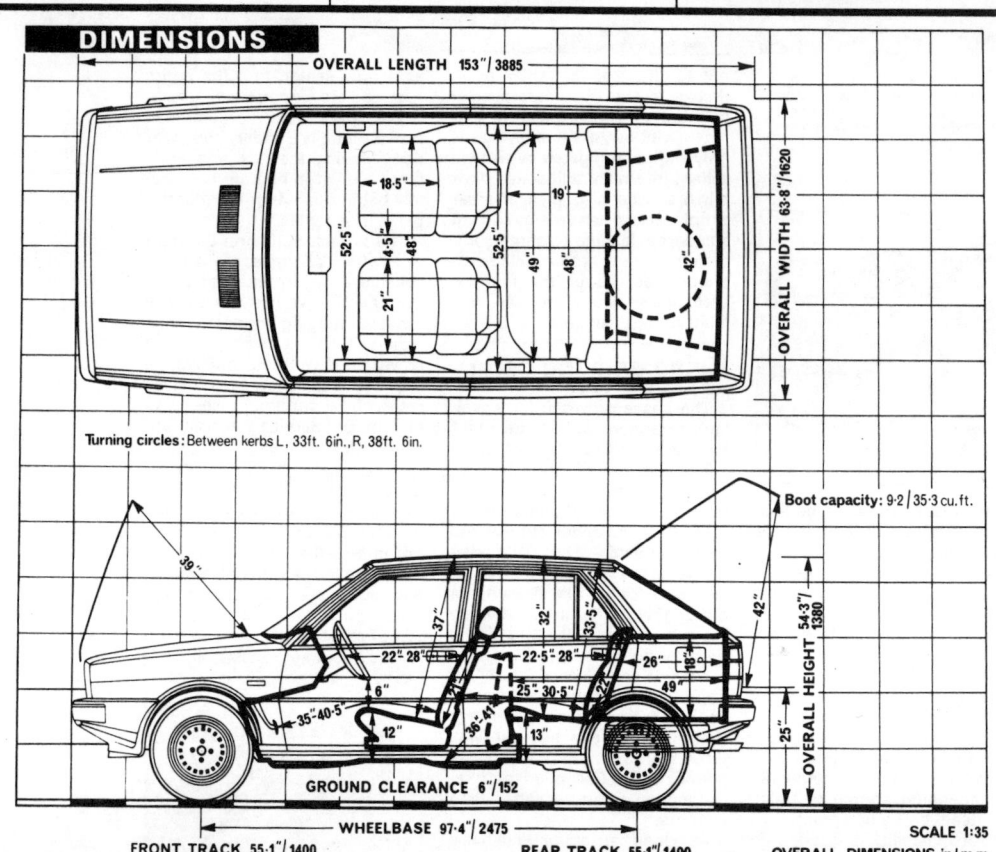

PRICES

Basic	£4,710.00
Special Car Tax	£392.50
VAT	£765.38
Total (in GB)	**£5,867.88**
Seat Belts	Standard
Licence	£80.00
Delivery charge (London)	£75.00
Number plates	£15.00
Total on the Road (exc. insurance)	**£6,037.88**

EXTRAS (inc. VAT)
*Alloy wheels £179.40
Metallic paint £83.47
Stereo radio cassette £100.70
*Fitted to test car

TOTAL AS TESTED ON THE ROAD £6,217.28
Insurance Group 5

SERVICE & PARTS

Change	Interval 6,000	12,000	24,000
Engine oil	Yes	Yes	Yes
Oil filter	Yes	Yes	Yes
Gearbox oil	No	No	No
Spark plugs	Yes	Yes	Yes
Air cleaner	Yes	Yes	Yes
Total cost	£27.25	£93.12	£108.62

(Assuming labour at £15.50/hour inc. VAT)

PARTS COST (including VAT)
Brake pads (2 wheels)—front £18.70
Brake shoes (2 wheels)—rear £14.88
Exhaust complete £168.44
Tyre—each (typical) £44.25
Windscreen (laminated) £92.26
Headlamp unit £49.36
Front wing £56.88
Rear bumper £123.33

WARRANTY
12 months' unlimited mileage

SPECIFICATION

ENGINE
Front, front-wheel drive
Head/block All alloy/cast iron
Cylinders 4, bored block
Main bearings 5
Cooling Water
Fan Electric
Bore, mm (in.) 86.4 (3.40)
Stroke, mm (in.) 63.9 (2.52)
Capacity, cc (in.³) 1,498 (91.41)
Valve gear Ohc
Camshaft drive Toothed belt
Compression ratio 9.2-to-1
Ignition Breakerless
Carburettor Weber 34 DAT
Max power 85 bhp (DIN) at 6,200 rpm
Max torque 90 lb ft at 3,500 rpm

TRANSMISSION
Type Three-speed automatic, with torque converter

Gear	Ratio	mph/1000rpm
Top	0.83	17.6
2nd	1.16	12.6
1st	1.94	7.5

Final drive gear Helical spur
Ratio 4.38 to 1

SUSPENSION
Front—location Independent MacPherson struts
—springs Coil
—dampers Telescopic
—anti-roll bar Yes
Rear—location Independent, MacPherson struts, transverse and trailing links
—springs Coil
—dampers Telescopic
—anti-roll bar Yes

STEERING
Type Rack and pinion
Power assistance No
Wheel diameter 15¼ in.
Turns lock to lock 3.8

BRAKES
Circuits Twin diagonal split
Front 8.94 in. dia. disc
Rear 7.28 in. dia drum
Servo Vacuum
Handbrake Centre lever working on rear drums

WHEELS
Type All alloy
Rim width 5 in
Tyres—make Michelin XZX
—type Radial ply
—size 165/70-13
—pressures F 26, R 26 psi (normal driving)

EQUIPMENT
Battery 12V 45Ah
Alternator 45A
Headlamps 110/120W
Reversing lamp Standard
Electric fuses 14
Screen wipers 2-speed plus intermittent
Screen washer Electric
Interior heater Air blending
Air conditioning not available
Interior trim Cloth seats, cloth headlining
Floor covering Carpet
Jack Screw scissor
Jacking points One each side
Windscreen Laminated
Underbody protection Pvc, wax injected bo sections + Crylagard protection, 6-year anti-rust warranty subject to annual inspection

HOW THE LANCIA DELTA AUTOMATIC COMPARES

Lancia Delta Automatic £5,867
Front engine, front drive
Capacity 1,498 c.c.
Power 85 bhp (DIN) at 6,200 rpm
Weight 2,240 lb/1,017 kg
Autotest 10 July 1982

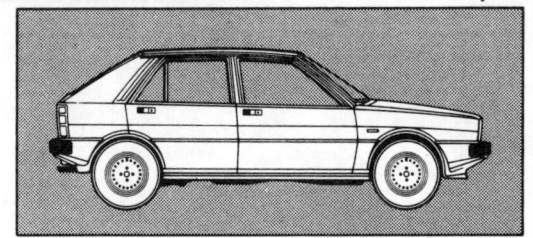

Citroen GSA Pallas (A) £5,251
Front engine, front drive
Capacity 1,299 c.c.
Power 65 bhp (DIN) at 5,500 rpm
Weight 2,184 lb/992 kg
Autotest GSA manual 5 January 1980

Honda Accord 3-door Hondamatic £5,500
Front engine, front drive
Capacity 1,602 c.c.
Power 80 bhp (DIN) at 5,300 rpm
Weight 2,093 lb/949 kg
Autotest Honda Accord EX 4-door 20 February 1982

Talbot Horizon 1.5GL (A) £5,414
Front engine, front drive
Capacity 1,442 c.c.
Power 83 bhp (DIN) at 5,600 rpm
Weight 2,240 lb/1,017 kg
Autotest 9 January 1980 Horizon SX

Vauxhall Astra 1.3SGL (A) £5,770
Front engine, front drive
Capacity 1,297 c.c.
Power 75 bhp (DIN) at 5,800 rpm
Weight 2,166 lb/985 kg
Autotest Astra L Estate manual April 1980

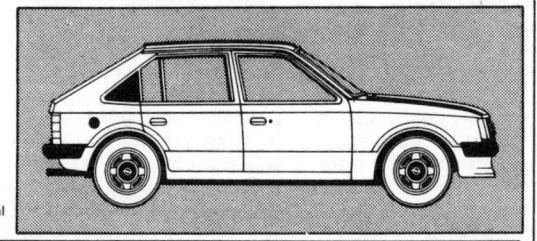

Volvo 345GL £5,446
Front engine, rear drive
Capacity 1,397 c.c.
Power 70 bhp (DIN) at 5,500 rpm
Weight 2,233 lb/1,013 kg
Autotest 9 February 1977

MPH & MPG

Maximum speed (mph)
* Vauxhall Astra 1.3SGL	99
Talbot Horizon GL(A)	97
Lancia Delta (A)	96
* Citroen GSA Pallas	95
+ Honda Accord 3-door (A)	91
Volvo 345GL (A)	87

Acceleration 0-60 (sec)
* Vauxhall Astra 1.3SGL	13.2
Lancia Delta (A)	14.0
Talbot Horizon GL (A)	14.1
* Citroen GSA Pallas	14.9
+ Honda Accord 3-door (A)	15.0
Volvo 345GL (A)	17.0

Overall mpg
* Citroen GSA Pallas	28.8
+ Honda Accord 3-door (A)	27.1
* Vauxhall Astra 1.3SGL	26.7
Lancia Delta (A)	26.6
Talbot Horizon GL (A)	26.5
Volvo 345GL (A)	26.0

* Test data for manual versions
+ Test data for Honda EX 4-door

Other small automatic hatchbacks worthy of comparison in a mainly manual transmission dominated sector of the market, would be the Mazda 323 1500 (£5,049), Colt 1400GLX (£5,509), Fiat Strada Super 85 (£5,468) and Golf 1500 (£5,824).

As it is the lack of test data has meant including figures for a couple of manual cars.

In fact the Delta stands up well in this company (particularly in relation to the manual transmission Citroen — the least powerful car here). Of the automatics the Italian is a point faster to 60 mph than the Horizon, if marginally slower in top speed. Indeed the only performance figures that stand out as being significantly worse than the rest are those for the Honda and Volvo, the latter in particular lacks top speed and acceleration thanks to its relatively poor power to weight ratio, nor is the Swede renowned for its fuel consumption, whereas the Honda does at least do somewhat better than average in this important respect.

ON THE ROAD

A pretty viceless bunch — ultimately — but not without their quirks. The Horizon for example has irritatingly low geared steering — a criticism that applies to a lesser extent on the Lancia, while similarly geared systems on the Vauxhall and Volvo are significantly more responsive about the straight ahead. Steering apart the Delta and Astra are likely to appeal most to the enthusiast since by comparison with the other fwd cars mentioned here, understeer and cornering roll is least evident on these two, and much less than one encounters in the Citroen.

All here are naturally straight line stable (though the Volvo is rather side wind sensitive), but the Lancia and Citroen are particularly so. The latter's hydropneumatic suspension easily gives it the best ride quality, and of course the most useful self levelling capability. Lancia ride has a tendency towards nobbliness on some surfaces (the Italian is rather worse than even the Volvo or Honda in this aspect of the ride), but this is perhaps balanced by its excellent braking capabilities — a quality shared by the Honda. By contrast, Horizon and Volvo brakes are lighter to work, and have rather more tendency to front lock in emergency. For heating and ventilation, top honours should go to the Honda and Vauxhall, whose air blending systems are generally more effective and versatile than the Delta's (in spite of the supposed Saab influence). Similar systems in the Citroen and Horizon lack crisp temperature control. Lancia, Volvo and Horizon air blenders need rather better fresh airflow.

SIZE & SPACE

Legroom front/rear (in.)
(seats fully back)
Volvo 345GL (A)	41/38
Vauxhall Astra 1.3SGL	39.5/37.5
Lancia Delta (A)	40.5/36
Talbot Horizon SX (A)	40/36
Citroen GSA Pallas	39/35.5
Honda Accord 3-door (A)	41/32

Not only does the Honda have the least rear seat legroom, but the Accord is only available as a three-door so those looking for a five-door Honda will have to opt for the Quintet. Knee and headroom are also lacking in the Delta, but for most purposes, it and the rest provide quite tolerable accommodation for four persons and their luggage, with the Volvo and Astra clearly the most spacious.

The Citroen, Volvo, Horizon and Vauxhall have most behind rear seat space, but as a load carrier it is the Vauxhall that wins thanks to its floor height rear sill. On the other hand this car does not have the Honda or Lancia's split rear seat back facility. Not so good is the Lancia's rear visibility as a result of its shallow hatch window.

VERDICT

The Citroen has superb ride quality (quite unaffected by the load carried) and self levelling suspension plus good steering and stability. But if top quality roll free ultimate behaviour is a priority (surely unlikely for those who have opted for an automatic in the first place) then the Lancia has much to recommend it. It is also above averagely refined thanks to low levels of road generated noise. However we would still rate the Astra as having equal handling, better ride and more interior room. Honda's "Hondamatic" semi automatic clutchless transmission works well in practice — a car with no obvious faults if somewhat less room than the opposition. The Volvo is roomy and. comfortable, and for those who prefer rear drive it is an obvious choice, though its "stepless" transmission means high revving rather fussy urban driving. Taking that into account the choice seems to lie between the Lancia, Vauxhall and Citroen. If accleration is important the Lancia and Vauxhall are the main protagonists. And of those two the Vauxhall (or Opel equivalent) would seem to have more than the Delta to recommend it.

Road & Track Impressions

LANCIA D

Ital by design... Fiat by nature! The Italian job that tries to be different...

DARING to be different, particularly when it comes to styling, has always been the Lancia hallmark. Similar to Alfa Romeo, some of their claim to fame has also been earned through motor sport. Consequently, when the new Lancia Delta arrived in Poole, we were possibly expecting a little too much from this 1.5 litre hatchback.

The sporting image has mellowed to become a middle of the road, slightly up-market, medium-sized car incorporating an uprated FIAT Strada 1500cc, four-cylinder engine in the now traditional, transverse, front wheel drive configuration.

Styled by the legendary Giugiaro of Ital Design — the phrase used in the press handout — there is little to show that he made any special effort at the drawing board to produce anything other than a pleasant five-door hatchback. Admittedly, some of the trimmings of opulence are included in the Delta LX specification with sliding steel sunroof, electric front windows, headlamp washers and tinted windows as standard, but we were looking for something more — that touch of class and refinement to lift the Lancia above the run of the mill, mid-range hatchbacks.

Unfortunately, after more than a week of road and track testing, we could find nothing to shout about. The Lancia Delta LX is simply an extremely pleasant, modern car.

The 1.5-litre, overhead camshaft, Fiat power pack, uprated by the fitting of a twin-choke Weber carburettor, plus electronic ignition, develops 85 brake-horse-power at 5,800rpm and bustles its way up to just on 100 miles-an-hour. It takes all five gears in the slick-shifting gearbox to do so, with the 6,500rpm red line being reached in the first four ratios.

The engine sounds too busy and gives the feeling that a sixth gear, as an overdrive, would be useful to improve fuel consumption and reduce the amount of noise in the passenger compartment. From 70mph up to maximum speed, engine noise becomes increasingly obtrusive and both annoying and tiring on long motorway journeys. There was slight wind noise, caused mainly by the internally adjustable, door-mounted rear view mirrors.

First impressions are probably the most important in convincing a potential buyer that a vehicle is 'his' or 'her' type of car. Here the Lancia Delta should score quite well, as the driver settles into the firm but comfortable, velour upholstered seat. An impressive, easily read instrument panel is clearly seen above the spokes of the adjustable steering wheel. All-round vision is reasonable with the long, sloping windscreen pillars intruding slightly into what the wife called 'good parkability'.

which switch

A survey of the bank of push-button control switches set centrally on the facia had us searching in the owner's manual for the method of turning on the lights. We soon discovered that the steering column mounted protrusion had to be rotated to obtain both parking and headlights and *not* pushed, pulled or flicked up and down. The control switches were at the bottom of the list of favourites, especially the rear window wash/wipe which only operated when its button was pushed. There didn't appear any way of obtaining intermittent or even constant rear wiper operation — a real nuisance on wet motorways.

Anyway, with instruments and controls checked out, a twist of the ignition key and the motor burst into life. The cable-operated clutch proved light and smooth to use, as was the gear lever. However, a conscious effort had to be made to release the gear lever to allow it to self-centralise to select third and fourth gears, or else a second to fifth gear movement was easily and mistakenly made.

Throttle pressure was light, but there was a surprisingly long movement required on the brake pedal to operate the servo-assisted brakes. For size seven feet-plus, the 'suspended' control pedals proved no problem, but the little woman at home complained about having to keep her right foot in the air to operate both brake and accelerator pedals. However, she did compensate by saying that she liked the easy steering and all-round vision for parking, plus the smart styling.

Further praise was lavished on the interior trim and the many, useful oddment trays inside the car. When it came to doing the shopping, there was minor criticism about the size of the boot and the height of the tailgate; but the divided, folding rear seat gave extra luggage capacity, providing there wasn't a full complement of passengers.

Two of the finer points about the Delta are its suspension and handling. Four MacPherson struts provide all-round, independent suspension that copes exceedingly well with all normal road surfaces and, coupled with the wider than average 5in. rims shod with low-profile tyres, gives excellent cornering with a minimum roll.

The confidence inspired by the steering tended to make our test track drivers try that little bit harder on the corners and they came back with the report: 'Very stable when cornering at speed — no appreciable under- or over-steer.' Praise indeed.

A couple of minor complaints came to light during our test; the first being the method of draining rain from the sunroof. With the drainholes at the front, water becomes trapped at the rear when the car is parked facing uphill. This we discovered after hearing water sloshing about above our heads when driving off. On the first brake application, the water surged forward and breeched the gulley to pour into the car. The second point needing attention is the weak rear parcel shelf bracket. The test car's sheared.

The fact that suspension and steering encouraged brisk driving probably accounts for the overall 30mpg fuel consumption, although even at steady speeds of 30 and 70mph, the consumptions of 69.1 and 31.3mpg respectively, are not particularly frugal for a 1500cc motor.

Finally, great emphasis is made of the corrosion protection now being given to the Delta, where structural adhesive and plastic sealant are being applied to all flanged joints to stop moisture infiltration.

anti-corrosion

Then there's 'Zincrometal' pre-treated steel body panels, underside and wheel arch PVC coating with all closed box sections being sprayed with Cryla-Gard using no less than 32 different spray heads to reach the parts that other spray heads can't reach! What does all this mean? Simply a six year anti-corrosion warranty for the Delta.

Overall impressions of this Lancia is that it is an average 1.5 litre hatchback in terms of performance, but can bring out the beast in many by providing better than average roadholding and steering. The price of £5,965.05p. including tax, accommodates numerous luxuries including electric windows and sunroof, without outstripping the competition.

SPECIFICATION

Engine: Transverse mounted four-cylinder, 1498cc, belt driven ohc, water cooled. Max power 85bhp (DIN) at 5800rpm. Max torque 90.4 ft lb at 3500rpm. Compression ratio 9.2:1 (four star petrol). Weber 34 DAT 8 carburettor. Fuel tank capacity 10 gallons.
Transmission: Five-speed, all synchromesh gearbox. Cable operated, diaphragm spring clutch. Front wheel drive.
Suspension: All independent, MacPherson struts and anti-roll bars front and rear. Telescopic hydraulic dampers.
Brakes: Servo-assisted, 8.9in discs front and 7.3in drums rear with handbrake operating on the rear. Dual circuits.
Steering: Rack-and-pinion with 3¾ turns lock-to-lock. Turning circle 34.5ft between kerbs.
Wheels and Tyres: Pressed steel 5in rims shod with Michelin 165/70 SR 13 ZX radial ply tyres.
Electrics: Battery 45amp/hr. Alternator 45amp. 16 fuses. Two-speed wipers with intermittent wipe action.
Dimensions: Length 12ft 9in; width 5ft 3¾in; height 4ft 6¼in.

PERFORMANCE

SPEEDOMETER CORRECTION (MPH):

Ind	30	40	50	60	70	80
True	28	38	47	57	65	74

ACCELERATION FROM REST:

0-30mph	3.8sec
0-40mph	6.0sec
0-50mph	8.5sec
0-60mph	11.9sec
0-70mph	16.9sec

ACCELERATION ON THE MOVE:

	3rd	4th	5th
30-50	6.0sec	8.8sec	11.5sec
40-60	6.3sec	8.8sec	11.7sec
50-70	8.2sec	10.4sec	13.3sec

MAXIMUM SPEED: (Mean) 98.6mph. Best ¼, 100mph

MAXIMUM GEAR SPEEDS:

1st	30mph
2nd	49mph
3rd	69mph
4th	90mph

FUEL CONSUMPTION: (STEADY SPEEDS)

	4th	5th
30mph	58.9	69.1
40mph	47.5	52.3
50mph	38.9	43.8
60mph	33.1	36.8
70mph	28.4	31.3

Overall 30.0mpg

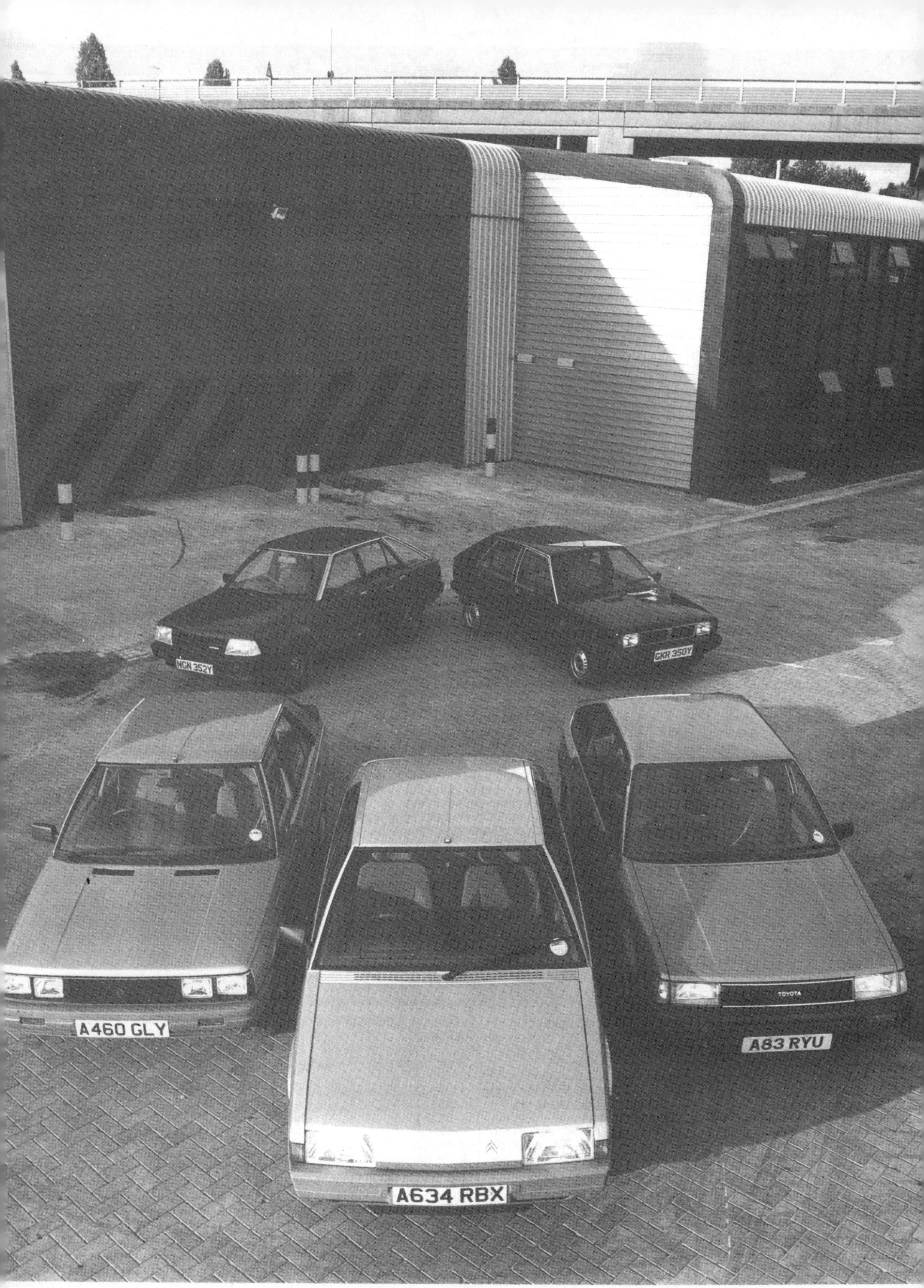
NGN 352Y
GKR 350Y
A460 GLY
A83 RYU
TOYOTA
A634 RBX

GROUP TEST: MID RANGE HATCHBACKS

Counter attack

Belated converts to front wheel drive, Toyota's new Corolla shows how fast the Japanese giants are catching up. But can they beat Europe's best?

Any car that sells well over ten million units in well under two decades just has to be taken seriously – by buyers, builders and commentators alike. But in the case of the Toyota Corolla – for several years the global best seller and hot on the heels of the VW Beetle and the Ford Model T to become the most successful model of all time – there has been a strange absence of universal recognition for the car, its qualities or its identity.

Perhaps the fault lies with the family Toyota's frequent changes of body style over the years: each of the four wholly different Corolla shapes since 1966 has lasted an average of under five years on sale, whereas rival eight-figure sellers have kept one single identity over a far longer time span. Yet despite the continual changes of style, each successive set of Corolla clothes has concealed identically depressing and conservative rear-drive mechanical elements – until now, and the fifth generation Corolla, that is.

With the latest Corolla Toyota have at last caught up with the field and have belatedly seen the sense of the space-saving front-wheel-drive layout that has been universal practice on rival models for a generation or more. The arrangement chosen is the conventional transverse engine installation with the five-speed gearbox to one side; a similar system is used on the bigger Toyota Camry as well as on scores of other small cars. Engine capacity remains at the ultra-competitive 1300 cc mark, and with an overall length well under 14ft the new Corolla is up against very much the same band of rivals as the old model; the difference is that with a much smarter body and an up-to-date technical layout the front-wheel-drive Corolla's appeal is at last likely to extend beyond the predominantly conservative customers who were the main buyers of the old model.

Two new Corollas are offered: a boxily shaped four-door saloon, plus the rather more attractive five-door hatchback which we here compare with similarly-priced imported models. Commercially the most significant of these is the Nissan, née Datsun, Stanza which regularly appears in the monthly sales top ten and is compared here in its £5995 1600cc five-door form, some £700 more than the £5307 Corolla GL; also in the bigger sales league is the Renault 11 which, though relatively recently launched, has just been fitted with a bigger 1700cc engine to form the GTX version. Surprisingly at £5500 this new version is actually cheaper than the top-specification 1400cc TSE model we tested in November, equipment differences accounting for the £500-odd price gap between the two versions.

The new BX marks something of a mainstream departure for the traditionally eccentric Citroen company: the £5451 14RE tested comes right in the middle of the five-car range and, with an engine capacity of 1360cc, is the model designed for maximum fuel economy. Less well known is our fifth contestant, the Lancia Delta: since the summer this attractively styled Italian former Car of the Year winner has been expanded into a three-model range topped by the rapid twincam GT and supported by the £4950 1300 which we include here.

Like the bigger Deltas the 1300 has all-round independent MacPherson strut suspension and rack and pinion steering: the cheaper car's powerplant, an adaptation of a Fiat unit, gives 78 bhp at 5800 rpm from its 1301 cc displacement and is mounted transversely driving the front wheels through a five-speed gearbox.

This arrangement is in fact shared by all four rivals, too: most powerful of the four is the 1600 cc, 81 bhp Datsun Stanza, again with all-round independent Macpherson strut suspension; a carbon-copy layout is followed by the new Toyota, albeit with a smaller 1295 cc engine giving 65 bhp at 5400 rpm. Both also have rack and pinion steering, now increasingly adopted by Japanese manufacturers.

The new Renault engine gives 80 bhp from its 1721 cc on a high 10.0:1 compression ratio; again with five-speed transmission, rack and pinion steering and servo disc-drum brakes, the 11GTX differs slightly in the suspension department where trailing arms and torsion bars at the rear complement the front MacPherson struts. More different still is the Citroen BX which, though conventionally equipped with transverse engine, five-speed gearbox and front-wheel-drive, uses a high-pressure hydraulic system to power its all independent hydropneumatic suspension and its four-wheel disc brakes. Power from the 1360 cc overhead cam engine, borrowed from the Peugeot and Talbot Samba ranges, is 72 bhp at 5750 rpm – enough to give the relatively large-bodied car an official fuel consumption figure of 51.4 mpg at a steady 56 mph, aided no doubt by the good air drag figure of 0.34.

The Datsun is longer still, yet despite a wind resistance coefficient of 0.38 achieves an identical 56 mph consumption figure to the Citroen's: this time a very high top gear is an important contributor. Smallest of all is the Delta, yet with rather low gearing it can manage only 48.7 mpg; the Renault, next smallest, has the biggest engine yet achieves the best official 56 mph fuel figure – 54.5 mpg, slightly ahead of the mid-sized Toyota's 53.3.

In terms of weight there's little to choose between the bulky Datsun at 18.7 cwt and the opposite extreme of the compact 17.6 cwt Renault: the performance and economy contests appear therefore to be wide open but, as we shall see, this is far from being the case in practice.

PERFORMANCE

RENAULT	●●●●
LANCIA	●●●
TOYOTA	●●●
DATSUN	●●●
CITROEN	●●

This is not exactly the class where one would expect to find a genuine performance hatchback, but the new Renault engine – the first overhead cam unit of their own design – is a good one and is responsible for transforming the 11 from a hard-worked also-ran to a pleasantly lively and enjoyable little car. The Delta gives surprisingly good results from its lively, but unpleasantly noisy little engine, beating the bigger and smoother Datsun by some margin; some way behind are the Toyota and Citroen with their small engines.

What makes the Renault such a quick car on the road is the smoothness and instant response available from its powerful and very torquey engine – illustrating perfectly the benefits of a big engine in a relatively small body. Even at under 25 mph in fourth (1500 rpm approx) the GTX pulls strongly, obviating any neeed to change down to overtake slower traffic; from 3000 rpm onwards there is a stronger rush of power for quicker acceleration still. All the while it is a silky smooth performer, an edge of bominess appearing only once the 4500 rpm mark is passed.

High engine speeds are strictly unnecessary on the 11, however, for as our performance figures so clearly show, flexibility at low rpm is very much better than the opposition. Yet using the full 6000 rpm red-lined on the rev counter the GTX will reach 60 mph from standstill in exactly 11 seconds and can run to 102 mph in fourth gear – a higher top speed than its rivals but still not quite as fast as one would expect a 1700 cc car to be.

Impressive on-paper performance is provided by the Delta's small but high-revving powerplant, yet in practice the car has to be driven uncomfortably hard to achieve these results and is disappointingly unrefined and noisy in the process. Low gearing allows a reasonable degree of flex-

GROUP TEST: MID RANGE HATCHBACKS

ibility in the intermediate gears, and the five ratios are well spaced for the kind of hard driving and frequent gearshifting that the car's character encourages. But, even making the appropriate allowances, it is still frustrating to have a second gear that won't even reach 50 mph and a third that will only just push the car past 70.

The Corolla's engine is not the last word in refinement, but at least it is reasonably flexible as well as smooth at high revs. There's still the throbbiness and boomy mid-range resonances typical of many Japanese engines, but the Corolla's is a decent, well-made unit designed for honest hard work and ease of operation. Around town the Corolla is responsive and likely to drive, aided in no small measure by its smooth driveline and outstandingly light five-speed gearbox. On motorways and fast A-roads it is not quite so much in its element, a downshift into fourth sometimes being necessary to help the small engine cope with upgrades or headwinds.

At the legal speed limit the Corolla is noticeably noisier than the Datsun – the quietest of the five – and, unlike the Citroen and Renault, becomes very much noisier beyond 85 mph.

The Stanza's 1600 cc engine pulls from low speed and revs satisfyingly and smoothly to deliver useful top-end acceleration. Overall it's a refined unit, with only some high-speed boominess to cause irritation. Extremely high gearing is a mixed blessing: on the one hand it makes for peaceful and economical cruising, on the other it takes the edge off the car's flexibility and makes necessary frequent use of the poor gearchange that is the weak link in the chain.

It is perhaps asking too much of a tiny 1360 cc engine to impart powerful performance in a big car such as the BX but, tuned to give 72 bhp and a decent spread of torque, Citroen's Peugeot-derived unit does quite well.

In gentle driving the BX appears pleasantly flexible, but for more urgent progress the engine has to be worked hard – often painfully so – to make the car move quickly enough. Relatively low gearing – as on the Delta – helps low-speed response, but a constant irritation and a reminder of the engine's 104 origins is the ever-present gearbox whine and the large amount of jerkiness and backlash in the transmission system.

Cruising at around 50-60 mph in the BX is acceptably quiet (though still noisier than the Datsun); at 70 mph is is a good deal noisier and, though sound levels do not increase very much above this speed, one would still wish for a higher fifth gear to provide a more restful regime.

Top speeds range from the easy hundred of the Renault and the eventual 100 of the Citroen, to the bare 95 attainable by the Japanese duo – in fourth, rather than top gear. Gearboxes, likewise, span the great divide between the Corolla's amazingly light and quick cable operated change (whose only fault is an annoying clunk at every shift) and the hit and miss mechanism of the Datsun which allowed one to become 'lost' in the gate to the right of fifth gear, and which crunched whenever a shift was taken quickly.

HANDLING AND RIDE

TOYOTA ●●●
RENAULT ●●●
CITROEN ●●●
LANCIA ●●
DATSUN ●

Although the overall results under this heading are on the disappointing side considering the strides made by smaller cars such as the Peugeot 205 and Fiat Uno, it is unusual to see a Japanese car at the top of the list.

The Corolla is not there entirely by default, either: its all-independent suspension, front-wheel-drive and rack and pinion steering give it surefooted, predictable and enjoyable handling as well as a ride that is good over most surfaces and which is certainly no worse than average under tougher conditions.

The type of surface which catches the Corolla out is one with sudden, isolated ridges such as a jointed concrete carriageway; these bumps jar through the car's structure surprisingly readily. Under most other conditions it is smooth-riding and comfortable, however, and even under hard cornering retains its composure.

Grip from the narrow tyres is better than one would expect, the firm rack and pinion steering and the roll-free cornering making the Corolla a responsive car to handle quickly. Breakaway, when it does occur, takes the form of a gentle sliding of both front and rear ends, though the tail can slide out more quickly under determined provocation.

Less satisfactory is the Corolla's straightline stability: though the steering retains its firm precision at all speeds, the body is badly deflected by sidewinds and truck slipstreams.

The Renault, too, suffers from some instability at speed but we found the 11GTX a much more comfortable proposition than the hardriding TSE we tested in August. The same speed and accuracy of handling are there, but for some reason the bigger-engined car succeeds in eradicating many more of the road's irregularities; travel, though still distinctly firm, is far from being rough.

One factor that the bigger engine and extra power do tend to exaggerate is the lightness of the 11's rear end: pushed hard into a corner and then backed off the throttle, the Renault can suffer a sudden rear-end slide – especially in the wet.

Hydropneumatic suspension assures the Citroen of a superbly

TOYOTA COROLLA, CITROEN BX, DATSUN STANZA, LANCIA DELTA, RENAULT 11GTX

CITROEN BX 14RE
Smaller engined Citroen has plentiful roadholding and superb ride but steering is very heavy and engine noisy when worked hard. Interior space is best of group but low build gives poor visibility and eccentric design offers no practical advantages. Hatchback boot is excellent, with low sill and clear floor to aid the loading of bulky and awkward items

DATSUN STANZA 1.6GL
Stanza is one of few modern cars with unsatisfactory handling, as shown by the sudden lurch and tailslide (above). Ride is reasonable but steering too heavy. Spacious interior is spoilt by poor front seats, though rear comfort is good; vague gearchange mars flexible and economical 1600cc power unit but high gearing at least makes for refined noise levels

TOYOTA COROLLA LIFTBACK GL
Front-drive Corolla's modern shape is attractive apart from ugly rear end: handling and roadholding are up to European standards, though smooth and economical engine needs more power for refined cruising. Gearbox is superb, interior design sober, and space adequate for four. Boot has low sill but seats don't fold flat, restricting luggage capacity

smooth and level ride over all manner of obstacles – almost as if the bumps weren't even there – but absorption of town-speed road mends, white lines and manhole covers is still not as effective as on some conventionally-suspended cars. High-speed stability is outstanding, with the solid steering providing tremendous precision – albeit at the price of extreme heaviness for cornering and, at the other end of the scale, parking; this heaviness is unacceptable on a medium-sized car and will see the Citroen struck off many a buyer's shortlist.

Where the BX represents a notable improvement over older Citroens is in its handling: no longer is there such exaggerated body roll on bends and, if one has the muscle to wrestle with the heavy steering, roadholding levels are high and handling responses rapid.

Just as our Renault unaccountably proved more refined than our previous test model, our 1300 Delta did just the reverse with a harsher ride, steering with less feel, and less taut road behaviour. Even so, the Delta remains an entertaining car to drive hard and corner quickly – despite steering that is too low-geared and vague around the centre point and an unforgiving firmness to the suspension system.

The Datsun, finally, is one of the very few modern cars whose handling we regard as unsatisfactory. Driven sedately, the Stanza sends the driver reassuring messages, with a smooth ride, firm steering and seemingly quick responses to the wheel. But in fact this impression of responsiveness is an indication of how quick the Stanza is – not to obey the driver's commands but to swerve, lurch, and ultimately slide out of control. What happens is briefly this: movement of the steering wheel turns the car into the corner fairly quickly – so far, so good – and the body takes on a couple of degrees of roll. But if the cornering speed is anywhere near moderate the car swerves, body roll turns into a nasty sideways lurch and the rear wheels lose their grip. If the driver is on the ball he or she may be able to correct the skid (at the risk of a similar swerve in the other direction); if caught unawares, a spin could result.

Even under normal conditions the Stanza's steering is unacceptably heavy, and knowledge of the way the handling can bite back with a vengeance takes away any confidence the owner might once have had in the rest of the car.

The brakes on our Renault were difficult to apply smoothly: they had poor response for gentle braking but could grab sharply when more pressure was applied to the pedal. On the Datsun, braking in the middle of a corner could cause the car to skid – even at modest speeds.

GROUP TEST: MID RANGE HATCHBACKS

The Citroen's power-assisted brakes are of course superbly responsive, but demand care.

ACCOMMODATION

CITROEN	●●●●●
DATSUN	●●●●
TOYOTA	●●●
RENAULT	●●
LANCIA	●●

For once a clear-cut contest: the two 'big' cars – Citroen and Datsun – offer enough clear advantage in interior space over the small contestants – Renault and Lancia – as well as the halfway-house Toyota to make them feasible five-person carriers rather than compact four seaters.

Best in relation to its overall length is the Citroen BX. Long and low, it provides its passengers with soft and supportive seating, excellent legroom front and rear, plus a reasonable amount of headroom for all four or five occupants. If there has to be a drawback it must be that the front seat backrests (conveniently adjusted via a knurled wheel on the front of the seat base) do not go upright enough for some preferences; the cheaper 14RE does not feature the convenient tilt adjustment found on the pricier 16TRS and clearance between the driver's legs and the steering wheel rim can thus be rather limited. In the rear the BX is quite simply excellent for its price.

The Stanza also provides a decent amount of space – albeit on a platform that is considerably longer than its opponents'. It is at the rear that the Datsun does best: the seat is well shaped and provides a good shoulder support, and both headroom and legroom and more than adequate for the class. In the front the Stanza is less satisfactory with uncomfortable seats devoid of thigh support, though fitted with lumbar adjustment. Headroom is good, but the seat does not slide back far enough.

The Corolla provides just enough interior space to accommodate a family of four – but little more. In the front the seats can accommodate a six-footer who likes a straight-arm driving style, there's very good headroom and the generous width of the cabin promotes a feeling of spaciousness. The seats themselves have the strong support in the centre of the back that has come to be typical of Toyota: they feel uncomfortable at first, but one soon begins to appreciate their shaping on longer runs. In the rear there is more room than the figures suggest, but the seat backrest is not as well shaped as the Datsun's or Citroen's, shoulder support is lacking and headroom is on the tight side.

As in our previous test, both Renault and Lancia have to be relegated to the compact class. The Renault, particularly in three-door form, feels very cramped indeed: its cabin is both narrow and short, though to some extent the unusual 'monotrace' front seat mountings help increase rear legroom by more than shows in our published kneeroom figures.

Front seats are quite comfortable but have limited rearward travel as well as Renault's peculiar "dentist's chair" tilt adjustment that does not suit everyone; the rear seat, too, is comfortable and well upholstered, but again lacks space. In the Lancia it is much the same story but with still poorer space in the rear, especially as regards headroom, plus uncomfortably hard backrests and poorly shaped bases to the front seats. An adjustable steering column is of limited consolation.

LIVING WITH THE CARS

TOYOTA	●●●●●
RENAULT	●●●●
LANCIA	●●●
CITROEN	●●
DATSUN	●●

In the face of the heavy-to-drive Datsun and Citroen and the numerous design irritations of the other two Europeans, the hundred percent hassle free Corolla is by a large margin the easiest car to live with in everyday service.

The Toyota interior is on the whole pleasantly trimmed, with attractive coarse cloth seats and door panelling plus a simple and conveniently laid out two-tone grey plastic dashboard. Instruments are confined to the bare minimum of speedometer, fuel and temperature gauges but are clearly presented; the bi-level heating and ventilation is simple but very effective, warm-up is quick and the driver's face level vents, mounted high up either side of the instrument dials, are a good idea.

The low-silled hatchback boot area scores well on practicality but for two complaints: the suspension towers intrude noticeably on the compartment's width and, because only the seat backrests fold down, the resultant load platform slopes upwards towards the front unless the whole seat base is removed from the car. Rear seat belts are a standard fitment, and the split seat backrest is of course a nowadays essential feature. The built-in FM mono radio performed poorly.

Automatic choke gives the Corolla easy starting and, unlike all four of its rivals, it drives smoothly from the word go. Gearchange, clutch, and brakes are all light in action, but the steering effort is higher than on many Japanese cars. Vision all round is much better than on the Citroen, making the Corolla a convenient town car. It is not always quiet, of course, but the honest, clean-running sounds of the engine are unworrying except at prolonged high speeds.

Renault have come a long way since the oddball ergonomics of their earlier cars, and the 11 must count as an utterly straightforward vehicle to operate. Dials, stalks and other controls are conventional, clear and simple, driving position is good and the heating/ventilation averagely effective. All that is missing is a sense of flair to the design. Plus points include a headlamp levelling control and a rev counter, but the rear window wiper has no wash facility and only operates while the switch is held down. The new GTX engine takes a surprisingly long time to warm up, too, and we were disappointed that the bigger-engined car was in fact less quiet at speed than the 1.4 TSE.

The Delta's interior is stylishly laid out and trimmed, but in everyday use the seats prove uncomfortable, the lack of rear seat room is frustrating and the high noise levels become strenuous. The boot, though small, is easily reached through the full-depth tailgate, and again split rear seats are a feature. An annoyance in town driving is the big steering wheel and its unnecessarily low gearing. The gearchange, too, is poor compared with the Corolla's.

Great efforts have been made by Citroen to woo more middle-of-the-road customers to the BX, and the company have certainly succeeded in making the new car unintimidating (if rather heavy) to drive. In its steering, gearchange and handling it is fairly simple: the only noticeable Citroen excesses come in the form of the powerful, sharp acting brakes and the outlandish minor switchgear which, though easy enough to fathom and operate, offers no real advantage over conventional stalks and switches. The revolving drum speedometer is rather lost in the predominantly blank instrument block.

Apart from its lack of a split rear seat the BX's loading area is excel-

TOYOTA COROLLA, CITROEN BX, DATSUN STANZA, LANCIA DELTA, RENAULT 11GTX

LANCIA DELTA 1300
Delta has smart styling and sharp handling but in 1300 form has little refinement to accompany lively performance: steering is low geared and suspension hard. Cabin is stylish but lacks space for tall passengers, especially in rear. Front seats are poor, noise levels high but small boot is convenient in use. Unnecessarily low gearing is a nuisance, too

RENAULT 11GTX
New, bigger engine for Renault 11 is smooth running, flexible and gives car good performance. High speed noise is disappointing but ride is better than earlier models. Monotrace seats ease cramped but plush interior: dashboard is plain and easily understood; boot space is adequate rather than generous and loading sill is inconveniently high

theory that a lazier big engine can give better economy than a heavily taxed smaller one; perhaps a GTX with more miles on the clock will do better. Yet supporting the big-engine philosophy, our 1300 Delta did less well at 28.0 mpg than the much faster twin cam 1600 which in October returned 28.7.

By some clever means Toyota have managed to have the Corolla classified group three for insurance purposes: this gives it an immediate advantage over the group four Citroen and a bigger bonus still over the group five Datsun and Lancia. In terms of servicing it is the Citroen, with its accent on minimal maintenance costs, that sets the pace with a mere 87 minutes workshop time required for each annual 12,000 mile service. The Datsun and Lancia require a major service every 12,000 miles and an oil change at the halfway stage: for the Renault and Toyota the mileages are 5000 and 10,000 miles.

In terms of initial value for money the BX emerges as a lot of car and equipment for the price, featuring not only electric windows but central door locking and a laminated screen, too: at £5500 the Renault 11GTX provides a big engine in place of the luxury equipment of the smaller TSE. We would judge the Toyota better value than the cheaper Lancia; the Datsun is quite a big car but is more expensive and little better equipped.

The provision of numerous plastic body parts on the Citroen may well be an added insurance against premature rusting: in any case the BX has a six-year corrosion warranty, as does the Lancia. The Renault has five years, the Japanese cars no cover at all. Datsun and Renault have over 400 dealers: Citroen and Toyota have over 200, Lancia under 150.

lent, with a low sill and a long load platform that is both tough, flat, and wholly unobstructed by any suspension intrusion. Self-levelling suspension ensures that the powerful BX lights remain correctly aligned even when there's a heavy load aboard.

In town the BX is not at its best. The steering is very heavy indeed (though the turning circle is quite moderate); the indicators don't self cancel and all-round vision is appalling.

Heavy steering also spoils the otherwise bland and inoffensive Datsun. Around town effort is intolerably high, and the poor gearshift quality makes matters worse. Instruments are big and plain, controls simple and storage space generous: the wipers even feature a variable delay intermittent setting.

Heater controls are over simplified into four pushbuttons and fan and temperature controls, resulting in insufficient versatility of settings; a single floor-mounted lever opens the boot when pulled and the fuel flap when pushed. The boot is big (though Datsun don't quote any claimed capacity) but is spoilt by a very high loading sill and side panelling which robs it of much useful width.

COSTS

TOYOTA	●●●●
CITROEN	●●●
DATSUN	●●●
RENAULT	●●
LANCIA	●●

The three 'modern' cars in this comparison all make impressive showings in the official fuel consumption tests, each averaging well over 41 mpg in the three-cycle aggregate. Yet in our admittedly impatient hands neither French car could better 30 mpg overall, and our consumption returns for all but the two Japanese models were frankly disappointing.

And surprisingly, it is the bigger of the two Japanese which tops the table, the Stanza's very tall gearing (and perhaps the timid driving its handling encourages) helping it to a good 35.4 mpg overall. The Toyota needed to be driven harder but even then averaged 34.2 and gave a peak of 35.7 mpg on a motorway run, signalling very good economy potential in restrained driving.

Only tenths of an mpg separated our 1300 cc BX 14's consumption from the bigger and faster 1600 car's 30.1, reflecting the need to drive the smaller car harder. Noticeable was the great difference between motorway returns of 35.7 mpg and town figures of under 24.

Particularly disappointing was the 27.6 mpg recorded by the Renault as this to some extent disproves the

VERDICT

TOYOTA	●●●●
RENAULT	●●●
CITROEN	●●●
LANCIA	●●
DATSUN	●

The Toyota is perhaps not quite as exciting as its attractive styling suggests, but for our money it is the best car in this group. Illustrating just how far Toyota have progresed of late, the Corolla is pleasant, practical, easy to drive and one hundred per cent hassle-free – some would say to the point of boredom: add to this list of virtues its fuel economy, low servicing demands and Toyota's good reputation for reliability and it becomes a keen candidate for top honours in a broader market where these passive qualities are high priorities for buyers.

The Renault and Citroen are poles apart in style but equally deserving of the customer's money. We had expected the 11GTX to offer more of a performance advantage over the

GROUP TEST: MID RANGE HATCHBACKS

CITROEN BX 14RE

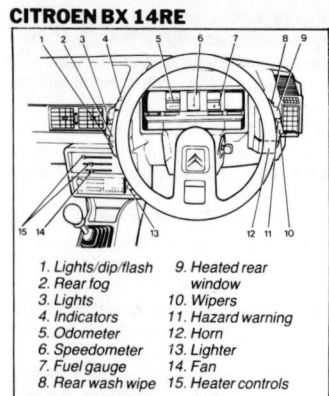

1. Lights/dip/flash
2. Rear fog
3. Lights
4. Indicators
5. Odometer
6. Speedometer
7. Fuel gauge
8. Rear wash wipe
9. Heated rear window
10. Wipers
11. Hazard warning
12. Horn
13. Lighter
14. Fan
15. Heater controls

DATSUN STANZA 1.6GL

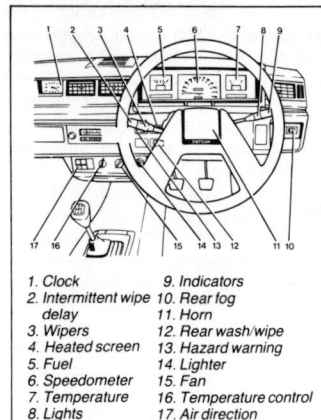

1. Clock
2. Intermittent wipe delay
3. Wipers
4. Heated screen
5. Fuel
6. Speedometer
7. Temperature
8. Lights
9. Indicators
10. Rear fog
11. Horn
12. Rear wash/wipe
13. Hazard warning
14. Lighter
15. Fan
16. Temperature control
17. Air direction

TOYOTA COROLLA GL

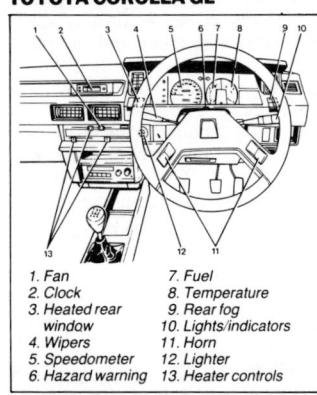

1. Fan
2. Clock
3. Heated rear window
4. Wipers
5. Speedometer
6. Hazard warning
7. Fuel
8. Temperature
9. Rear fog
10. Lights/indicators
11. Horn
12. Lighter
13. Heater controls

LANCIA DELTA

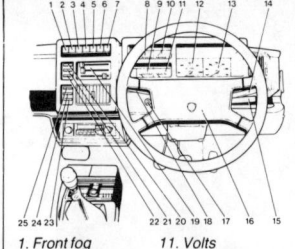

1. Front fog
2. Rear fog
3. Rear wash/wipe
4. Heated rear window
5. Headlamp wash
6. Hazard warning
7. Panel check
8. Economy
9. Fuel
10. Oil pressure
11. Volts
12. Speedometer
13. Rev counter
14. Lights
15. Indicators
16. Horn
17. Trip re-set
18. Wipers
19. Heater controls
20. Rheostat
21–25. Air vent controls

RENAULT 11 GTX

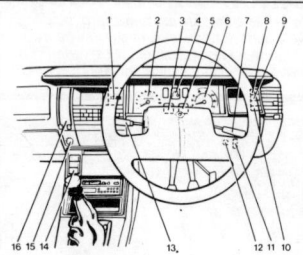

1. Digital clock
2. Speedometer
3. Temperature
4. Fuel gauge
5. Oil level
6. Rev counter
7. Wipers
8. Heated rear window
9. Rear fog lights
10. Rear wiper
11. Rheostat
12. Choke
13. Indicators/horn/dip/flash
14. Hazard warning
15. Heater fan
16. Heater controls

CAR	Citroen BX 14RE	Datsun Stanza 1.6GL	Lancia Delta 1300	Renault 11 GTX	Toyota Corolla GL
PRICE	£5451	£5995	£4950	£5500	£5307
Other models	3 hatchbacks	1 saloon / 3 hatchbacks	2 hatchbacks	6 hatchbacks	1 saloon / 1 estate
Price span	£4790-£6100	£5819-£6695	£4950-£5990	£4350-£5900	£5133-£5307
PERFORMANCE					
Max in 5th (mph)	100	93	99	100	91
Max in 4th (mph)	90	95	88	102	95
Max in 3rd (mph)	72	82	75	86	75
Max in 2nd (mph)	48	57	47	55	56
Max in 1st (mph)	29	30	28	30	29
0-30 (sec)	4.0	3.8	3.4	3.9	4.1
0-40 (sec)	6.6	5.9	5.2	6.0	6.0
0-50 (sec)	9.4	7.9	7.3	8.0	8.9
0-60 (sec)	13.5	11.5	10.9	11.0	12.6
0-70 (sec)	18.1	15.8	14.2	13.2	16.4
0-80 (sec)	25.8	21.6	18.8	17.8	23.9
0-400 metres (sec)	19.5	18.3	17.7	17.4	18.4
Terminal speed (mph)	73	72	75	76	70
30-50 in 3rd/4th/5th (sec)	6.3/10.0/15.6	6.2/11.1/14.8	5.5/9.7/12.5	4.6/7.8/10.8	6.7/12.5/12.0
40-60 in 3rd/4th/5th (sec)	6.9/8.9/12.9	6.8/12.2/17.1	6.0/9.8/15.2	4.9/7.5/19.1	6.8/10.5/13.0
50-70 in 3rd/4th/5th (sec)	7.0/9.9/13.4	8.2/13.0/19.6	6.7/10.6/15.1	5.8/8.3/10.7	7.4/10.4/14.7
60-80 in 3rd/4th/5th (sec)	—/11.6/16.5	9.8/13.9/24.7	8.2/11.4/17.0	7.7/9.8/14.0	10.1/12.6/17.7
SPECIFICATIONS					
Cylinders/capacity (cc)	4/1360	4/1598	4/1301	4/1721	4/1295
Bore x stroke (mm)	75x77	78x84	86x56	81x83	76x71
Valve gear	ohc	ohc	ohc	ohc	ohc
Compression ratio	9.3:1	9.0:1	9.5:1	10.0:1	9.3:1
Power/rpm (bhp)	72/5750	81/5200	78/5800	80/5000	65/5400
Torque/rpm (lbs/ft)	80/3000	96/3200	77/3400	100/3250	72/3000
Steering	rack/pin	rack/pin	rack/pin	rack/pin	rack/pin
Turns lock to lock	3.7	3.7	3.8	4.0	4.0
Turning circle (ft)	33	36	35	33	31
Brakes	P/Di/Di	S/Di/Dr	S/Di/Dr	S/Di/Dr	S/Di/Di
Suspension front	I/Wi/HP	I/McP	I/McP	I/McP	I/McP
rear	I/TA/HP	I/McP	I/McP	I/TA/TOR	I/McP
COSTS					
Test mpg	23.9-35.7	29.7-36.7	22.5-28.9	25.5-33.1	33.3-35.7
Govt mpg City/56/75	33.2/51.4/39.8	29.1/51.4/38.2	31.7/48.7/35.8	32.1/54.3/42.4	34.9/53.3/38.7
Tank galls (grade)	9.7(4)	11.9(2)	10.8(4)	10.3(4)	11(2)
Major service miles (hours)	12,000 (1.45)	6000 (2.0)	12,000 (3.3)	10,000 (1.5)	10,000 (1.8)
Parts costs (fitting hours)					
Front wing	£36.00 (0.7)	£66.39 (0.8)	£62.40	£36.40 (1.8)	£44.91 (1.3)
Front bumper	£49.00 (0.6)	£62.30 (0.7)	£138.84	£56.20 (0.7)	£49.41 (1.0)
Headlamp unit	£30.00 (0.4)	£69.73 (0.8)	£77.79	£21.50 (0.4)	£38.74 (0.5)
Rear light lens	£15.00 (0.4)	£41.58 (0.4)	£6.06	£14.50 (0.3)	£23.16 (0.3)
Front brake pads	£18.00 (0.4)	£13.28 (0.6)	£17.07	£16.90 (0.4)	£11.16 (1.0)
Shock absorber	£25.00 (0.4)	£27.24 (1.2)	£37.79	£33.20 (—)	£55.21 (1.9)
Windscreen	£49.00 (3.5)	£76.58 (2.5)	£104.83	£62.00 (1.0)	£57.70 (3.0)
Exhaust system	£79.00 (0.4)	£84.62 (1.4)	£159.70	£65.20 (1.3)	£104.39 (0.6)
Clutch unit	£49.38 (3.9)	£57.84 (3.2)	£49.00	£66.32 (3.8)	£49.95 (3.5)
Alternator	£49.00 (0.4)	£100.85 (0.6)	£82.26	£47.90 (0.5)	£61.37 (0.7)
Insurance group	4†	5	5†	4†	3
Warranty	12/UL	12/UL	12/UL	12/UL	12/UL
Anti rust	6 yrs	none	6 yrs	5 yrs	none
EQUIPMENT					
Alloy wheels	£203	no	no	yes	no
Automatic transmission	n/a	£1035*	n/a	£370*	£301
Five-speed gearbox	yes	yes	yes	yes	yes
Central locking system	yes	no	yes	no	no
Load-adjusting headlights	yes	no	no	no	no
Electric windows	front	no	no	no	no
Tinted glass	£98	yes	yes	no	no
Rev counter	no	yes	yes	yes	no
Adjustable steering column	no	no	yes	yes	no
Sound system	pre-radio	AM radio	none	pre-radio	FM radio
Seat height adjustment	no	no	yes	tilt	no
Rear seat belts	no	yes	yes	yes	yes
Split rear seat	no	no	yes	no	yes
Rear wash-wipe	yes	yes	yes	wipe	yes
DIMENSIONS					
Front headroom (ins)	36	36.5	35	35-36.5	38
Front legroom (ins)	35-43.5	34-40	37-43	32-39	35-42
Steering-wheel-seat (ins)	11-20	12-18	15-22	14-22	12-20
Rear headroom (ins)	34	34.5	32	34	34
Rear kneeroom (ins)	28-37	29.5-35.5	25.5-32	26-32	24.5-32.5
Length (ins)	166.5	169	153	156.4	163
Wheelbase (ins)	104.5	91.2	97.3	97.8	95.7
Height (ins)	53.8	55	54.3	55.5	54.5
Boot Load height (ins)	27	33	25	32	24.5
Boot depth (ins)	35/58	38/67	26/49	33/157	37/56
Boot height (ins)	17.5	14.5	17	16	15
Boot width (ins)	65.3	66	63.8	64.2	64.4
Overall width (ins)	65.3	66	63.8	64.2	64.4
Int. width (ins)	55	53.5	52.5	51	54
Weight (cwt)	17.7	18.3	18.4	17.6	17.5
Towing weight (cwt)	19.7	21.7	18.4	15.8	19.7
Boot capacity (cu. ft)	16/52	—/34	9/35	11/42	14/32

KEY. Valve gear: ohc, overhead camshaft. **Steering:** rack/pin, rack and pinion. **Brakes:** Di, discs; Dr, drums; S, servo assistance; P, power assistance. **Suspension:** I, independent; C, coil springs; Tor, torsion bar springs; Hp, Hydropnumatic. Wi, Wishbone. McP, MacPherson struts; TA, trailing arm. * auto model. † estimate.

1400 cc TSE, but even so, despite its limitations of space, it is definitely the driver's car of the group.

If the Renault is for the driver, the BX is without doubt the passenger's car – supremely comfortable, as well as practical for large loads. It is also at last convincingly cheap to run: for those who appreciate its high specification and are keen to steer clear of the traditional tin-box image it is very good value and could well be the answer. But we believe that the bigger 1600 engine is worth the extra £150, and even then many buyers will be put off by its heavy steering, poor vision and high noise levels.

We must have fallen too heavily for the charms of the Delta GT and its fabulous twincam engine tested in October, for the 1300 model allows only a limited quantity of Lancia character to filter through. In reality the 1300, though a good performer for its capacity, is disappointingly noisy and unrefined and, although quite cheap to buy, has little other than a classic name to distinguish it from mass-produced rivals.

Nor can we recommend the Datsun. Something is disastrously wrong in its suspension design, yet even without this inexcusable fault and the resultant poor handling we would have judged the car's good fuel economy an insufficient advantage to make up for its heavy steering, high price and poor boot.

Prisma

While the Delta had been designed as a hatchback to take advantage of the increasing popularity of this configuration, there was still a healthy demand for conventional "three-box" saloons of medium size. Lancia moved to meet this demand by producing a saloon of this type based on the Delta. Unlike its immediate predecessors, the new model was not given a Greek letter as its name but was christened Prisma. However, it did follow the pattern of the Delta in using a modified version of the Fiat Regata floorpan and engines based on fiat power units.

The Prisma was introduced in 1982 in three versions, identified as 1600, 1500 and 1300 models. The smaller engines were the same as those used in the equivalent Delta Models while the 1600 was powered by a 1585cc twin-cam unit similar to those used in Lancia's Beta range. In the Prisma the twin-cam engine produced 105bhp (DIN) at 5,800rpm with a maximum torque of 13.8kgm at 3,300 rpm. Equipped with a five-speed gearbox the 1600 had a top speed of 110mph. The Prisma was launched in the UK in July 1983 in three versions: 1600, 1500 manual and 1500 automatic. The Prisma 1300 was never to appear in right-hand drive form.

Early in 1985 both 1500 models were discontinued in the UK though they stayed in production in LHD throughout the life of the Prisma. The standard 1600 was endowed with a high level of

Prisma 1500

equipment, equal to that of the Delta, and in the UK was joined by an even better equipped version, the 1600LX, later in 1985. The following year the improved front suspension and slight external styling changes provided for the "facelifted" Deltas were also applied to all versions of the Prisma. More importantly, both versions of the 1600 were enhanced by the provision of electronic fuel injection, distinguished by the letters "ie" following the model number. While giving only a marginal increase in maximum power, 108bhp at 5,900rpm, the revised engine enabled the Prisma 1600ie to reach a top speed of 115mph. From this time only the fuel-injected cars were imported into the UK, but both carburetter and fuel-injected versions were built in LHD. The 1600ie models continued in the UK until 1989 when the LX version was replaced by the Prisma Symbol. This was mechanically the same as both the LX and standard 1600ie, but incorporated a number of "extras" in its specification. These included electric windows front and rear, Alcantara upholstery, sunroof, alloy wheels, rear sun blinds, and metallic paint.

Prisma 1600

The Prismas sold in the UK represented only a part of the range available in other markets. One of the most significant of these was the 1900 DS, a diesel-powered model introduced in 1984; the first passenger car of the marque to use a diesel engine although

Lancia's commercial vehicles had used diesel power for many years. The reason why Lancia held back from using a diesel engine in its cars was that the company believed that the noise and vibration inherent in this type of power unit was not in keeping with the character of the cars. However, improvements incorporated in a newly-designed engine overcame these objections. The Prisma power plant was a four-cylinder unit of 1,929cc capacity producing 65bhp at 4,600 rpm with a maximum torque of 12.1mkg at 2,000 rpm. At 223kg the engine weighed only 16kg more than the 1600 petrol engine and was mounted on specially-designed mountings to reduce the transmission of vibrations to the body. Great attention was paid to sound-proofing to achieve noise levels inside the car lower than those of an equivalent petrol-engined car.

The normally-aspirated diesel was joined by a turbodiesel-powered model in 1985 - though production of this car actually started at the end of 1984 - the two versions being produced in parallel until Prisma production ended in 1989. Designated the Prisma 1900 TDS, its turbocharged engine gave 80bhp at 4,200rpm with maximum torque increased considerably to 17.5kgm at 2,400rpm. Maximum speed of the turbocharged car rose to 106mph compared with 98mph for the original version. Two years after the turbodiesel Prisma was launched the same engine was fitted to the Delta. Identified as the 1900 TDS, the turbodiesel Delta had the same performance as its Primsa counterpart, while the bodywork and trim were the same as that of the Delta HF. Production of the Delta turbodiesel ended in 1991.

Another Prisma variant not seen in the UK was the 2000 4WD, a four-wheel drive model introduced in 1986. This was powered by the 1,995cc fuel-injected twin-cam engine previously used in the Trevi but restricted to 115bhp at 5,400rpm (112bhp at 5,500rpm with catalyst). The Prisma 4WD incorporated a permanent four-wheel drive system which was simpler than that of the Delta, described later, though it employed a viscous coupling for the centre differential and a limited slip differential at the rear. While not having the sporting character of the Delta, the Prisma 4WD had quite a lively performance with a maximum speed of 114mph. The model name changed to Prisma integrale in 1987, but no mechanical or trim changes were made.

All the variants of the Prisma remained in production from their different introduction dates until the model was phased out after the launch of the Dedra in 1989.

Models covered in this section: 1300, 1500, 1600,/1600LX, 1600ie, 1600ieLX Symbol, 1900 DS and 1900 TDS.

Production Data

Model	Introduction from	to	No. built
1300	1982	1989	82,818
1500/1500LX	1982	1989	67,761
1600	1982	1989	99,746
1600ie	1986	1989	50,690
1900 DS	1984	1989	38,909
1900 TDS	1984	1989	42,251
2000 4WD/integ	1986	1989	4,522

Prisma 1900 DS Lancia's first diesel-engined car

OnTheRoad

Bridging the gap?

Roger Bell finds the new Lancia Prisma to be quite lively, attractive and able. But will it sell against bigger rivals in an important sector of the market?

The Prisma is longer than the hatchback Delta, but the wheelbase is the same

ON THE outside, it's a Delta with a booted tail. On the inside, it is something else — and in most respects rather better, as well it should be, for the new Lancia Prisma (*Motor*, w/e December 18) is to be sold from a higher rung of the market ladder. Lancia project the Prisma as a prestige middleweight that bridges the gap between the Delta and the Trevi. In terms of overall length, it *is* a larger car than the hatchback Delta. But the wheelbase — and therefore the cabin accommodation — is just the same, hence the different interior styling. Change the decor and style and, hey Primsa — a new model.

As our recent description outlined, the Delta and Prisma are mechanically identical. Just to recap, both ranges include 1300 and 1500 ohc models and a 1600 twin-cam, all mated in standard form to five-speed gearboxes. All have independent all-strut suspension. Both ranges are also strongly related to the Fiat Strada by their familiar running gear. Where the Prisma most differs from its Fiat-Lancia stablemates is inside.

The first thing we noticed was the smell — a curious "Italian Plastic" odour, quite pungent and not particularly pleasant. The second was that neither seat belt was usable, as the tongues didn't mate with the anchors, which were evidently of a different make. While the rest of the Press convoy swept into the thick of Rome's traffic (some crews unbelted — and apparently unconcerned — for the same reason) we dallied: I feel undressed without belts, especially in Rome! Another car was presented for appraisal — a 1500, rather than the 1600 we'd originally selected. Never mind. At least it had belts that buckled.

The first positive thoughts were good ones. Lancia have finally banished their "Latin Ape" model and placed the major controls so that six-foot Anglo-Saxons can drive without acutely bent knees and extended arms. In fact the driving position behind the thick-rimmed, low-set steering wheel is excellent. You get a marvellous view out through deep windows with very low sills which make the front of the cabin feel very airy and spacious. Despite its smell, the trim is little more plasticky than that, say, of a small BMW, and far less so than the Delta's. Cloth-

Mechanically, the Prisma and Delta are identical

covered seats and doors add to the air of luxury.

Ahead is a very neat and comprehensive cluster of colourful instruments and warning lights. Apart from the column stalks, the minor switchgear is rather poorly arranged on a low central panel, along with the easy-action rotary heating and ventilation controls. Regrettably, all fresh air vents are interconnected with the heater so you can't have both working together efficiently. It is all very easy on the eye, if not ergonomically perfect, and far less contrived than the "blocked" theme of the Delta's dash.

The 1500 ran sweetly out of Rome, good clutch and throttle actions making it easy to drive smoothly in stop-start traffic. The spongey brakes are characteristically over-servoed, though, and less progressive on release than on application. Situation normal. Brake *performance*, however, is excellent. Steering is particularly light and easy — so easy that it might be mistaken for a powered set-up. Minimising castor has certainly relieved the driver of all the heave-ho parking effort demanded by most Fiat-Lancia products of the past, but it's also robbed the steering of feel. Unless you're cornering hard, the Prisma is rather dead at the helm.

Our test route included a tediously long stretch of *autostrada*, which the 1500 tackled well. With a final drive giving 20.7 mph/1,000 rpm, it cruised easily at 90 mph with a modest 4,350 rpm on the tacho. Wind noise was well suppressed at speed, engine roar pleasantly muted.

The boring *autostrada* run soon highlighted one problem area: the seats. They are so short in cushion length that half your thighs overhang without support. Side location is also poor, both at hip and shoulder level, though the firm backrest pushes into your lumbar regions well. Legroom in the back is cramped unless the front seats are pushed forward. Headroom is also inadequate for a six-footer.

Like the Delta, the Prisma handles well. Its stability and manners are impressive, so deliberate provocation is needed to make the car do anything untoward. On the all-too-short twisty section of the route, we swapped 1500 for 1600 power — and were slightly disappointed: it didn't feel like another 20 bhp, aided by lower gearing. Even so, the top Prisma has quite a wallop, particularly if you stir the gears. Although Lancia make much of their "floating" gearchange anchorage, lever movement is not outstandingly light or precise.

Again we found ourselves with useless seat belts, not to mention a trip computer that didn't work. But we did achieve a speedo reading of over 180 kph (112 mph), which is what Lancia claim as a maximum.

Ignoring the problems, which we'll attribute to pre-production hiccups, the Prisma is quite a lively, attractive and able car, if not an especially roomy one. Whether it will sell against bigger rivals in the prestige middleweight sector remains to be seen. Much will depend on the price, not yet announced in Britain.

AutoTEST Lancia Prisma
Reflections through the looking glass

Lancia Prisma
First new offering from Lancia in this country since marketing of the company's range passed into the hands of the Heron Corporation-owned Lancar. Mid-range three-box saloon looks very much like "booted" version of front-wheel drive Delta hatchback and is mechanically similar in most respects. UK market will receive single overhead cam 1500 manual and automatic models and 1600 manual fitted with familiar 105 bhp Fiat-Lancia twin-cam unit. Fits into existing range above Delta and below Trevi.

PRODUCED BY:
Lancia & Co.,
Fabbrica Automobili S.p.A.,
Via Vicenzo Lancia 27,
Turin 10141,
Italy.

SOLD IN THE UK BY:
Lancar Ltd,
Henwood Industrial Estate,
Ashford,
Kent TN24 8DH.

AS JETTA is to Golf, as Orion is to Escort, so Prisma is to Delta. The manufacturers may wish to avoid the dismissive catch-phrase of "just another booted hatchback", but superficially at least, that is perhaps the easiest way to describe the models mentioned. Nor is it in any way to demean them. There is no doubt that the front-wheel drive hatchback explosion of the 1970s produced vehicles with a very useful combination of virtues — good handling, ride and roadholding plus, with the rear seats folded down, vast amounts of cargo space. But with the concept firmly accepted by the car buying public, manufacturers then wisely looked to plugging the gap created by a dearth of conventional mid-range three-box saloons.

Simply to describe the resultant cars as with a boot stuck on the back is to dismiss the amount of work necessary to create an integrated design, especially in these days when aerodynamic efficiency is close to the top of a manufacturer's list of priorities for any model. And so it is that Lancia's latest offering, the Prisma, bears more than a passing resemblance to the hatchback Delta. Yet the now de rigueur high boot line and larger curved rear window contrive to give the car much more of a wedge-shape than its smaller brother.

Specifically, Prisma is a new bodyshell which is said to use a different floorpan from the Delta, but has an identical wheelbase and is fitted with the same doors. Suspension arrangements are essentially the same as equivalent Delta models, incorporating MacPherson struts front and rear with offset telescopic dampers and coil springs. Front struts are located by what are best described as "boomerang-shaped" lower wishbones and forward facing links — an arrangement which is good for absorbing longitudinal loads — while the rear struts are controlled by two parallel transverse links mounted through rubber bushes to a crossmember bolted to the bodyshell. This suspension layout, which also incorporates an anti-roll bar, is said to keep rear wheel toe-in nearly constant, thus reducing rear wheel steer effects to a minimum during hard cornering manoeuvres.

Steering is rack and pinion with a fairly low-geared 3.8 turns required to get from lock to lock, while brakes are four-wheel discs on the 1600 Prisma (front discs and rear drums on the 1500 models) with servo assistance and connected by the now normal twin diagonal circuits. The 1600 is also fitted with 14in. diameter wheels carrying low rolling resistance 165/65-14 Pirelli P8 tyres.

The two engine options for UK Prismas (the 78 bhp 1300 model is not being imported) are either the single overhead cam 1500 (85 bhp at 5,800 rpm), as seen in the Fiat Strada range, or the twin cam 1600 (105 bhp at 5,800 rpm) used in the Lancia Trevi, Coupé and HP Executive models as well as the Fiat Strada 105TC. The twin cam is fitted with a 34DAT downdraught Weber carburettor and differs from the 1600 installed in other Lancias by having a higher 9.5-to-1 compression ratio, modified inlet and exhaust port shapes to improve torque and fuel consumption, bimetal exhaust valves and the Digiplex electronic spark advance control system. This is activated by electro-magnetic sensors on the crankshaft and alters the spark advance to keep it at the optimum setting for the load under which the engine is being placed. The five-speed gearbox fitted to the 1600 Prisma has what could be described as a set of "sports" ratios, with top speed achieved in fifth rather than the now more common fourth gear. There is no automatic available as an option in keeping with the 1600's more sporting image.

In terms of price, Prisma slots into the Lancia range above Delta and below Trevi and is intended to strengthen the company's position in what they term the "upper middle class segment" of the market. A tough task in a class occupied with stiff opposition from the other big European car manufacturers, but one in which Lancia must succeed.

Performance
Class competitive

The Prisma gave an initial impression to some testers that it performed very well, a feeling which, with familiarity, turned to one of slight disappointment that it did not go a little better than it did. In actual fact, a perusal of the pertinent figures discloses that it goes very well in comparison with its theoretical rivals. With a fairly lusty 105 bhp produced by the free-revving twin cam engine having to propel a relatively light 2,218lb vehicle, the result is predictable enough — a healthy 108 mph top speed, just 2 mph down on Lancia's claim of 110 mph, 0-60 mph in 10.8 sec, and the quarter mile traversed in 17.8 sec with a terminal speed of 77 mph. This level of performance puts the Prisma near the top of its class in almost every respect.

Its willing engine is, of course, aided by a good choice of gear ratios, which are both relatively closely spaced and, in the case of fifth, permit top speed to be achieved with the engine turning over at 5,700 rpm, right at the peak of the power curve. This, incidentally, is different from the five-speed gearbox fitted to the 1500 Prisma. That is closer to a 4+E box where top speed is achieved in fourth and fifth is an overdrive-type cruising gear and aimed at improving fuel economy.

Final drive in the 1600 is 3.59-to-1 (3.77 in the 1500) which produces mph per 1,000 rpm figures of 18.7, fifth; 15.5, fourth; 11.6, third; 8.1, second, and 5.0, first, which correspond to maximum speeds of 97 mph in fourth, 74 mph in third, 52 mph in second and 32 mph in first, these being arrived at with the needle of the electric rev counter just touching the red zone, which starts at 6,500 rpm. It should be noted that although not recommended procedure, the eager 1600 engine winds up into the red zone with contemptuous ease, which means that should it prove advantageous to hold on to a gear between corners on a twisty bit of road, it is possible to do so, albeit to the accompaniment of a fair bit of not unpleasant engine noise. This characteristic seems to bear out that old adage about Italian engines thriving on revs, although less charitable individuals might classify the power unit as exhibiting a certain amount of Italianate fussiness.

Power delivery is generally smooth, although a slight initial flatspot did manifest itself on the test car; that is, if accelerating hard from rest, the revs would sag slightly before the power came in strongly. Once under way, the in-gear incremental times proved to be commendably consistent, especially in the important mid-range speeds.

Economy
A bit thirsty

It's the old story of a relatively high performance engine combined with lowish overall gearing — fuel economy suffers as soon as one starts to tap the power available. Our overall figure of 26.5 mpg covers a best of 28.3 mpg, after a session of high speed, fifth gear cruising, and 25.5 mpg after our figure-taking session at MIRA. Our computed average consumption of 30.5 mpg should therefore be well within the realms of possibility for the normal owner, providing he or she curbs any instinct to exploit the Prisma's performance too frequently. Working on the 26.5 mpg average, the Prisma's 9.9 gallon petrol tank would endow it with a range of about 250 miles, which could rise to around the 300 mile mark if a 30 mpg average were attained.

Squeezing in the last gallon of fuel petrol is a messy rather than time consuming operation since, although the tank trickle brims in about 90 sec after the initial cut-off of the pump, it spits back rather a lot through the overflow pipe, which is located at the top of the filler neck, as air is expelled from the expansion space. Access to the filler is gained through a locking flap which is released with the door key — a nice security touch.

The Fiat-Lancia twin-cam engines tend to be fairly heavy users of oil, perhaps because of their willingness to rev, so our 1,800 miles per litre figure is not altogether surprising. It could conceivably have been better, though, if the engine of the test car had had a few more miles on it.

Noise
Not overbearing

The Prisma, in common with many Italian vehicles, is not an overly quiet car, and the twin-cam engines have always had a reputation for being mechanically busy. Having said that, engine noise is only marginally greater than wind and road noise when the Prisma is cruising at motorway speeds. In fact, the chief noise generator in our test car above 50 mph was the optional steel sunroof, a point which was easily demonstrated by pushing up on the trailing edge of the unit — wind noise was immediately reduced by at least 50 per cent. The remainder was apparently coming from around the door seals, the windscreen pillars and the fairly large, internally adjustable, door mirrors.

The engine, as one would naturally expect, was loudest un-

Rear view of the Prisma emphasises the high boot line and resultant wedge shape. Although there is a distinct similarity between Prisma and Delta, even including shared doors, everything behind the C-pillar is new. Much attention has been paid to making this new Lancia as aerodynamically slippery as possible with bumpers faired into the bodywork and external projections kept to a minimum

The dohc engine with its single twin-choke downdraught carburettor fills the engine compartment quite effectively, although dipstick, reservoirs and battery can be reached easily enough. Radiator is small, slim unit sitting immediately behind the grille and assisted by a thermostatically operated electric fan

der hard acceleration and accompanied by a healthy — but not unpleasant — bark from the exhaust. Induction roar from the twin-choke downdraught Weber carburettor quietened to the "merely present" level once cruising speeds had been reached.

Road noise is well suppressed, there being less than average bump-thump transmitted from the Pirelli P8s through to the occupants, and there was no obvious body boom. The test car did exhibit a distinct heterodyning from even relatively low speeds when the sunroof was fully open.

During the course of the test, the weather was very warm, and even with the top fan speed selected to boost ventilation, the fan was surprisingly quiet; since the unit works well enough at slower speeds for most purposes, it can really be discounted as a significant noise source.

Road behaviour
Generally good

Steering is like that on the Delta — low geared at 3.8 turns from lock to lock and, combined with overly large minimum turning circles of 35 ft to the left and 37 ft 1in. to the right, the two factors lead to considerable twirling of the wheel when negotiating right angle bends in town driving. The steering is probably well weighted for most male drivers, although some women might consider it to be on the heavy side, particularly at low speeds when carrying out parking manoeuvres.

Once travelling more quickly, though, it is possessed of good feel and precision with a reassuringly steady feel to it, the good weighting and fairly strong self-centring action combined with the car's natural lack of roll encouraging enthusiastic driving on winding country roads. Two slight criticisms of the steering concern the detectable amount of torque steer, particularly when accelerating from rest with lock applied, plus what can best be described as a high speed instability which makes its presence felt in an occasional weaving motion when travelling in an apparently straight line. It may be what tyre engineers refer to as "white lining", which is the effect created when the front tyres in particular are deflected by small longitudinal imperfections in a road surface, such as dips and ridges as well as cats eyes and road markings, from whence the name derives.

When pushed hard in corners, Prisma displays typical front-wheel drive understeer, but this is at higher speeds than might normally be expected. If the throttle is lifted off in mid-corner, the result is the normal tuck-in, or tightening of line, which is the built-in safety factor associated with most front-wheel drive cars. The back end is easily retrieved should a bend taken too quickly require the driver to back off the power for more than an instant, the overall effect being speed killing tyre scrub.

Interestingly, Prisma exhibits a ride quirk which it shares with the Delta. Specifically, you are aware of a small amplitude, high frequency shake which occurs throughout the speed range in reaction to small, sharpish unevenesses in the road surface. It tends to make for a somewhat busy primary ride at times and may, since it seems to come more from the front than the back of the car, be the engine-gearbox unit moving up and down slightly on its mountings. These mountings, it should be added, do an excellent job of damping out "shunt" — the driveline snatch which is caused, particularly on front-wheel drive cars, when the driver puts his foot on and off the throttle at low speeds in a low gear.

The servo-assisted brakes work well enough in normal driving conditions, being light and responsive, 10-20lb of pedal pressure being easily enough to generate the sort of "check-braking" required around town. For four-wheel disc brakes, they were only able to generate a slightly disappointing 0.9g of stopping power with 60lb pressure applied to the pedal — less than the 0.96g at 80lb achieved by the Delta's disc-drum combination.

Our tough 10-stop fade test, in this case from 77 mph, revealed a fairly normal pattern with the pressure required to maintain a 0.5g stop building as the friction material heated up to the high working temperature being required of it and then levelling out from the seventh stop onwards. There was considerable smoke from the front pads by the tenth stop, but the brakes returned to normal once they had a chance to cool.

The handbrake generated 0.27g of stopping power and, surprisingly for a front-wheel drive car, proved capable of holding the Prisma both in the upward *and* downward directions on the 1 in 3 test hills. This is possibly the result of a slightly less front-heavy weight distribution (61-39) than some front-wheel drive cars, such vehicles with rear operating handbrakes usually sliding down the 1 in 3 hill with the back wheels locked.

Behind the wheel
Well thought out

Considerable time has obviously been spent on the interior layout of the Prisma. The driver is not forced to adopt an overly "Italian ape" driving position since the steering column is adjustable for rake, and although the driver's seat didn't slip far enough back on its runners to suit the leg length of some of our taller testers, it was generally quite acceptable for someone of average build, with a good pedals-wheel-seat relationship and relatively good support under the thighs. The seats themselves are supportive enough in a vertical direction, but the flat cushions lack lateral location once one

The rev counter is on the left, speedometer on the right with a resettable odometer, and battery condition, fuel contents and engine temperature along the bottom. A non-calibrated econometer is situated above the petrol gauge with groups of warning lights on either side. To the left of the instrument cluster is a separate check panel. Centre console below contains rotary controls for heating and ancillary switches. Steering column stalks look after washers and wipers (right), turn indicators, headlamps and headlamp flasher (left). There are two separate horn buttons on the steering wheel spokes

starts to corner at all enthusiastically. The now common back rake adjusting wheels live on the inside edges of the seats.

The "vee-shaped" spokes of the steering wheel permit an uninterrupted view of the clearly laid out rectangular instrument binnacle, the only request here being to replace the rather half-hearted econometer with its red, yellow and green segments with something far more useful, such as an oil pressure gauge. Otherwise, the Prisma is nicely instrumented and has well laid out controls. The horn is operated, Japanese-style, from two little press pads on the spokes (too easy to touch when twirling the wheel), an arrangement which is not absolutely ideal — in an emergency, a centre-push is much preferable — but certainly better than the occasionally hard-to-find stalk switch.

The gearchange is reasonable enough, with a relatively smooth action, although as with Fiat's front-wheel drive cars, the change cannot be rushed or it becomes decidedly baulky. One particular annoyance is the positioning of the handbrake in relation to the gear lever — it's too far forward, which means that when the driver selects reverse while the handbrake is on, he can hit his hand rather hard on the handle.

A particularly nice point is the warning light check system with its light emitting diodes indicating the condition of all front and rear light bulbs plus warning lights for brake pad life, coolant level, battery condition, oil level, washer reservoirs, gearbox oil and engine temperature.

The heating and ventilation system is very good indeed. It is of the air blending variety with good throughput and controllability, and the one notable problem with the Delta's system — the lack of flow from the side vents — has been effectively solved. Full marks to the Lancia engineers for the work they've obviously done in this area.

Forward visibility is good, aided in wet weather by windscreen wipers which are properly oriented for right-hand drive, and the screen pillars are of a reasonable thickness. Rear vision is better than on the Delta, courtesy mainly of the larger, curved window, the only real intrusion being from the overly large rear seat head restraints. At least these can be removed easily enough, though, which improves the situation instantly.

Living with the Lancia Prisma

We liked the Delta very much when it arrived on the scene back in 1980, finding it generally highly acceptable in almost every respect. The Prisma is generally possessed of the same good qualities, thus making it a highly agreeable car. The driver and passengers are well catered for where headroom and legroom are concerned and, although there are no door pockets, interior storage space is excellent. This is mainly due to the deep, bucket-like tray which extends the width of the car beneath the facia, the only worthwhile addition perhaps being some sort of anti-skid mat at the bottom of the tray to prevent lighter items flying about during hard cornering. The glovebox, a low-slung item above the passenger's knees operated by a lockable rotating knob to the right of the lid, is augmented by small oddment bins attached to the inner wheel arches.

Two very useful features on the 1600 Prisma were the central locking system, which receives full marks for working off the driver's as well as the passenger's door locks — very useful if loading parcels from the nearside while the car is parked at the kerb — and the electrically operated front windows which, although a little slow to lift at 3.7 sec, work efficiently enough.

There are several good detail points as well. The door mirrors are reasonably sized and have a smooth-operating mechanical remote control mechanism, while grab handles in the roof over the three passenger doors are spring-loaded and thus remain out of the way when not needed. Also at head level, in a small panel above the windscreen, is a large interior light as well as rotating high intensity map reading light, while rear seat passengers are catered for with a large courtesy light of their own. The front light is on a 15 sec time delay once the doors have been closed.

A feature more frequently seen in hatchbacks, the Prisma has a 50-50 split rear seat, the back of which is released by two small levers. This allows access to the boot from inside the car and also helps accommodate long or awkward-shaped loads. One point which engendered a fair amount of discussion among testers was the two-tone check cloth, designed exclusively for Lancia by Ermenegildo Zegra, and used for the seats, door trim and headliner. Some thought it exceptionally smart, while others found it oppressive and particularly irritating on bare skin, the order of the day during the recent spell of hot weather.

The twin cam engine and its ancillary equipment, reached through the forward-hinging bonnet, fills the engine compartment, although access to regular service points is reasonable enough. The only niggle is the use by Lancia of a Ford-type dipstick with an electonic sensor wired to the top. This type of unit may be good for warning the forgetful owner that the oil level is marginal, but it has two annoying drawbacks: engine vibration can cause a break in the circuit, thus illuminating the warning light when there is still sufficient oil in the sump — something which happened twice during the test period — and because the wire leading from the sensor is short, it is all too easy to drop oil into the engine compartment when checking the level.

The bonnet release has been moved to the right side of the car, near the steering column, which although arguably a small point, shows the time and thought taken by Lancia in producing the right-hand drive version of the Prisma for the British market.

The Prisma range

Three versions of Prisma are to come to the UK: a 1500, 1500 automatic and 1600, the subject of this week's autotest. Prices are £5,550, £5,989 and £6,150 respectively. Metallic paint is optional on all models at £99, while a sunroof is available on the top-of-the-range 1600 at £250.

Far left: Relatively wide opening doors give easy access to interior. Seats are supportive enough in a vertical direction but the flat cushions do not give good sideways location. Close proximity of the gear lever and handbrake can cause problems, particularly when selecting reverse

Left: Although not obvious on paper, there is a definite improvement in rear seat legroom in the Prisma when compared with the Delta. Seat belts are provided for three rear seat passengers

A separate panel in the boot floor lifts to reveal the spare wheel which is stored horizontally. A removeable panel on the right-hand side of the boot contains a small but usefully equippped tool kit. Low sill to the boot means the loading of awkwardly-shaped or heavy objects is a relatively straightforward exercise. There is some intrusion into the usable space by the inner wheel arches

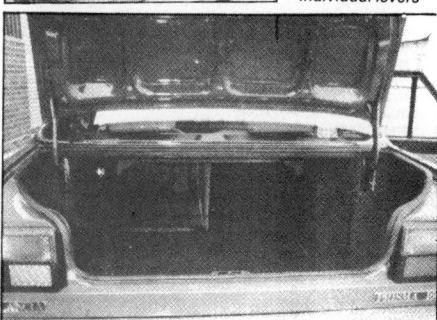

A very useful feature on the Prisma is the 50-50 split of the rear seats, which not only permits the carrying of longer objects than would otherwise be possible, but also allow access to the boot from inside the car. Seat backs are released by small individual levers

53

HOW THE LANCIA PRISMA 1600 PERFORMS

Figures taken at 1,010 miles by our own staff at the Motor Industry Research Association proving ground at Nuneaton. All Autocar test results are subject to world copyright and may not be reproduced in whole or part without the Editor's written permission.

TEST CONDITIONS:
- Wind: 5-8 mph
- Temperature: 26 deg C (79 deg F)
- Barometer: 29.8in. Hg (1,011 mbar)
- Humidity: 64 per cent
- Surface: dry asphalt and concrete
- Test distance: 681 miles

ACCELERATION

FROM REST

True mph	Time (sec)	Speedo mph
30	3.2	32
40	5.6	43
50	7.6	54
60	10.8	64
70	14.5	76
80	19.5	86
90	26.9	96
100	38.3	107

Standing ¼-mile: 17.8 sec, 77 mph
Standing km: 33.1 sec, 97 mph

IN EACH GEAR

mph	Top	4th	3rd	2nd
10-30	—	11.2	6.9	4.7
20-40	12.5	9.4	6.0	4.2
30-50	11.0	8.3	5.6	4.3
40-60	11.2	8.0	5.8	—
50-70	11.5	8.3	6.6	—
60-80	12.7	10.0	—	—
70-90	16.9	12.7	—	—
80-100	27.7	—	—	—

MAXIMUM SPEEDS

Gear	mph	kph	rpm
Top (mean)	108	174	5,700
(best)	109	176	5,800
4th	97	156	6,500
3rd	74	119	6,500
2nd	52	84	6,500
1st	32	52	6,500

FUEL CONSUMPTION

Overall mpg: 26.5 (10.7 litres/100km)
5.84 mpl

Constant speed

mph	mpg	mpl	mph	mpg	mpl
30	54.4	12.0	70	31.5	6.9
40	50.8	11.2	80	28.5	6.3
50	43.9	9.7	90	22.7	5.0
60	38.0	8.4	100	18.2	4.0

Autocar formula: Hard 24.9mpg, Average 30.5mpg, Gentle 36.1mpg
Grade of fuel: Premium, 4-star (97 RM)
Fuel tank: 9.9 Imp. galls (45 litres)
Mileage recorder reads: 1.1 per cent short

Official fuel consumption figures
(ECE laboratory test conditions; not necessarily related to Autocar figures)
- Urban cycle: 27.7 mpg
- Steady 56 mph: 44.1 mpg
- Steady 75 mph: 33.6 mpg

OIL CONSUMPTION
(SAE 20W/50) 900 miles/pint

BRAKING

Fade (from 77 mph in neutral)
Pedal load for 0.5g stops in lb

	start/end		start/end
1	24-10	6	34-88
2	20-30	7	40-94
3	24-44	8	40-66
4	24-52	9	44-64
5	28-74	10	38-76

Response (from 30 mph in neutral)

Load	g	Distance
10 lb	0.30	100 ft
20 lb	0.43	70 ft
30 lb	0.57	53 ft
50 lb	0.77	39 ft
60 lb	0.90	33 ft
Handbrake	0.27	111 ft

Max. gradient: 1 in 3

CLUTCH
Pedal 20 lb; Travel 5in.

WEIGHT
Kerb, 19.8 cwt/2,218 lb/1,006 kg
(Distribution F/R, 61.2/38.8)
Test, 23.1 cwt/2,590 lb/1,175 kg
Max. payload, 994 lb/450 kg

DIMENSIONS

- OVERALL LENGTH 164.6" / 4180
- OVERALL WIDTH 63.8" / 1620
- OVERALL HEIGHT 54.5" / 1385
- WHEELBASE 97.4" / 2475
- FRONT TRACK 55.1" / 1400
- REAR TRACK 55.1" / 1400
- GROUND CLEARANCE 6" / 152
- Boot capacity: 15.9 / 26 cu. ft.
- Turning circles: Between kerbs L, 35ft., R, 37ft. 1in.
- SCALE 1:35
- OVERALL DIMENSIONS in / mm

PRICES

Basic	£4,936.4
Special Car Tax	£411.3
VAT	£802.1
Total (in GB)	**£6,150.00**
Licence	£85.00
Delivery charge (London)	£75.00
Number plates	£20.00
Total on the Road (exc. insurance)	**£6,330.00**

EXTRAS (inc. VAT)
Metallic paint	£99.2
*Sunroof	£249.5

*Fitted to test car

TOTAL AS TESTED ON THE ROAD: £6,579.5
Insurance: Group n/a

SERVICE & PARTS

Change	Interval 5,000	10,000
Engine oil	Yes	Yes
Oil filter	Yes	Yes
Gearbox oil	Check	Yes
Spark plugs	Yes	Yes
Air cleaner	Yes	Yes
Total cost	**£64.88**	**£71.35**

(Assuming labour at £12.35/hour inc. VAT)

PARTS COST (including VAT)
Brake pads (2 wheels) – front	£27.1
Brake pads (2 wheels) – rear	£44.2
Exhaust complete	£176.1
Tyre – each (typical)	£52.9
Windscreen (laminated & tinted)	£120.5
Headlamp unit	£83.5
Front wing	£63.00
Rear bumper	£171.3

WARRANTY
12 months' unlimited mileage

SPECIFICATION

ENGINE
- Type: Transverse front wheel-drive
- Head/block: Al. alloy/cast iron
- Cylinders: 4, bored block
- Main bearings: 5
- Cooling: Water
- Fan: Electric
- Bore, mm (in.): 84.0 (3.44)
- Stroke, mm (in.): 71.5 (2.93)
- Capacity, cc (in.^{3}): 1,585
- Valve gear: Ohc
- Camshaft drive: Toothed belt
- Compression ratio: 9.3-to-1
- Ignition: Breakerless
- Carburettor: Weber 34DAT 13/250 twin choke
- Max power: 105 bhp (DIN) (77.2 kW ISO) at 5,800 rpm
- Max torque: 100 lb ft at 3,300 rpm

TRANSMISSION
- Type: Five-speed
- Clutch: Diaphragm spring, 8.2in. diameter

Gear	Ratio	mph/1000rpm
Top	0.96	18.7
4th	1.16	15.5
3rd	1.55	11.6
2nd	2.23	8.1
1st	3.58	5.0

- Final drive gear: Helical spur
- Ratio: 3.59

SUSPENSION
- Front – location: Ind. MacPherson strut
- – springs: Coil
- – dampers: Telescopic
- – anti-roll bar: Yes
- Rear – location: Ind. MacPherson strut, parallel transverse links
- – springs: Coil
- – dampers: Telescopic
- – anti-roll bar: Yes

STEERING
- Type: Rack and pinion
- Power assistance: No
- Wheel diameter: 15 in.
- Turns lock to lock: 3.8

BRAKES
- Circuits: Twin, diagonal split
- Front: 10.5 in. dia. disc
- Rear: 9.3 in. dia disc
- Servo: Vacuum
- Handbrake: Centre lever acting on rear discs

WHEELS
- Type: Alloy
- Rim width: 5.5 in
- Tyres – make: Various – Pirelli P8 fitted to test car
- – type: Radial tubeless
- – size: 165/65SR-14
- – pressures: F 31, R 31 psi (normal driving)

EQUIPMENT
- Battery: 12V 40Ah
- Alternator: 55A
- Headlamps: 120/110W
- Reversing lamp: Standard
- Electric fuses: 12
- Screen wipers: 2-speed plus intermittent
- Screen washer: Electric
- Interior heater: Air blending
- Interior trim: Cloth seats, cloth headlining
- Floor covering: Carpet
- Jack: Screw scissor
- Jacking points: One each side
- Windscreen: Laminated
- Rust protection warranty: To be decided

HOW THE LANCIA PRISMA 1600 COMPARES

Lancia Prisma 1600 — £6,150
Front engine, front drive
Capacity 1,585 c.c.
Power 105 bhp (DIN) at 5,800 rpm
Weight 2,218 lb/1,006 kg
Autotest 30 July 1983

Audi 80CL* — £6,645
Front engine, front drive
Capacity 1,588 c.c.
Power 85 bhp (DIN) at 5,600 rpm
Weight 2,050 lb/930 kg
Autotest 7 April 1980
*GLS tested

Renault 18 GTL* — £5,700
Front engine, front drive
Capacity 1,647 c.c.
Power 79 bhp (DIN) at 5,500 rpm
Weight 2,128 lb/966 kg
Autotest 17 February 1979
*GTS tested

Talbot Solara 1.6 GLS — £6,695
Front engine, front drive
Capacity 1,592 c.c.
Power 87 bhp (DIN) at 5,400 rpm
Weight 2,251 lb/1,146 kg
Autotest 14 March 1981

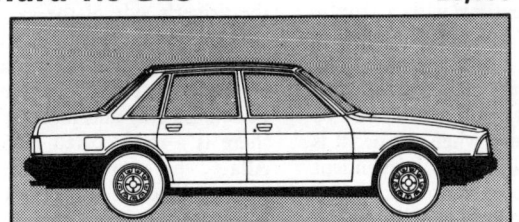

Vauxhall Cavalier 1.6 GLS* — £6,987
Front engine, front drive
Capacity 1,598 c.c.
Power 90 bhp (DIN) at 5,800 rpm
Weight 2,296 lb/1,044 kg
Autotest 18 December 1982
* 5-door tested

Mazda 1600 LX 4-door* — £5,349
Front engine, front drive
Capacity 1,587 c.c.
Power 81 bhp (DIN) at 5,500 rpm
Weight 2,304 lb/1,045 kg
Autotest May 1983
* Hatchback tested

MPH & MPG

Maximum speed (mph)
Lancia Prisma 1600	108
Vauxhall Cavalier 1.6 GLS	103
Mazda 1600 LX	102
Audi 80 CL	99
Renault 18 GTL	96
Talbot Solara 1.6 GLS	96

Acceleration 0-60 (sec)
Lancia Prisma 1600	10.8
Vauxhall Cavalier 1.6 GLS	10.8
Talbot Solara 1.6 GLS	12.0
Audi 80 CL	12.1
Mazda 1600 LX	12.3
Renault 18 GTL	13.4

Overall mpg
Vauxhall Cavalier 1.6 GLS	31.9
Mazda 1600 LX	30.6
Talbot Solara 1.6 GLS	29.0
Renault 18 GTL	28.8
Audi 80 CL	28.1
Lancia Prisma 1600	26.5

The 1600 Prisma's performance bias can be clearly seen when it is measured against what must be considered its prime opposition. The top speed of 108 mph is down slightly on the 110 mph claimed by Lancia, although that figure is probably close to a true maximum, bearing in mind tyre scrub created by the banking on MIRA's high speed circuit and the high ambient temperature in which the car was tested. On 0-to-60 mph acceleration, it ties with the 1.6 Cavalier at 10.8 sec, both cars being significantly quicker from a standing start than the rest of its rivals.

Interestingly, even if other possible contenders were taken into consideration — Alfa Giulietta 1.6 (103 mph, 12.2 sec), BMW 316 (102 mph, 12.1 sec), and Ford Sierra 1.6 (101 mph, 13.0 sec) — the Prisma 1600 still comes out at the top of the overall performance table.

On fuel consumption, the story is predictable enough, for with the level of performance offered by the Prisma, it is no surprise that it is the heaviest user of petrol in the group.

ON THE ROAD

All the cars in the group behave, within reason, in similarly predictable fashion. Being nose-heavy front-wheel drivers, they all display classic understeer which changes to tuck-in, or a tightening of line, when one lifts off the throttle in mid-corner.

The Audi probably offers the best overall package with its very refined road behaviour, the Renault and Lancia being its closest rivals. Specifically, the Audi offers a commendably smooth engine, an excellent gearchange and a firm ride. It handles very well, but can get caught out at times by badly surfaced roads. It is also arguably the quietest car of the group. The Renault is inherently stable and soaks up poorly surfaced roads with ease, but does not have quite the good seating or heating-ventilation of the others. It also has a gearchange that is only adequate at best. The Talbot Solara is similar to the Renault in terms of ride, its main drawback being its heavy steering and pronounced body roll when cornered quickly.

The Cavalier, by comparison, has a decidedly firm ride, but offers thoroughly chuckable handling in return, while the Mazda is mechanically quite refined, and has good neutral handling but suspect ride quality at low speeds.

Into the middle of this group comes the Prisma, which although suffering from a slightly busy ride and a fair amount of mechanical noise, is nevertheless an eminently competent all-round package.

SIZE & SPACE

Legroom front/rear (in.)
(seats fully back)
Mazda 1600 LX	40/39
Talbot Solara 1.6 GLS	40/38,5
Audi 80 CL	40/38
Vauxhall Cavalier 1.6 GLS	39/39
Lancia Prisma 1600	41/36
Renault 18 GTL	39/35

As a result of its slightly smaller overall dimensions than some of the other cars in the comparison group, (13.7ft long and 5.3ft wide), the Prisma is not terribly strong — at least on paper — on rear legroom, the two figures being virtually identical to those of the Delta tested previously by us (26 July 1980).

It must be said, though, that in the legroom category, the top five cars in the group are all very close in overall terms, and in fact with the front seat fully back, the Prisma offers an inch more legroom than its nearest competitors.

Where it scores particularly, though, is as one would reasonably expect, on the ample amount of boot space available — 15.9 cu. ft., to be precise. The Prisma also has a very useful feature more often associated with hatchbacks than three-box saloons — a 50-50 split rear seat.

VERDICT

Prisma is undoubtedly up against some very stiff opposition. In terms of the performance and economy available for the cash expended, the Mazda 1600 looks to offer very good value for money, as does the Lancia. The Talbot Solara offers impressive amounts of usable space, and if the price looks a bit high in comparison with the others, it must be remembered that Talbot always quote on-the-road prices.

The Audi is superbly finished, although it is quite costly — a problem the Vauxhall Cavalier suffers from as well. With the wide range of Cavaliers available, however, it would be possible to choose a version — say, the 1.6GL four-door — which would be more cost-competitive with the others. The Renault 18 has predictably safe, stable road behaviour, although the steering isn't quite as precise as it might be and gearchange quality is only average.

The Prisma is a keenly prices, coolly competent three-box saloon which, although not doing anything exceptionally well, offers good value and performance for the money.

ROAD IMPRESSIONS

SHARING a common origin with the very successful Delta, the Prisma boasts amongst other features one of the best instrument layouts available.

Lancia Prisma 1500

WHAT A useful and jolly little car is the latest version of the Lancia Prisma — and not so little either, because this four-door booted saloon comfortably takes four and their luggage. It follows the current line of four-cylinder transverse-engined, front-wheel-drive family cars but tries to impart a little extra. To one who remembers the great Lancia models of the past it is not a Lancia, but it is a nice car nevertheless. We should really have tried it in twin-cam, 105 bhp, 110 mph form. As it was, the 1500 engine with belt-driven single oh-camshaft, alloy head, and five-bearing crankshaft develops 85 bhp at 5,800 rpm, which pushes the Prisma along very well without undue noise or fuss. The claim that this is one of the lightest cars in its class, at 2,068 lb, seems justified, translated into willing pick-up. There is 12.5 kgm torque at 3,500 rpm and it all works out at 0 to 60 mph in rather under 12 seconds, while it is possible to push this 86.4 × 63.9 mm (1,498 cc) Prisma up to just over 100 mph. Its "Otto-4" engine, as Lancia have it, now has a c.r. of 9.2 to 1, and it has been given a high-turbulence head and breakerless electronic ignition.

So what is it like on the road? The latest Prisma can be summed up as having comfortable seats upholstered in sensible tweed, a five-speed gearbox with a stubby lever whose action is average-good, but a thought rubbery, so it would not have done for an Aprilia, disc/drum brakes which feel unreassuring at first but prove to be powerful and progressive when the pedal is firmly depressed, and all-round independent MacPherson-strut suspension which very effectively irons out the bumps, if jittery and rattly over bad surfaces. The engine runs commendably quietly and the steering (3½ turns lock-to-lock) is light. The instrumentation is effective, with the smaller dials consisting of a battery-meter, thermometer and accurate fuel-gauge with low-level light. The warning lights are small and neat, on the compact Veglia instrument panel, but the digital clock was sometimes difficult to read in daylight. Seat belts front and rear are standard equipment, as are courtesy lag interior lighting, a laminated windscreen, an adjustable steering column and two external rear-view mirrors. The triple stalk-controls are arranged in a pattern common to Italian cars, the lh one for the lamps longer and ahead of the turn-indicators' control, which is matched by the rh wipe/washers stalk. The external door handles are akin to those of an Alfa 6. The instruments are minutely calibrated, engine heat normally being 87 deg C and the tachometer needle going into the red at 6,500 rpm. The central console contains a row of switches, the heating and ventilation adjusted with three remarkably big knobs, with clear symbols above them, and, below, a Panasonic radio / stereo having a roof aerial.

The plastic fascia surround incorporates a useful shelf, properly angled to retain loose objects, unlike that of an Alfa 6, but the wells in the scuttle are irritatingly far forward, especially in this age of seat belts (door pockets would be preferable) and the cubby hole is too small and although lockable, this involves using the key to open it and then turning a knob to drop the lid. The Carello lamps gave a good beam once properly adjusted, which is easy to do. The petrol tank holds 9.9 gallons, its screw-type filler tap (a Fiat contribution?) being beneath a lockable panel on the near-side. Although the hand-out says "Pirelli low-profile tyres are used to reduce rolling resistance and for improved directional stability", in fact the test car was on our old friend the Michelin "X" (165 × 70 SR 13 ZX). It was rather disconcerting to find an exposed steering-column, the universal joint of which rubbed one's left foot when cornering!

Using this likeable Prisma for well over 1,000 miles, some of them hurried ones, such as driving back to Wales after the VSCC Cadwell Park races, fuel thirst worked out at 33.3 mpg, with up to a remarkable 36 mpg when merely cruising and although a flickering warning light caused me to consult the dip stick, no oil was required. The front-hinged bonnet is self-supporting, with plastic release lever on the driver's side of the car, but one has to remove the electrical connection before the awkward dip stick can be read.

The gear lever is gaitered, with a large lidded ash tray and inset lighter ahead of it, there is a roof-located map lamp as on an Alfa 6, the doors are of generous size and possess effective "keeps", and although the literature suggested that a check panel for services is only to be found on the Prisma 1500 Automatic and 1600 models, there was one, very compactly set in the left side of the instrument binnacle, on the test car. There was also an economy gauge, best ignored unless you are a "mimser". This single-cam 1500 proudly carried a reminder that a Lancia Rally *con compressore volumetrico volumex* won this year's Monte Carlo Rally outright, and I look forward to trying a normal Lancia with this engine-driven supercharger.

For the present, let it be said that the Prisma, which is built at the Chivasso plant and comes between the Delta and Trevi, and costs £5,550, is an attractive small car worth considering by those seeking family transport with a prestige badge. Lancia claim to have overcome their notorious rusting problems (those Betas with engines virtually falling out) by using various treatments, including 38 to 40 microns of primer on body panels compared to the standard 18 to 20 micron thickness. — W.B.

CAR TEST
LANCIA PRISMA 1600

WITH fluctuating exchange rates, the ever spiralling cost of the local content programme and the successful efforts being made by Western countries to make their industrial operations more cost effective, there has been an increasing interest in imported cars, despite the heavy import duties they involve, and one of the most affordable is the Lancia Prisma.

The modern but hardly original styling doesn't suggest anything very special, but in fact it's a front-drive thoroughbred.

KEY FIGURES	
Maximum speed	175,6 km/h
1 km sprint	32,47 seconds
Fuel tank capacity	45 litres
Litres/100 km at 80	5,59
Optimum fuel range at 80	805 km
*Fuel Index	7,27
Engine revs per km	1 988
National list price	R18 750
(*Consumption at 80, plus 30%)	

Its lines are clean and crisp, with a sense of balance and poise that few three-box sedans can rival, and its stance is sure-footed, masculine. It is practical, with a capacious trunk and plenty of utility space, thanks to its folding rear seats, and there is ample accommodation for five adults, a comfortable driving position and a very high level of appointment.

It is also bang up to date in the equipment field, with a check control panel to keep tabs on 13 different functions, central locking, electric windows in front, an interior light delay, multi-function clock, and optional air-conditioning. But above all, it is mechanically spirited in the best Italian tradition, equipped with an eager twin-cam engine, a five-speed manual gearbox and a chassis/suspension system that equates with the best available in any current medium size sedan. That neat, rather plain exterior can be misleading — the Prisma appeals in ways you are unlikely to grow tired of.

Lancia has always been an interesting carmaker, apart from the period immediately after it became part of the giant Fiat empire. It has now re-established its identity and the Lancia name is once more held in high regard by enthusiasts who enjoy innovative, honestly presented driver's cars.

The Prisma is certainly one of these — it's one of those rare machines with which the driver can form a perfect rapport, something that develops outwards from the basic feel.

It is sobering to see Lancia describe its chassis and front-drive suspension layout as "by now traditional, as we first laid out in 1960". This "traditional" layout includes such state of the art features as an intermediate bearing on the longer driveshaft to effectively create equal-length driveshafts, and front MacPherson struts with slightly offset coils and offset dampers to reduce stiction and obtain a variable springing rate.

Similarly the single lower "sickle" arms locate the lower side of the wheel uprights, carry the anti-roll bar and brakes, and impart an anti-dive effect, while the steering is rack and pinion and the scrub offset is zero.

The dampers are gas-filled and the rear suspension is fully independent, using MacPherson struts articulating on twin parallel transverse links and an anti-roll bar. The long links attach to a narrow cross member in the centre of the car and are set wide apart to secure consistent geometry and a greater resistance to roll.

Brakes are discs all round, with diagonally-split hydraulics, and the car is shod with low profile 165/55 SR 14 Pirelli P8s on handsome alloy wheels, providing prodigious grip and low rolling resistance.

A front-drive thoroughbred that appeals in ways you won't grow tired of...

A fold-down, split rear seat back increases versatility and provides a total of 940 dm³ utility space, in this medium sedan (below). Note the inertia-reel rear seatbelts. Instrumentation is clear and comprehensive (opposite page, left), and the twin-cam 1600 (right) is a thoroughbred.

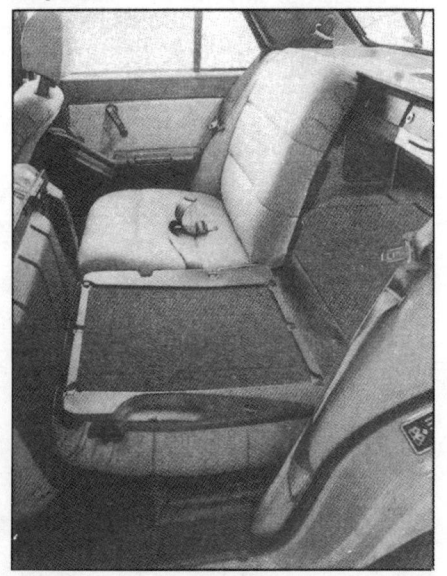

The 1 585 cm³ twin-cam transverse engine is the latest version of a unit that has powered other Lancia models as well as the Fiat 124 ST and 125. It is a willing and proven high performer which in Prisma form puts out 77 kW at 5 800 r/min and 135 N.m at 3 300 r/min. It pulls strongly through a wide rev range, although at speeds below about 2 000 r/min it will baulk if you are not in the right gear.

A single Weber twin-choke carb takes care of induction and the water-temperature-controlled automatic choke shuts off in stages as the engine warms up. Ignition is managed by a fully-mapped Marelli Digiplex electronic system and there is substantial valve overlap. The crisp crackle in the exhaust is distinctly Italian.

WELL EQUIPPED INTERIOR

The Prisma's interior is fortunately one of Lancia's successes, because their interiors tend to be either remarkably good or terrible. It is fairly simple, but also restful, functional and comfortable. The sweeping fascia with full-width parcel tray carries a large rectangular binnacle stretching across to above the centre console and containing all instruments and primary controls, while the large console carries the secondary controls.

Instruments are clear white on black and consist of speedometer and tachometer flanking fuel level, oil pressure and temperature gauges, plus an econometer, as well as a battery of warning lights laid out in clear graphic style.

To the right is the check panel and multi-function digital clock. On the console, three wheels with sliding cursor indicators control the heating/ventilating system — a typical example of successful Lancia individuality.

Air is distributed by adjustable vents at either end of the fascia and two in the centre and although it's effective, one cannot obtain a blend between cold fresh air and heated air — it's either one or the other. The short, "hockey-stick" gearshift has a large boot and there is a smallish, soft-grip steering wheel plus three column stalks, which is in our opinion one too many.

FLEXIBLE DRIVING POSITION

The Prisma has a good driving position which doesn't force you to sit in the "Italian" position, with pedals too close and wheel too far away. The interior is finished in a finely woven cloth and accurate mouldings, with cloth door panels, and shows none of the notorious Italian inattention to assembly detail.

There are inertia-reel seatbelts front and rear, plus a central lap belt in the back, and standard equipment includes twin door mirrors — the driver's adjustable manually from inside and the passenger's, electrically from the driver's seat. There is a manually operated optional steel sunroof (which leaked in hard rain), a roof console with a delay action interior light and a swivelling maplight, an adjustable steering column, spring-loaded grabhandles which fold flush against the roof when not in use and an electrically operated aerial for the optional stereo radio/tape deck.

SPECIFICATIONS

ENGINE:
- Cylinders . . . four in-line, transverse
- Fuel supply . . . . Weber 34 DAT 13 carburettor
- Bore/stroke . . . . . . 84/71,5 mm
- Cubic capacity. . . . . . 1 585 cm^3
- Compression ratio. . . . . . . 9,3 to 1
- Valve gear . . . . d-o-h-c
- Ignition. . . . . electronic
- Main bearings . . . . five
- Fuel requirement . . 98-octane Coast, 93-octane Reef
- Cooling. . . . . water

ENGINE OUTPUT:
- Max. power I.S.O. (kW) 77,2
- Power peak (r/min) 5 800
- Max. usable r/min. 6 500
- Max. torque (N.m) 135,4
- Torque peak (r/min) 3 300

TRANSMISSION:
- Forward speeds . . . . five
- Gearshift. . . . . console
- Low gear . . . . 3,583 to 1
- 2nd gear . . . . 2,235 to 1
- 3rd gear . . . . 1,550 to 1
- 4th gear . . . . 1,163 to 1
- Top gear . . . . 0,959 to 1
- Reverse gear . . . . 3,714 to 1
- Final drive . . . . 3,588 to 1
- Drive wheels . . . . front

WHEELS AND TYRES:
- Road wheels. . . . . .alloy
- Rim width 5,5J
- Tyres . . . . 165/65 SR 14
- Tyre pressures (front) . 200 to 220 kPa
- Tyre pressures (rear) . 200 to 220 kPa

BRAKES:
- Front. . . . . .discs
- Rear . . . . .discs
- Hydraulics . . dual circuit, diagonally split, load sensitive valve at rear
- Boosting . . . . vacuum
- Handbrake position. . .between seats

STEERING:
- Type . . . . rack and pinion
- Lock to lock. . . . . .3,75 turns
- Turning circle . . . . . .10,6 metres

MEASUREMENTS:
- Length overall . . . . . . .4 180 mm
- Width overall . . . . . . .1 620 mm
- Height overall . . . . . . .1 385 mm
- Wheelbase . . . . .2 475 mm
- Front track . . . . .1 400 mm
- Rear track . . . . .1 400 mm
- Ground clearance . . . . . .170 mm
- Licensing mass. . . . . .975 kg

SUSPENSION:
- Front. . . . . independent
- Type . .coil spring struts, lower wishbones, stabilizer bar
- Rear . . . . independent
- Type . . coil spring struts, transverse links, longitudinal reaction rods, stabilizer bar

CAPACITIES:
- Seating. 4/5
- Fuel tank. . . . . 45 litres
- Luggage trunk . . . . 465 dm^3
- Utility space . . . . 940 dm^3

WARRANTY:
12 months irrespective of distance.

TEST CAR FROM:
T.A.K. Motors.

In the boot, the (alloy) spare wheel is covered by a padded hardboard cover with a grip strap and there is a separate jack compartment, complete with toolkit, in the right strut housing. The sill is low with (functional) fog and reversing lights built into the rising portion.

The boot lid glides up on gas struts, and both luggage and engine compartments have lights. In the engine compartment finish and attention to detail are good and there is reasonable access for maintenance.

BEHIND THE WHEEL

In use, our major criticism of the interior was the shape of the rather overstuffed seats, which lack lumbar, thigh and shoulder support but are comfortable over long distances. We would replace the front seats with Recaros. We also did not like the odd glovebox release catch — a knob placed some distance away from the lid.

On the move, the Prisma doesn't show its full mettle until you use the gearbox and the loud pedal, but once clear of stop-start traffic, it comes alive, surging ahead in long strides of smooth power, never electrifying but certainly a cut above the average 2,0-litre car of this size. Unfortunately the gearbox is not one of the better features, for the changes are slow, vague and very notchy, and the test car had to visit the agents during the test for a new clutch, which improved matters but didn't cure them.

You learn to change smoothly with deliberation, but oh for a gearbox which would match the rest of the drivetrain. The engine is powerful, smooth and instantly responsive with a wide spread of torque, and at cruising speeds it copes with hills, slow traffic and long bursts of acceleration with a sense of urgency — a real thoroughbred.

It returns remarkably good figures for a sedan of this size with only 1 600 cm^3 under the bonnet — 0 to 100 km/h in just 11 seconds and a true top speed of 175,6 km/h.

It is the chassis which really impresses, with taut, crisp responses and rock steady roadholding. Balance and poise are remarkable, and the fat Pirellis hang on even in wet or greasy conditions. On a long, fast trip on a round route which included several mountain passes in the pouring rain, it never once felt like losing adhesion — even in hairpin corners under hard throttle.

The brakes are in a similar league, remarkable more for the way they obtain the average stopping times recorded (3,9 seconds) than for the times themselves. The all-disc setup is progressive, fade-free and with ample feel, with that satisfying initial bite discs provide to enable you to scrub off speed very quickly in response to only light pedal pressure. However, they are really set-up for use at speeds higher than 100 km/h and they tend to lock towards the end of a stop, despite the proportioning valve.

TEST SUMMARY

R18 750 may sound like a lot of money for a medium-to-small sedan powered by a 1 600 cm^3 engine. But when the car is compared with the BMW 3-Series, the VW Passat, the Alfa 159 and the Giulietta, the price isn't so out-of-line. And of course you get a measure of exclusivity that the others can't provide. . .

The front seats (above) are comfortable on a long trip but lack lateral support. They are finished in high quality cloth. A look inside the boot (below) reveals attention to detail — the (alloy) spare wheel cover is substantial and has a handle, there is a neat bin for jack and tools in the strut housing and gas struts make the big, low-silled lid easy to lift.

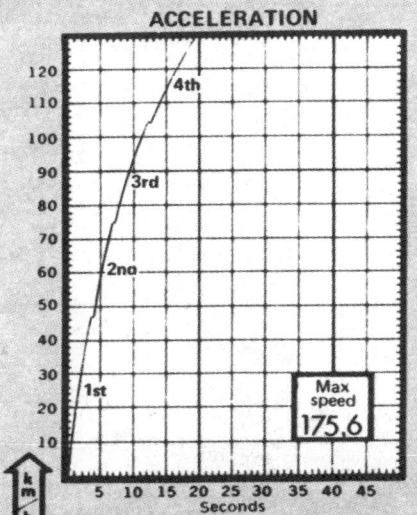

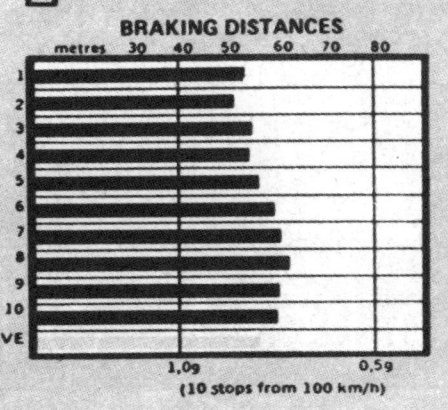

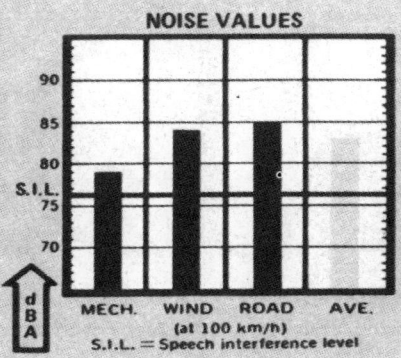

test — Lancia Prisma 1600

PERFORMANCE

PERFORMANCE FACTORS:
- Power/mass (W/kg) net 79,18
- Front area (m²) 2,24
- km/h per 1 000 r/min (top) .. 30,18

INTERIOR NOISE LEVELS:

	Mech.	Wind	Road
Idling	.50	—	—
60	.70	—	—
80	.75	80	83
100	.79	84	85
Average dBA at 100			82,7

ACCELERATION (seconds):
- 0-60 4,82
- 0-80 7,34
- 0-100 11,06
- 1 km sprint 32,47
- Terminal speed ..... 158,5 km/h

OVERTAKING ACCELERATION:

	3rd	4th	Top
40-60	3,72	5,52	7,44
60-80	3,48	5,40	7,32
80-100	4,08	5,20	8,28

MAXIMUM SPEED (km/h):
- True speed 175,6
- Speedometer reading 160
- Calibration:

Indicated:	60	70	80	90	100
True speed:	60,6	72,8	83,6	93	102

FUEL CONSUMPTION (litres/100 km):
- 60 4,90
- 70 5,19
- 80 5,59
- 90 6,12
- 100 6,93

BRAKING TEST:
From 100 km/h
- Best stop 3,64
- Worst stop 4,40
- Average 3,99

GRADIENTS IN GEARS:
- Low gear ..... 1 in 2,7
- 2nd gear ..... 1 in 3,6
- 3rd gear ..... 1 in 5,3
- 4th gear ..... 1 in 7,1
- Top gear ..... 1 in 10,8

(Tabulated from Tapley (x gravity) readings, car carrying test crew of two and standard test equipment)

GEARED SPEEDS (km/h):
- Low gear 47
- 2nd gear 75
- 3rd gear 108
- 4th gear 144
- Top gear 175

(Calculated at engine power peak — 5 800 r/min)

TEST CONDITIONS:
- Altitude ..... at sea level
- Weather ..... fine, light wind
- Fuel used 98-octane
- Test car's odometer ..... 8 560 km

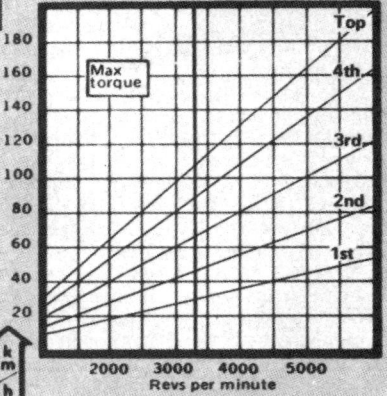

IMPERIAL DATA

ACCELERATION (seconds):
- 0-60 m-p-h 10,66

MAXIMUM SPEED (m-p-h):
- True speed 109,1

FUEL ECONOMY (m-p-g):
- 50 m-p-h 50,36
- 60 m-p-h 42,66

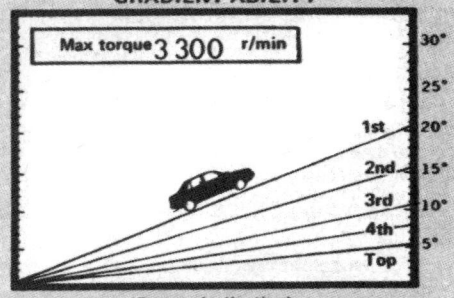

CRUISING AT 100

- Mech. noise level ..... 79 dBA
- 0-100 through gears ..... 11,06 seconds
- Litres/100 km at 100 6,93
- Optimum fuel range at 100 ..... 649 km
- Braking from 100 ..... 3,99 seconds
- Maximum gradient (top) ..... 1 in 10,8
- Speedometer error ..... 2 per cent under
- Speedo at true 100 98
- Tachometer error ..... 11 per cent over
- Odometer error ..... 0,6 per cent over
- Engine r/min at 100 ..... 3 313

Max torque 3 300 r/min

NEWCO

NEW PRISMAS SHOW PROMISE

Mild alterations to the Prisma do some good, as does the arrival of a 2.0litre 4wd version. But quality needs to improve/Georg Kacher

AFTER MANY DIFFICULT YEARS, Lancia is back on the road to success. The turnaround began in 1982, the year the company introduced the notchback Prisma. Since then, sales have rocketed from only 146,000 cars to 226,800 units in 1985. During the past 12months alone, the company's European market share increased by 14.2percent to 2.1percent. Although the most recent growth is due largely to the favourable acceptance of the Thema and the Y10, the Prisma is also more popular than ever. Over the past four years, sales of the compact notchback climbed from 59,000 vehicles in 1983 to a projected 85,000 in 1986.

To cement this positive trend, Lancia has now modified and extended the Prisma range. The new additions are a fuel-injected 108bhp 1.6litre version and the Prisma 4wd which features a permanent 4wd system and a 2.0litre 115bhp engine.

Externally, these mid-year '86 models differ from their predecessors in having revised front and rear bumpers, a bigger Thema-style grille flanked by modified headlamps. The car also has a new bonnet and flush hubcaps. According to Bruno Cena, Lancia's technical director, 'these changes are subtle enough not to hurt the resale value of older cars but big enough to improve the drag coefficient by 0.02 points.' Inside, the car has redesigned instruments, better shaped front seats mounted on longer runners, new trim materials and a more efficient heating and ventilation system.

Not surprisingly, the changes under the skin are more significant than the facelift. The 1.3 and 1.5litre engines, for instance, are now equipped with modified twin-choke carburettors which, together with an integrated fuel-feed cut-off device and electronic ignition, help to improve the fuel efficiency by up to six percent. The carb-fed 1.6litre powerplant, whose maximum output is down from 105 to 100bhp, but has a flatter torque curve, is now installed east-west instead of west-east for superior thermodynamic efficiency and better serviceability.

The top-of-the-line front-wheel drive Prisma is powered by twin-cam 1.6litre unit which boasts the same electronic Weber-Marelli engine management system as is used in the Thema and Croma 'ie' models. The Prisma 1600ie develops 108bhp at 5900rpm and 97.6lb ft of torque at 3500rpm. The same chip-controlled injection and ignition technology is fitted to the 2000ie unit of the Prisma 4wd. This engine, which has counter-rotating silencer shafts to cut noise and vibration, delivers 115bhp at 5400rpm and maximum torque is 118lb ft, available at 3250rpm.

All new Prisma models have four instead of three engine mounting points, additional sound deadening material, softer suspension bushings, uprated springs and dampers, allegedly more responsive steering systems and 27percent bigger (12.6gal) fuel tanks.

It takes only a few miles to discover that the revisions made to the suspension have done the Prisma a lot of good. The ride is now more supple and better balanced (the Thema could learn a lot in this respect from its little brother), the cabin is quite well insulated from tyre and road noise and driveline vibration.

Encounters with deep potholes and sharp ridges are no longer felt throughout the entire body. Fortunately, the car's handling and roadholding have barely suffered from the softer suspension settings. Body roll is kept in check by two fat anti-roll bars, the low-profile 165/65SR14 tyres hug the road as well as ever, and the optional power steering is light, precise and nicely weighted. The slow, vague and very heavy manual rack and pinion steering is to be avoided at all costs, though.

Prisma 4x4 has unusual 56/44 front to rear torque split; leads to car's fine balance. New to Prisma is 4wd's 115bhp injected 2.0litre first seen in the Thema

The Nouva Lancia Prisma is, again, unusually well equipped. Included in the price (still to be determined) are power windows and door locks. The cars also have very comprehensive instrumentation, complete with on-board computer, an adjustable steering column and much more. Still on the credit side, the Prisma has excellent all-round visibility, the well-spaced five-speed gearbox, powerful and progressive brakes and a roomy cabin.

But despite these virtues, the new Prisma is still not as good as it should be. The transmission, for example, which is as sticky and imprecise after four years as at day one. The build quality is still below par if the Italian registered test cars are anything to go by. Recurrent faults on test-cars included squeaking dashboards, carelessly fitted trim and upholstery, one loose head restraint and several stubborn door locks. Criticism is also drawn by the messy facia (confusing push buttons, impractical heater and ventilation knobs, awkward radio position, tiny glove box), the wipers (poor pattern, poor wiped area, lift at speed), plus excessive engine and wind noise. Performance and fuel economy are not bad, though. The 2189lb Prisma 1.6ie accelerates in 10.0sec from 0-62mph, tops 116mph and averages 35.3mpg. The cruising range thus works out at an astonishing 447miles.

Because of a 187lb weight penalty, plus greater friction and rolling resistance, the 2.0litre, 115bhp Prisma 4wd is not quite as lively as the 1.6litre 108bhp 2wd model. The four-wheel-drive car takes an extra 0.5sec to sprint from 0-62mph, its maximum speed is also marginally lower at 115mph, and the ECE fuel consumption works out at a rather less impressive 29.3mpg. On the road, however, the four-wheel-drive edition is the sharper handling, better balanced and much more competent car.

The 4wd hardware, which was developed with Steyr-Daimler-Puch (who also supplies the main components), consists of a centre differential which incorporates a Ferguson Formula viscous control unit, a two-part prop shaft, two rear driveshafts and a rear axle differential. Whereas the soon to be released Lancia Delta Mk2 has a self-locking Torsen rear axle diff, the Prisma must do with a pneumatically operated manual rear axle lock.

Other specific 4wd features include a modified rear suspension, quicker ratio power steering, a revised exhaust system, the relocation of the space-saving spare from under the boot floor to a position inside the luggage compartment (this reduces the boot volume by a full 20percent), fog lamps, sill extensions, light alloy wheels shod with wider 185/60HR14 tyres and redesigned instruments. What the Prisma 4wd still lacks, however, is anti-lock brakes, 'but we are working on an ABS system that is compatible with four-wheel drive,' says Bruno Cena.

Since Lancia has opted for an asymmetrical 56/44 front to rear torque split, the Prisma 4wd retains foolproof handling, but is capable of throttle-on oversteer. Normally, even ambitious drivers will be surprised by the Prisma's tidy and undramatic handling. When we chased it through a series of steep and wet hairpins, traction was always impeccable, and both ends of the car stayed obediently in line as if it were running on rails.

In 1986, Lancia will build a mere 3000 Prisma 4wds all of which will be sold in Italy, Germany, Switzerland and Austria. Production of the first right-hand-drive models should begin early next year, but at this time it is too early to tell when the car will be available in the UK (the new front-wheel-drive models should arrive in June) and at what price. According to a semi-official source, the Prisma 4wd will cost 35percent more than the current 1.6litre version which indicates a tag of just over £9700, some £2600 below the cheapest Quattro. This looks like fair – but only if Lancia manages to sort out the quality problems.

Steyr-Daimler-Puch developed and supply 4wd hardware; includes viscous centre diff and manually locked rear axle diff. The dash has clearer instruments

ROADTEST

The Prisma has had a tough time trying to make an impact in the mid-sized saloon market. Cut-throat marketing and high-profile advertising by the volume car makers has tended to keep the four-door Lancia off the shopping lists of many potential buyers.

The arrival of a substantially revised Prisma, with an aggressive price tag and significant changes to everything from the shape of the body to the position of the engine beneath the bonnet, should give the car a stronger footing on which to fight back.

Major changes include the adoption of Weber/Marelli electronic ignition on the 1.6-litre twin cam engine to improve performance and economy. The bodyshell gets a revised front end with new bumper, grille and bonnet to improve aerodynamics, while subtle changes have been made to the interior.

And as a measure of Lancia's determination to win a slice of this lucrative medium-sized saloon market, the new Prisma has gone on sale at a competitive £7400, undercutting its rivals by a good amount.

The familiar iron-block aluminium headed twin overhead camshaft four has been subtly but substantially revised. There is a new cylinder head and camshafts, new piston crown profiles and Weber Marelli IAW electronic indirect injection which operates in conjunction with a fully mapped electronic breakerless ignition control. The head has also been turned through 180 degrees which puts the exhaust manifold to the front, closer to the airstream, and the whole assembly is canted forward.

Power output is a healthy 108 bhp at 5900 rpm with 100 lb ft of torque at a usefully low 3500 rpm and this is enough to propel the 1017 kg Prisma to a top speed of 116 mph, faster than all our selected rivals with only the Green Cloverleaf Alfa 33 edging close at a whisker below 115 mph. The Lancia needs to be revved, however, to make this kind of progress, a smooth and willing process right to the rev limiter which intrudes at an indicated 7000 rpm, but even then the car's sprinting prowess is less good than some of the rivals. The 60 mph mark is reached in 10 seconds, four tenths slower than the Alfa, seven tenths adrift of the front-drive Vitesse and over a second from the Mazda's 8.8 sec best-of-the-bunch effort.

Despite very short 18.7 mph/1000 rpm overall gearing, the Lancia has to give best in the gears too. In fourth the Alfa, Mazda and Rover can lug ahead from 30 to 50 mph, and even the less sporting Peugeot 309's 7.7 sec effort performance is better than the Lancia's. For all that, the Italian twin-

LANCIA PRISMA 1600 ie

Lancia's booted middle-ranger has a redesigned engine and cleaned up body and offers excellent equipment and performance for the money. Handling and brakes, however, are an acquired taste

cam is pleasant to live with, and appears happy enough cruising on the motorway. High revs are accompanied by a pleasingly sporting rasp from under the bonnet and there is no boom or harshness anywhere in the rev range to deter full use of the engine.

Economy overall, is only fair, on a par with the rivals. An overall thirst of 28.3 mpg is better than the appetite of fellow Italian Alfa 33 and similar to that of the Mazda, but the Lancia's touring consumption reflects its short gearing. At 31.6 mpg, it's the greediest of the bunch. The Rover however is a clear category winner when the performance is taken into account.

In a car which encourages full use of the gears, a shift which is imprecise and rubbery tends to give a false impression of the performance available. The Prisma has good synchromesh, but the change is vague and the spring biasing inhibits gentle handling, particularly into fourth when the pattern has a curious path curving the lever away to the left. The clutch action is light and easy, and there is little transmission snatch to challenge the smooth driver.

The Prisma's handling is something of the proverbial curate's egg, rather depending on the driver's mood. The enthusiast will delight in the energetic turn-in, almost verging on instant oversteer, and this actually increases mid-corner even with power applied, so that tiresome front drive ploughing is never a problem. The flip side is that this characteristic translates to a feeling of slight straight-line instability and a smart outstep from the tail when the power is cut in mid-corner. The steering is low geared and lacking in sharpness and response, and although not particularly heavy once on the move it tended to feel dead, and at variance with the lively nature of the car. It's almost fun, but doesn't quite suit Lancia's image of drivability. Entertaining certainly, restful, no.

Ride however is good. It's firm and a little jittery at low speeds, but the body control is good and the suspension deals with heavy potholes without crashing, and severe undulations without wallowing. The suspicion of waywardness when cruising tends to give the feeling of a soggy ride at high speeds, but the disturbance is lateral rather than vertical and is less disturbing for the driver who has the reassurance of the steering wheel, than the passengers.

The brakes were a disappointment. The pedal action is heavily servo-assisted, but there is some travel before the brakes come on so the pedal is a fair way to the floor even before heavy braking starts. This makes heel-and-toeing extremely difficult and there is a lack of feel to the brakes in general. They did however prove effective in normal use and could be made to lock the wheels even from high speed.

The noise test meter says that the Prisma is noisy at high speeds; certainly the engine makes its presence felt when spinning hard and this is the major, although not particularly unpleasant disturbance. Wind and road noise are quite well suppressed, however.

The Prisma is well made, and representative of the strenuous efforts that the company are making to improve the build quality of their products. There is still some evidence of style's triumph over practicality, and the minor controls, in common with Fiat, use more stalks for wipers, indicators and headlights than do most other manufacturers. The instruments have benefited from a clean-up and are clear and easily read. Heating and ventilation is new and features the programmable automatic system seen throughout the Lancia range. It can seem a little gimmicky at first and needs a little time to learn how to use it effectively, but throughput is good from large face level vents and there is good stratification of cool and warm air, to the face and feet. The seats too have been redesigned and are comfortable and supportive, and there is a new larger range of seat adjustment to cater for even the tallest of drivers, while the steering wheel is adjustable for height. Legroom in the front is an ample 41.8 in and this, combined with headroom of 37.5 in, gives the Prisma class topping front accommodation. The Rover can better the Lancia's front legroom by a fraction, but then its rear legroom is a good two and a half inches less. As the only hatch amongst our rivals, the Mazda can boast more leg and headroom in the rear, but naturally has to give best to the Lancia's boot space. On balance the Prisma is the roomiest of the bunch, but within an overall length of 164 in (4.16 m), which is slightly longer than all but the Volvo.

For the price, equipment is generous. Electric front windows, central locking, tinted glass and rear seat belts all come as standard equipment, as does a split rear seat which expands the rear load area from an already generous and class competitive 0.36 m^3.

The Prisma is very good value for money. £7400 is cheaper than all the selected rivals bar the Mazda which – although quicker and better composed in the chassis department – offers less equipment for its £7349 price tag. The Lancia is a well-equipped, good-value performer with a good top

MOTOR ROAD TEST
LANCIA PRISMA 1.6 ie

Facia layout is new and features automatic heating and ventilating system, visible just above the radio (above)

Seat covering is also new and accommodation is generous, particularly in the front (above). The engine is a tight fit under the bonnet (below) but the revised cylinder head allows the injection to feed neatly from the rear

speed. It lacks a little flexibility perhaps, and the chassis lacks the class that the car's exterior undoubtedly promises. For all that, it's distinctive, likeable and a little different from the mainstream. **M**

Make: Lancia **Model:** Prisma 1.6ie **Country of Origin:** Italy
Maker: Lancia SpA, Turin, Italy
UK Concessionaire: Lancia Ltd, 46-62 Gatwick Road, Crawley, West Sussex RH10 2XF.
Tel: 0293 518933

Total price: £7400 **Options:** None **Extras** fitted to test car: Sun roof, alloy wheels, rear electric windows, metallic paint (LX pack, £895) **Price** as tested: £8,295

PERFORMANCE

WEATHER CONDITIONS
Wind	18 mph
Temperature	59 deg F/ 15 deg C
Barometer	29.8 in Hg/ 1010 mbar
Surface	Dry tarmacadam

MAXIMUM SPEEDS
	mph	kph
Best ¼ mile (5th gear)	188.8	191.1
Terminal speeds:		
at ¼ mile	80	128.7
at kilometre	101	162.5
Speeds in gears (at 6500 rpm):		
1st	32.6	52.5
2nd	52.2	84.0
3rd	75.4	121.3
4th	100.7	162.1

ACCELERATION FROM REST
mph	sec	kph	sec
0-30	3.6	0-40	3.0
0-40	5.1	0-60	4.8
0-50	7.4	0-80	7.3
0-60	10.0	0-100	10.7
0-70	13.3	0-120	15.0
0-80	17.4	0-140	21.6
0-90	23.4	0-160	31.2
0-100	31.4		
Stand'g ¼	17.6	Stand'g km	32.5

ACCELERATION IN TOP
mph	sec	kph	sec
20-40	10.8	40-60	6.8
30-50	11.7	60-80	7.1
40-60	11.5	80-100	7.3
50-70	11.8	100-120	8.0
60-80	13.1	120-140	8.8
70-90	13.6	140-160	11.5
80-100	16.8		

ACCELERATION IN 4TH
mph	sec	kph	sec
20-40	8.4	40-60	5.0
30-50	8.2	60-80	5.1
40-60	8.1	80-100	5.1
50-70	8.4	100-120	5.9
60-80	9.7	120-140	7.4
70-90	11.9	140-160	12.4
80-100	16.8		

FUEL CONSUMPTION
Overall	28.3 mpg
	10.0 litres/100 km
Touring*	31.6 mpg
	8.9 litres/100 km
Govt tests	28.8 mpg (urban)
	45.6 mpg (56 mph)
	33.8 mpg (75 mph)
Fuel grade	97 octane
	4 star rating
Tank capacity	57 litres
	12.5 galls
Max range*	395 miles
	636 km
Test distance	703 miles
	1132 km

*Based on official fuel economy figures – 50 per cent of urban cycle, plus 25 per cent of each of 56/75 mph consumptions.

STEERING
Turning circle	10.5 m 34.4 ft
Lock to lock	3.8 turns

NOISE
	dBA
30 mph	67
50 mph	72
70 mph	78
Maximum†	85

†Peak noise level under full-throttle acceleration in 2nd

SPEEDOMETER (MPH)
True mph	30	40	50	60	70	80	90	100
Speedo.	31	41	51	63	74	84	94	104

Distance recorder: 1.7 per cent fast

WEIGHT
	kg	cwt
Unladen weight*	1017	20.0
Weight as tested	1210	23.8

*No fuel

Performance tests carried out by Motor's staff at the Motor Industry Research Association proving ground, Lindley, and Millbrook proving ground, near Ampthill.

Test Data: World Copyright reserved. No reproduction in whole or part without written permission.

GENERAL SPECIFICATION

ENGINE
Cylinders	4 in line
Capacity	1585 cc
Bore/stroke	84/71.5 mm
Max power	108 bhp (80.5 kW) at 5900 rpm (DIN)
Max torque	99.8 lb ft (135.3 Nm) at 3500 rpm (DIN)
Block	Cast iron
Head	Aluminium alloy
Cooling	Water
Valve gear	Double overhead camshafts driven by toothed belt
Fuel system	Weber IAW electronic injection.
Ignition	Weber IAW, fully mapped.
Bearings	5 main

TRANSMISSION
Drive	To front wheels
Type	5-speed, synchromesh
Internal ratios and mph/1000 rpm	
Top	0.96/18.7
4th	1.16/15.5
3rd	1.55/11.6
2nd	2.24/8.0
1st	3.58/5.0
Rev	3.7141
Final drive	3.714

AERODYNAMICS
Cd	N/A

SUSPENSION
Front	MacPherson struts, lower wishbones, anti-roll bar
Rear	MacPherson struts, transverse links, longtitudinal reaction rods, anti-roll bar.

STEERING
Type	Rack and pinion
Assistance	None

BRAKES
Front	Ventilated discs 25.7 cm dia
Rear	Discs, 22.7 cm dia
Servo	Yes
Circuit	Diagonal safety split
Rear valve	Yes

WHEELS/TYRES
Type	Pressed steel 5½J x 14 in dia
Tyres	165/65 SR 14
Pressures F/R	
(normal)	29/29 psi 2.0/2.0 bar
(full load)	32/32 psi 2.2/2.2 bar

ELECTRICAL
Battery	12 V, 45 Ah
Alternator	65 Amp
Fuses	14
Headlights	
type	Quartz halogen
dip	110 W total
main	120 W total

GUARANTEE
Duration	3 years unlimited mechanical
Rust warranty	6 years anti-corrosion.

MAINTENANCE
Major service	6000 miles

THE RIVALS

BMW 316 4dr, £8445; Citroën BX16 TRS, £7122; Renault 11 TXE, £6995; Ford Orion 1.6 GL, £7481; Vauxhall Belmont 1.6 GLS, £7962; VW Jetta 1.6 TX, £7149

LANCIA PRISMA 1.6ie

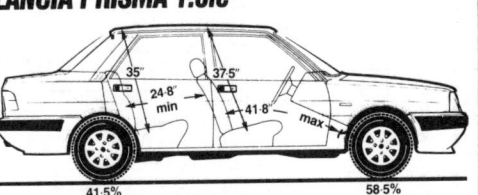

Length 4·19m (164·5") Width 1·62m (63·7") Front track 1·40m (55·1")
Wheelbase 2·47m (97") Height 1·38m (54·5") Rear track 1·40m (55·2")

£7400

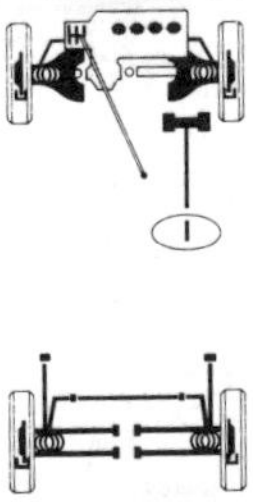

Capacity	1585
Power bhp/rpm	108/5900
Torque lb ft/rpm	99.8/3500
Max speed, mph	116
0-60 mph, sec	10.6
30-50 mph in 4th, sec	8.2
mph/1000 rpm	18.7
Overall mpg	28.3
Touring mpg	31.6
Weight kg	1017
Drag coefficient Cd	N/A
Boot capacity m³	0.35

Revised version of Lancia's middleweight Prisma gets fuel injection and a price tag that undercuts most of the opposition. Good performance with average. Balanced handling tending towards tail happiness, but steering low-geared and lacks sharpness. Ride however is well controlled and taut. Brakes are powerful though heavily overservoed. Overall roomy, comfortable and well equipped. Build quality is good too.

ALFA ROMEO 33 GREEN CLOVER LEAF

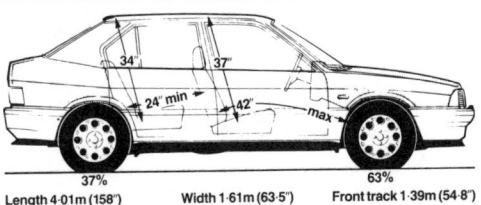

Length 4·01m (158") Width 1·61m (63·5") Front track 1·39m (54·8")
Wheelbase 2·45m (96·5") Height 1·30m (51·3") Rear track 1·36m (53·5")

£7769

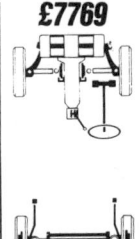

Capacity	1490
Power bhp/rpm	105/6000
Torque lb ft/rpm	98.3/4000
Max speed, mph	114.7
0-60 mph, sec	9.6
30-50 mph in 4th, sec	7.9
mph/1000 rpm	18.8
Overall mpg	26.7
Touring mpg	33.3
Weight kg	899
Drag coefficient Cd	N/A
Boot capacity m³	0.35

The most powerful 33 yet, with the 105 bhp "boxer" engine, alloy wheels and low-profile tyres. Top speed is competitive, as is its mid-range punch and sprinting ability, but economy is below par. Good grip and handling, but a little nervous on bumpy surfaces under power. Firm yet comfortable ride, excellent accommodation and supportive seats. Well equipped, for the price, but lacks the ultimate finesse and charm of the early 'Sud Ti.

MAZDA 323 1.6i

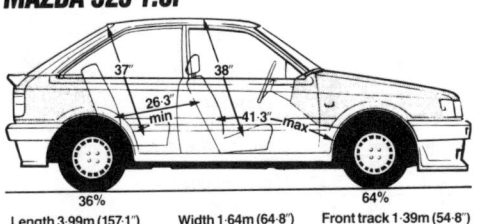

Length 3·99m (157·1") Width 1·64m (64·8") Front track 1·39m (54·8")
Wheelbase 2·40m (94·5") Height 1·39m (54·7") Rear track 1·42m (55·8")

£7349

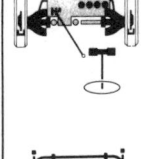

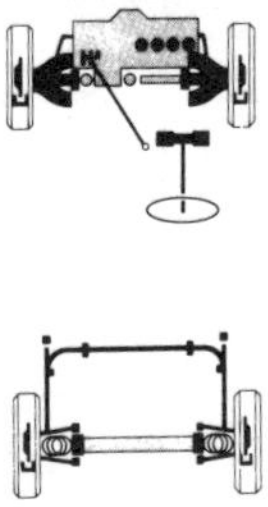

Capacity	1597
Power bhp/rpm	104/6000
Torque lb ft/rpm	101/4200
Max speed, mph	112.2
0-60 mph, sec	8.8
30-50 mph in 4th, sec	7.1
mph/1000 rpm	19.5
Overall mpg	26.6
Touring mpg	31.7
Weight kg	960
Drag coefficient Cd	0.37
Boot capacity m³	0.25

Sporting derivative of latest 323 is powered by new 1600 cc fuel injected engine. Reasonable outright performance, good flexibility but economy and refinement are disappointing. Able chassis gives crisp handling and taut well-controlled ride. Good accommodation, heating and ventilation and equipment levels but bland unimaginative interior. Understated appearance could thwart sales, but overall, fine value for money.

PEUGEOT 309SR

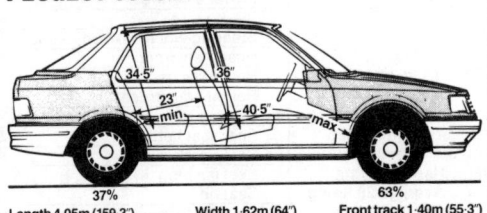

Length 4·05m (159·3") Width 1·62m (64") Front track 1·40m (55·3")
Wheelbase 2·46m (97·3") Height 1·38m (54·3") Rear track 1·37m (54")

£7495

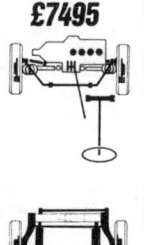

Capacity	1580
Power bhp/rpm	80/5600
Torque lb ft/rpm	98/2800
Max speed, mph	102.3
0-60 mph, sec	10.7
30-50 mph in 4th, sec	7.7
mph/1000 rpm	22.3
Overall mpg	30.2
Touring mpg	41.0
Weight kg	860
Drag coefficient Cd	0.33
Boot capacity m³	0.39

Important British-built Escort/Astra rival from Peugeot – the 309 is a grown up Peugeot 205 to replace the Talbot Horizon. Only average performance and economy from 1580 cc XU engine, but excellent flexibility and smoothness. Fine ride and handling but sensitive brakes. Roomy low drag body has a generous boot and is tidily trimmed. An able car with no serious vices. SR is top of range and has good equipment level.

ROVER 216 VITESSE

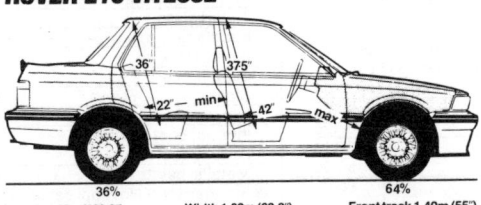

Length 4·16m (163·8") Width 1·62m (63·8") Front track 1·40m (55")
Wheelbase 2·46m (97") Height 1·37m (54") Rear track 1·44m (56·8")

£8727

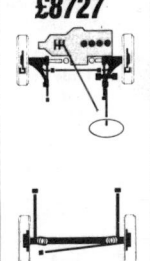

Capacity	1598
Power bhp/rpm	103/6000
Torque lb ft/rpm	102/3500
Max speed, mph	109.6
0-60 mph, sec	9.3
30-50 mph in 4th, sec	8.0
mph/1000 rpm	19.7
Overall mpg	31.9
Touring mpg	39.2
Weight kg	926
Drag coefficient Cd	0.40
Boot capacity m³	0.42

The fuel-injected S-series engine gives the Vitesse lively, if rather thrashy, performance with its strength lying in the mid-range. Economy is competitive, too. Good grip and taut responses are two ingredients of the Rover's fluid handling, and the ride is good given the short wheel travel. Super slick gearchange, neat instruments and good visibility are other virtues. Smart alloy wheels and plushly-trimmed interior complete the picture.

VOLVO 340GLE 1.7 4dr

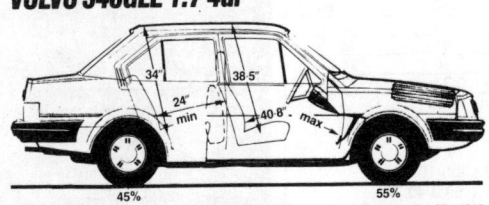

Length 4·41m (173·8") Width 1·65m (65") Front track 1·37m (54")
Wheelbase 2·39m (94·3") Height 1·39m (54·8") Rear track 1·34m (53")

£7500

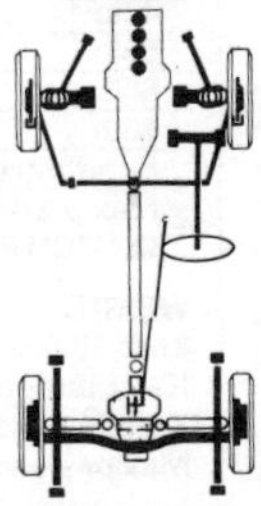

Capacity	1721
Power bhp/rpm	80/5400
Torque lb ft/rpm	97/3000
Max speed, mph	100.2
0-60 mph, sec	11.7
30-50 mph in 4th, sec	9.3
mph/1000 rpm	23.0
Overall mpg	27.8
Touring mpg	39.4
Weight kg	1025
Drag coefficient Cd	0.37
Boot capacity m³	0.38

Strong-selling 340 series improved by the adoption of new 1.7 litre engine. Performance and economy adequate, though still inferior to most rivals, but flexibility and smoothness are good. Good ride, but potentially good handling marred by poor traction. Traditional shortcomings of poor packaging and an indifferent gearshift remain, but the Volvo is a well equipped, solidly built car at a competitive price.

TEST UPDATE

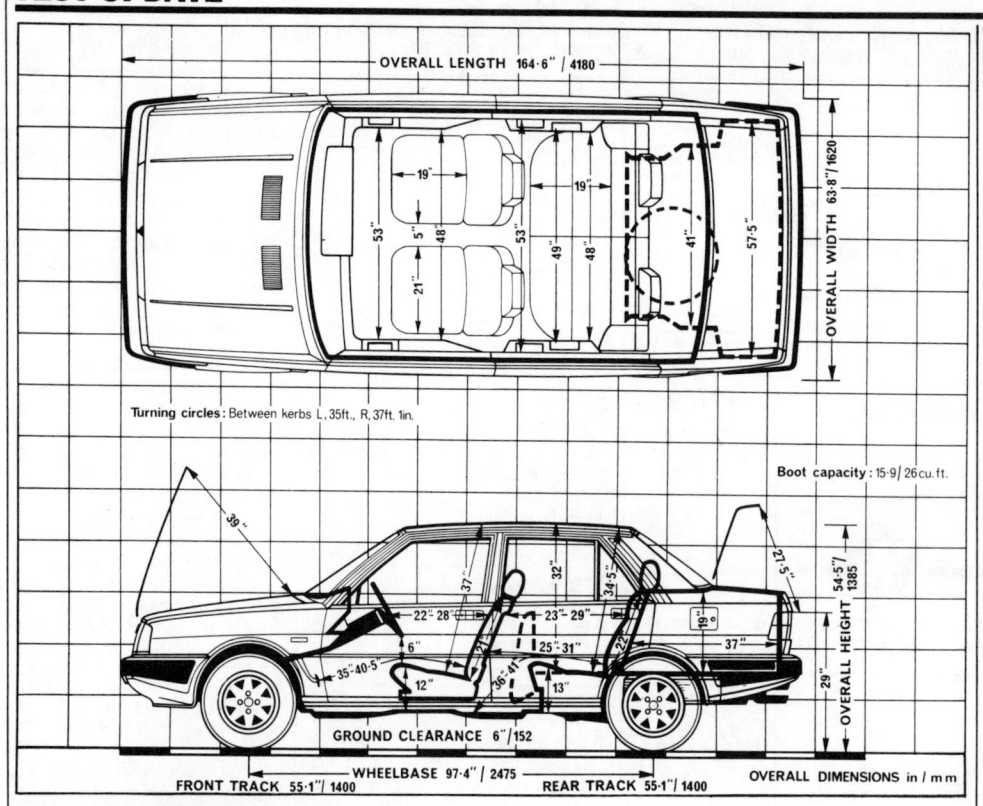

Overall length 164·6" / 4180
Overall width 63·8" / 1620
Overall height 54·5" / 1385
Front track 55·1" / 1400
Rear track 55·1" / 1400
Wheelbase 97·4" / 2475
Ground clearance 6·0" / 152
Boot capacity: 15·9 / 26 cu. ft.
Turning circles: Between kerbs L, 35ft., R, 37ft. 1in.

Overall dimensions in / mm

MODEL

LANCIA PRISMA 1600ie LX
PRODUCED BY:
Lancia, Via V, Lancia 27, 10141 Torino, Italy.

SOLD IN THE UK BY:
Lancar Ltd., 46-62 Gatwick Road, Crawley, W. Sussex RH10 2XF

SPECIFICATION

ENGINE
Transverse front, front-wheel drive. Head/block al. alloy/cast iron. 4 cylinders in line, bored block, 5 main bearings. Water cooled, electric fan.
Bore 84mm (3.31in), **stroke** 71.5mm (2.81in), **capacity** 1585cc (96.8 cu in).
Valve gear 2 ohc, 2 valves per cylinder, toothed belt camshaft drive. **Compression ratio** 9.7 to 1. Electronic ignition, electronic Weber IAW fuel injection.
Max power 108bhp (PS-DIN) (79.4kW ISO) at 5900rpm. **Max torque** 135.4lb ft at 3500rpm.

TRANSMISSION
5-speed manual. Single dry plate clutch 7.9in dia.

Gear	Ratio	mph/1000rpm
Top	0.959	18.69
4th	1.163	15.41
3rd	1.550	11.56
2nd	2.235	8.02
1st	3.583	5.00

Final drive: helical spur, ratio 3.588.

SUSPENSION
Front, independent, MacPherson strut, lower wishbones, coil springs, telescopic dampers, anti-roll bar.
Rear, independent, struts, parallel trailing arms, longitudinal links, coil springs, telescopic dampers, anti-roll bar.

STEERING
Rack and pinion. Steering wheel diameter 14.5in, 3.8 turns lock to lock.

BRAKES
Dual circuits, split diagonally. **Front** 10.1in (257mm) dia discs. **Rear** 8.9in (227mm) dia discs. Vacuum servo. Handbrake, centre lever acting on rear discs.

WHEELS
Alloy, 5.5in rims. Radial tubeless tyres (Goodyear Grand Prix on test car), size 165/65R14, pressures F29 R29 psi (normal driving).

EQUIPMENT
Battery 12V, 40Ah. Alternator 65A. Headlamps 110/120W. Reversing lamp standard. 14 electric fuses. 2-speed, plus intermittent wipers. Electric screen washer. Air blending interior heater; Cloth seats, cloth headlining. Carpet with heel mat floor covering. Scissor jack; 2 jacking points each side. Laminated windscreen.

PERFORMANCE

MAXIMUM SPEEDS

Gear	mph	kph	rpm
Top (Mean)	116	187	6200
(Best)	119	192	6300
4th	103	166	6700
3rd	77	124	6700
2nd	54	145	6700
1st	33	90	6700

ACCELERATION FROM REST

True mph	Time (sec)	Speedo mph
30	3.2	32
40	4.9	41
50	6.9	50
60	9.5	60
70	12.7	72
80	16.7	82
90	22.8	92
100	31.0	103

Standing ¼-mile: 17.3sec, 81mph
Standing km: 31.7sec, 102mph

IN EACH GEAR

mph	Top	4th	3rd	2nd
10-30	—	—	7.3	4.3
20-40	12.7	9.4	6.2	3.9
30-50	12.7	9.0	5.8	4.0
40-60	11.9	8.5	5.9	—
50-70	11.9	8.0	6.1	—
60-80	14.1	8.8	7.2	—
70-90	16.5	10.9	—	—
80-100	20.7	14.0	—	—

CONSUMPTION

FUEL
Overall mpg: 27.1 (10.4 litres/100km) 6.0mpl
Autocar constant speed fuel consumption measuring equipment incompatible with fuel injection.
Autocar formula: Hard 24.4mpg
Driving Average 29.8mpg
and conditions Gentle 35.2mpg
Grade of fuel: Premium, 4-star (97 RM)
Fuel tank: 12.5 Imp galls (57 litres)
Mileage recorder: 2.3 per cent long
Oil: (SAE 15W/40) negligible

BRAKING
Fade (from 81mph in neutral)
Pedal load for 0.5g stops, in lb

start/end		start/end
1 40-15	6	50-85
2 40-25	7	60-125
3 40-25	8	60-110
4 40-35	9	60-85
5 40-65	10	60-85

Response (from 30mph in neutral)

Load	g	Distance
20lb	0.30	100.3ft
30lb	0.46	65.4ft
40lb	0.58	51.9ft
50lb	0.79	38.1ft
60lb	0.91	33.1ft
70lb	0.96	31.4ft
Handbrake	0.31	97.1ft

Max gradient: 1 in 3
CLUTCH Pedal 20lb; Travel 6in

WEIGHT
Kerb 20.9cwt/2338lb/1058kg
(Distribution F/R, 59.3/40.7)
Test 24.5cwt/2748lb/1243kg
Max payload 992lb/450kg

COSTS

Prices
Basic	£6658.19
Special Car Tax	£554.84
VAT	£1081.96
Total (in GB)	**£8294.99**
Licence	£100.00
Delivery charge (London)	£200.00
Number plates	£20.00
Total on the Road	**£8614.99**
(excluding insurance)	
Insurance group	7

EXTRAS (fitted to test car)
Stereo radio/cassette and manual aerial £83.00
Total as tested on the road **£8697.99**

SERVICE & PARTS

	Interval		
Change	6000	12,000	18,000
Engine oil	Yes	Yes	Yes
Oil filter	Yes	Yes	Yes
Gearbox oil	No	No	No
Spark plugs	Yes	Yes	Yes
Air cleaner	Yes	Yes	Yes
Total cost	£46.45	£81.40	£46.45

(Assuming labour at £18.40 an hour inc VAT)

PARTS COST (inc VAT)
Brake pads (2 wheels) front	£35.49
Brake pads (2 wheels) rear	£35.49
Exhaust complete	£226.41
Tyre — each (typical)	£51.11
Windscreen	£126.59
Headlamp unit	£86.11
Front wing	£92.56
Rear bumper	£168.90

WARRANTY
36 months/unlimited mileage, 6-year anti-corrosion

EQUIPMENT

Automatic	N/A
Economy gauge	●
Electronic ignition	●
Five speed	●
Limited slip differential	N/A
Power steering	N/A
Rev counter	●
Steering wheel rake adjustment	●
Headrests front	●
Heated seats	N/A
Height adjustment	N/A
Lumbar adjustment	N/A
Rear seat belts	●
Seat back recline	●
Seat cushion tilt	N/A
Seat tilt	N/A
Split rear seats	●
Electric windows	●
Heated rear window	●
Interior adjustable headlamps	N/A
Sunroof (manual)	●
Tinted glass	●
Headlamp wash/wipe	N/A
Central locking	●
Cigar lighter	●
Clock	●
Fog lamps	●
Internal boot release	N/A
Locking fuel cap	●
Metallic paint	●

● Standard N/A Not applicable DO Dealer option

TEST CONDITIONS
Wind: 13mph
Temperature: 10deg C (50deg F)
Barometer: 29.7in Hg (1006mbar)
Humidity: 69per cent
Surface: dry asphalt and concrete
Test distance: 798miles

Figures taken at 60/78 miles by our own staff at the Motor Industry Research Association proving ground at Nuneaton.
All *Autocar* test results are subject to world copyright and may not be reproduced in whole or part without the Editor's written permission.

DESIGN BY DEGREE

FOR:
PERFORMANCE
HANDLING
AGAINST:
STEERNG
VISIBILITY

The Lancia Prisma has made the transition from hatchback to booted saloon very successfully while retaining its sporting character. We tested the new engine configuration in the recently introducd LX model

A number of well-known hatchbacks have made the transition to booted, three-box saloons — the Ford Escort-Orion, Volkswagen Golf-Jetta, and Fiat Strada-Regata ranges to name but three examples — but for some reason the Lancia equivalents, the Delta and Prisma, sometimes seem to be overlooked in comparisons of such cars. Both models deserve rather better, since they are certainly competitive against their obvious rivals, both in terms of performance, practicality, and dynamic behaviour.

At first glance, the latest top-of-the-range Prisma, the LX, powered by a fuel-injected version of Fiat-Lancia's well-known 1585cc four-cylinder, twin-cam engine, appears little changed from the Prisma 1600 tested by *Autocar* (30 July 1983). Granted, there is a new, more aerodynamic front bumper, pronounced chin spoiler, new bonnet with raised section on the leading edge echoing the frontal treatment of the Thema, and revisions to the windscreen strips, window frames (now of anodised aluminium) and drip channels, but these are hardly of earth-shattering significance.

Rather more important is what has taken place under the bonnet, for the 1.6-litre engine has been turned through 180deg and inclined forwards at an angle of 18deg. Furthermore, the engine block itself has been lowered by 46mm to improve the car's centre of gravity, and it is secured in the engine bay with four rather than three engine mounts to reduce driveline 'shunt' and help damp out vibration.

In this configuration, the exhaust manifold is now routed down the front of the engine rather than the back, which means the exhaust is directly exposed to the air flow, thereby aiding cooling and maintaining a lower under-bonnet temperature. The intake manifold is therefore behind the engine, a position which apparently helps prevent icing effects in carburettor-equipped Prismas as well as aiding fuel efficiency. Numerous internal engine components have been modified to improve the unit's thermodynamic efficiency, including the cylinder head, piston crown profile and camshafts, while other changes include a new design of camshaft cover and timing belt casing.

Arguably the most interesting development as far as this new Prisma is concerned, however, is the fitment of the Weber Marelli IAW integral fuel injection and ignition system previously seen on the Fiat Croma and Ford Sierra RS Cosworth. This controls the fuel injection and ignition timing with a comprehensive electronic 'map' as well as engine start-up and slow running. Sensors feeding a central microprocessor monitor air intake pressure and temperature and the ECU in turn controls slow running, provides for the sequential and phased control of the individual injectors and controls the spark advance on the basis of engine speed and load. The primary gains from this system are said to be in the areas of fuel economy and exhaust emissions.

In this latest form, the power output of the 1.6-litre twin cam is 108bhp at 5900rpm (previously 105bhp at 5800rpm) while the torque remains the same as the previous Prisma at 100lb ft generated at 3500rpm. The five-speed gearbox contains the same intermediate and final drive ratios as the Prisma 1600 tested in 1983, while tyre size also remains identical at 165/65SR14.

The suspension layout is unchanged in basic specification, although there have been a number of detail alterations, notably to the geometry and spring and damper rates. One worthwhile alteration has been to the fuel tank, which has ▶

Prisma's 1.6 *twin cam has been turned through 180deg*

New seats *and heating systems are included in the interior revisions*

Spare wheel *is stowed neatly under capacious boot floor*

Heating: *now press-button*

grown in size by a healthy 2.6 gallons to 12.5, thus endowing the Prisma with a useful increase in range. Steering remains of the non-power assisted rack and pinion variety with 3.8 turns required to get from lock to lock.

There are a number of interior refinements, again with obvious Thema inspiration. The seats, for instance, are of a new design with stronger foam padding and springing for improved support, while fore and aft movement of the front seats is facilitated by the use of ball bearings in the runners.

In addition, new upholstery fabrics with improved wear characteristics have been incorporated in the Prisma interior. There is also a new heating and ventilation arrangement, the air blending heater being fed from a larger air intake with distribution and air flow notably improved.

Clearly the power and torque gains for the Prisma's power unit in this latest, fuel-injected guise do not fall into the 'startling' category; nonetheless our figures reveal that the Lancia's performance is usefully improved.

Although the 0-30mph time for the old and new 1.6-litre models is the same at 3.2secs — indicating reasonable traction from a standing start — thereafter the 1600 LX opens out an impressive lead of 1.3secs by the 60mph mark (9.5secs compared with 10.8secs) and and 7.3secs by the time 100mph is reached (31secs versus 38.3secs), and that with good test conditions in each case.

Interestingly, there doesn't seem to be a similar level of disparity in the fifth gear incremental figures. The injected Prisma is in fact 1.7secs slower than its carburettored predecessor over the 30 to 50mph increment, 0.4secs slower from 50 to 70mph and then reverses the tables by a similar amount from 70 to 90mph, possibly an indication of the revised engine's improved breathing at higher revs.

Despite the fact that the overall gearing remains unchanged at 18.69mph per 1000rpm, there has been quite an impressive improvement in the Lancia's maximum speed. The mean figure achieved by the Prisma LX around MIRA's high-speed circuit worked out at

Steering column *is adjustable*

Systems failure *is monitored*

116mph with the engine turning over at 6200rpm, while the previous car could only manage 108mph, corresponding to 5700rpm. Once again, this is an indication that the twin-cam power unit is now pulling rather more strongly at higher engine speeds with the result that the maximum speed is achieved 300rpm above the stated power peak of 5900rpm.

Subjectively the Prisma in LX guise feels quite an eager performer, although in the best Italian sporting tradition, the engine note makes its presence very clearly heard the moment the driver becomes at all enthusiastic with the accelerator pedal. The fuel-injected power unit's cold start characteristics are good, the engine firing into life after just a couple of cranks of the starter but, funnily enough, power delivery is notably hesitant until the twin cam has started to build up some heat after approximately a mile.

One might also reasonably expect a car from a manufacturer with the heritage of Lancia to handle well — and the Prisma LX does. With front-wheel drive and a 59.3-40.7 nose-heavy weight distribution, it is also a fair prediction that understeer will be the dominant handling trait.

Push the Lancia hard through a corner and certainly it does understeer, although not in as strong a fashion as one might suspect. In fact, when the cornering limit is reached, the Prisma simply spins an inside front wheel grudgingly, which in turn slows the car down and allows the front wheels to regain traction. Lifting off the power in mid-corner causes little attitude change, the front end merely 'tucking in' slightly.

One area with which we were less than impressed was the steering. This rack and pinion system, which requires 3.8 turns of the steering wheel to get from lock to lock, is quite heavy at low speeds, making parking manoeuvres rather hard work. Yet despite this attribute, which is often associated with high-geared steering, normal urban driving requires far more wheel 'twirling' than one might expect. It is also quite noticeable that there are several inches of free play either side of the straight-ahead position which renders the steering 'rubbery' at higher speeds and therefore not as reassuring as it might be.

Refinement, as has already been partially suggested, is good rather than outstanding, the growl from the engine being the prime contributor to the overall noise picture at urban speeds, although this is augmented by a fair amount of gear whine in the intermediate ratios. When cruising at the 70mph limit on a steady throttle, wind noise enters the picture, with a slight roar detectable from around the A-posts. Road noise is generally well controlled throughout the speed range.

Fuel economy turned out to be another area showing improvement over the previous Prisma model. The LX recorded an overall figure of 27.1mpg, which is an improvement of 0.6mpg over that achieved by the previous version.

On climbing behind the steering wheel of the Lancia, the first impression is that little has changed. The instrument binnacle contains a straightforward, clearly marked analogue speedometer (marked to 120mph) and rev counter (the solid red segment begins at 6700rpm) plus gauges for battery condition, fuel contents, and engine coolant temperature. There are the usual clusters of warning lights in this area, as well as a vehicle schematic to warn of any major system failures. Also situated in the centre of the facia are the controls for Lancia's automatic heating and ventilation system. These consist of rocker-type switches which control the volume of fresh air, and the temperature. There is also an 'Auto' button which allows the cabin temperature to be preset and then maintained.

The most frequently used controls are looked after by the usual Fiat-Lancia trio of steering column stalks, while the column itself is adjustable for rake. The driving position is good, with a reasonable steering wheel-pedals-seat relationship, the brake and accelerator pedals being well positioned for heel and toe down changes. The seats seem to offer a little more support in the critical areas than those fitted to earlier Prismas, although some testers observed that they could still do with more lumbar and side support and that the cushions could be a little longer. Forward visibility is good thanks to the slim windscreen pillars, while the over-the-shoulder view poses few problems. Rearward visibility is another matter, however; the thick C-posts and bulky rear seat headrests obscure the view out of the back window and conspire to make reversing manoeuvres more a case of guesswork than they should be.

Detailed criticisms aside, though, the Prisma has made the transition from hatchback to booted saloon rather more successfully than one or two of its obvious rivals. It has also managed to retain the Italian breed's sporting nature, thereby endowing it with a certain amount of character in a market segment which is perhaps noted more for its blandness than anything. Now if only the Lancia engineers could do something about that steering, they really would be on to a winner with the Prisma. ∎

Large rear headrests *obstruct view, as do C-pillars*

Front bumper/spoiler *distinguishes the new Lancia Prisma LX*

Delta Hot Hatches

While the medium size hatchback was growing in popularity as a family car, there was also a demand for higher performance versions of these cars combining a sporting character with everyday practicality. Several manufacturers had entered this market with three-door models, but the five-door Delta offered greater comfort and carrying capacity. So it was that three years after the original Delta was introduced Lancia launched their first "hot hatch" in 1982.

Delta 1600 GT carburetted model

Like their competitors Lancia adopted the "GT" designation for the new car even though none of these modified family cars bore any real relationship to true Grand Touring cars. Powered by the 1585cc twin-cam engine developed for the Beta range, the new Delta was known as the 1600 GT. The twin-cam power unit gave 105bhp (DIN) at 5800rpm with a maximum torque of 13.8mkg at 3300rpm. The resulting performance was well up to class standards with a top speed of 112mph and a 0-60mph time of less than 10 seconds. the 1600 GT was distinguished from the smaller-capacity models by some minor exterior styling changes, most visibly black window frames and door handles in place of the bright metal of other versions. The car was well-equipped with Digiplex electronic ignition, disc brakes all round, alloy wheels, tinted glass, and electrically-operated front windows.

Right-hand drive versions of the 1600 GT arrived in the UK in June 1983 while later that year Lancia launched an even hotter version of the Delta. This was the 1600 HF Turbo, which had a turbocharged version of the same twin-cam engine. Lancia had built up considerable experience of turbocharging the twin-cam engine in the Beta Montecarlo Turbo with which they had won the World Sportscar Championship in 1980 and 1981. A prototype of the turbocharged Delta was shown at the Turin motor show in April 1982 so it was no suprise that a production model should appear during the following year.

The 1600 HF Turbo was claimed to be the "ultimate hot hatchback" when it was introduced and with 130bhp giving it a top speed of 121mph and 0-60mph time of 8.6 seconds the claim was well justified. A Garrett T3 turbocharger with air-to-air intercooler provided the boost, while other features included Microplex digitally-controlled ignition, sodium-cooled exhaust valves, and a ZF five-speed gearbox. Interior equipment was the same as that of the 1600 GT with the addition of Recaro front seats as standard in the UK. One unusual feature of the HF Turbo was that the road wheels were of 340mm diameter, limiting the choice of tyres. In 1985 more conventional 14inch wheels were fitted.

The HF Turbo did not replace the 1600 GT, both models continuing in parallel. Like the other Delta models, both the 1600 versions received the suspension improvements and styling "facelift" in 1986. In addition, the power units were upgraded by the incorporation of electronic fuel injection, identified by the letters "ie" in the model names. The effect on the GT version was marginal; power output increased by 3bhp to 108bhp giving a maximum speed of 115mph and 0-60mph time of 9.7 seconds. The change to fuel injection was more noticeable in the HF Turbo with power output rising to 140bhp and maximum torque, with overboost, of 21mkg at 3750rpm. Top speed was now

Delta 1600 HF Turbo

126mph and the 0-60mph time was reduced to 8.4 seconds.

Versions of both models equipped with catalytic converters for the exhaust were added to the range in 1987. The impact of the change is illustrated by the reduction in power output to 90bhp and 132bhp for the GT and HF Turbo, respectively, while their maximum speeds were lowered to 109mph and 123mph in common with the rest of the Delta range, the 1600 models underwent a final "facelift" in 1991. There were no mechanical modifications to the cars, but paired headlights of the type fitted to the early four-wheel drive models were an obvious external change. Internally the changes were limited to the materials used for the upholstery and trim. Production ended when the replacement model was introduced in April 1993.

Models covered in this section: 1600 GT, 1600ie GT, 1600 HF Turbo and 1600ie HF Turbo.

Production Data

Model	In production from	to	No. built
1600 GT	1982	1986	25,330
1600 HF Turbo	1983	1986	9,711
1600ie GT	1986	1991	29,441
1600ie HF Turbo	1986	1991	21,668
1600ie GT	1991	1993	8,151
1600ie HF Turbo	1991	1993	4,372

OnTheRoad
FANCIER LANCIA

The Delta — 1980's Car of the Year — faces the future with a selection of engine and transmission changes, while the new twin-cam GT1600 takes aim at Europe's hot hatchbacks. Russell Bulgin reports.

STOP PRESS: Lancia have *not* restyled the Delta. Given the discreetly urbane shaping of Giorgetto Giugario's original concept for the 1980 Car of the Year, any radical reworking of the razor-edged wedge was deemed unnecessary from an aesthetic standpoint. No, any revisions to the Delta — three seasons into its role as Lancia's mid-size hatchback contender — are concerned with wider motives than mere cosmetic enhancement.

Between 1980 and 1983, Lancia are scheduled to spend $418 million to introduce new models and update existing production facilities. Sales at home in Italy are increasing — the Delta finally slid into the Top Ten best sellers on the domestic market, for the first time, in September — and both production and exports are rising steadily. Given that huge investment, new models are on the way, and the launch of the upcoming three-box Prisma appears to have much to do with the restructuring of the Delta range. That still-secret saloon will plug the wide gap which now exists between the Delta and Trevi models.

Although only the Lancia Delta 1500 has been available in Britain, a 1300 variant is marketed in most European territories, and that version gets an uprated 78bhp higher-compression engine giving, Lancia claim, both increased performance and economy.

The existing 85bhp 1500 is now available only with 3-speed automatic transmission (except in Sweden and Switzerland); and a new GT1600 model is introduced for 1983.

Predictably, Lancia have worked on reducing both weight and aerodynamic drag in the revised cars. But, in developing the twin-cam GT1600 to give 105 bhp at 5,800 rpm, and reducing the car's bulk, Lancia found they were pursuing conflicting objectives; the new variant required some strengthening of the main structure to reduce noise transmission to the passenger compartment. Thus a number of peripheral components were rethought, and finally 88 lb was skimmed off the kerb weight. Windows, dashboard, battery, radiator, and the thermoplastic bumpers were all slimmed down to fight the flab.

As well as shaving 22 lb off the bumpers, the integral front valance was reshaped to form a modest chin spoiler as part of the drag-reducing campaign. Perhaps emphasising the technical independence of Fiat and Lancia, while the recently revised Strada has lost its neatly raised trailing edge roof the Delta receives a subtle add-on lip, disguised in body colour. Lancia claim a 15 per cent reduction in drag — and a two or three mph top speed increase — for these small changes. Perhaps significantly, given the Delta's bluff-fronted profile, an overall drag coefficient is not quoted.

In Lancia's strangely American argot, the GT1600 is "a personal car; a sport-GT machine in the Lancia tradition". The company cite the car's qualities as offering good comfort, equipment and engine characteristics, without extremes of performance; you must assume the order as quoted to be important, given the

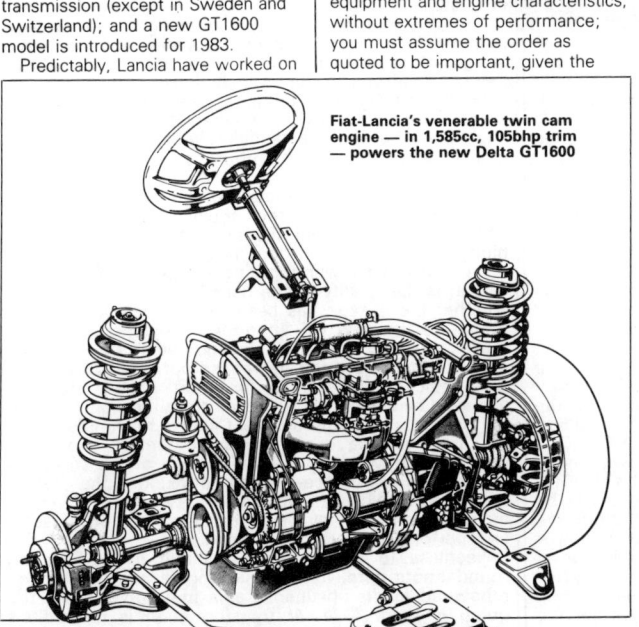

Fiat-Lancia's venerable twin cam engine — in 1,585cc, 105bhp trim — powers the new Delta GT1600

Spring 1983 introduction of the 4wd turbo model to clean-up in the all-out performance stakes.

Lancia explain that the GT1600 motor is the pre-Volumex Trevi unit, mildly modified for its new installation. Displacing 1,585cc, the five-bearing, alloy headed twin cam unit has had the ports reworked to improve torque and fuel consumption, new-material exhaust valves specified, a wide bore exhaust fitted and the adoption of the Digiplex inductive discharge ignition system as standard. Optimum engine performance is — naturally — the aim, with the spark advance curve programmed by a mini-computer responsive to a number of criteria in a service-free system.

With Lancia pitching the GT1600 at luxury or sporting hatchback models from Ford, Volkswagen, Opel and Volvo, the omission of fuel injection from the engine specification may seem odd; Lancia stayed with a single Weber carburetter and rejected fuel injection on grounds of cost, and technical and servicing complexity.

The five-speed gearbox gets new ratios, and the front MacPherson struts are redesigned. They now feature new damper and spring rates to cope with the revised demands of the new model. Low rolling-resistance Pirelli P8 65-series tyres are mounted on one-inch larger, 14in diameter wheels, and the Delta becomes one of the few 1600cc cars to use disc brakes all round.

Italian clothes and fabric designer Ermenegildo Zegna drew up the upholstery material for the new seats, the front pair being redesigned to give greater rear leg-room; there is greater noise-deadening material on the front bulkhead; and full dashboard instrumentation is standard. Emphasising the car's upmarket target, options include a six-function trip computer, air-conditioning, central locking, and a sunroof.

Driving impressions

So, the GT1600's specification implies that this latest Delta should be a nimble, comfortable yet sprightly machine. Is it? A test drive indicated that the GT1600 follows Lancia custom in offering a fine ride-handling compromise biased towards driver enjoyment. The ride gets a little jiggly at town speeds, but fast progress on a winding road begins to bring out the best in the car, allowing the light and direct steering to be enjoyed and the impressive new braking system to be exploited. Lancia timing gives the GT1600 a 0-62 mpg time of 10.2 seconds and a top speed in fifth of around 112 mph, with better-than-previous-model fuel economy.

On the debit side, the gearchange is a little rubbery and occasionally tetchy over its long-throw gate, and the new ratios seem a little short for totally relaxed motorway cruising. Taller gearing would compromise the expected Lancia brio without markedly improving fuel consumption, say Lancia; that may be so, but combined with a much needed reduction in front-pillar wind noise at motorway speeds it would go some way towards matching the easy-on-the-eye visual sophistication with its aural equivalent. Although the engine is claimed to be built to substantially different specifications to the similar unit mounted in its Fiat Strada 105TC cousin, it still seems too harsh and noisy when worked hard.

Unusually for an Italian car, the driving position — complete with adjustable tilt steering column — suits occupants who are taller rather than short, and the subtly-patterned seats seem hardbacked on first acquaintance but comfortable and supportive as the miles roll by.

Next spring is a likely launch date. In Italy, the GT1600 costs just five per cent more than its 1500 predecessor, while the 1300, by comparison, is cheaper by a similar percentage.

Stop Press: Lancia in new car price drop shock . . .?

PUTTING IT IN PERSPECTIVE
Type: Lancia Delta 1300, 1500 and GT1600
When available: Early 1983
Prices: UK prices unavailable at present
Engine: 1,301cc ohc four-cyl; 1,498cc ohc four-cyl; 1,585cc dohc four-cyl
Transmissions: 5 speed manual on 1300 and GT1600 models; 1500 available only with 3 speed automatic
Dimensions: 151in length; 62in width; 53in height

Lancia Delta 1600GT

As it was always meant to be

It is always interesting going back to a fondly remembered car after an interval of a couple of years of more; particularly if it is a new, more tempting version. In view of the current crop of advances evident in the newest designs, can it be as impressive second time around? It is especially interesting when the car concerned is an ex-Car of the Year (COTY) winner.

...the 1600GT delivers its promise, with no small thanks to the fitment of a Fiat/Lancia Digiplex ignition system which serves to smooth-out the twin cam to a remarkable degree...

I remember a week of varied motoring in Scotland with the original Delta 1500, including leisurely touring, and other times behind the wheel in a more self-indulgent mood. The overriding memories are of the concept: compact, very highly specified and with an extremely competent and attractively proportioned body designed by Giugiaro.

Less impressive were some of the trim details and, principally, there rests in the memorybank the frustration of a wild and woolly gearchange with a throw which could at best be described as 'variable' and a temperament which was awkward to the point of complete non co-operation.

In fairness, it must be added that the road test was carried out *before* the UK launch date on a car which had been sampled by several press persons on a one-day-thrash basis.

It turned out that the 1600GT sampled here *was* more impressive on a second acquaintance, even if it was a more fully equipped and more powerful extension of the range. Of most interest is the installation of the 1585cc, 105bhp twin-cam engine from the Fiat Strada 105TC. It gives the Delta the urge and eagerness it has always been able to handle, and with the forthcoming turbo HF (*CCC* February issue) the emphasis on performance will be taken a logical stage further. But good though the 1600GT's performance is, and the HF's is (with 130bhp – see Feb '84 **CCC**) perhaps the ultimate personification of the concept would be a Delta fitted with the 1995cc, fuel injected, 122bhp engine from the Lancia Coupe/HPE. *That* would be a car-and-a-half. With suitably raised gearing such a Delta would really represent the philosophy of the small car with a large engine, endowing it with the kind of easy-going performance which that 'Gran Turismo' nomenclature deserves.

Certainly, the Delta's taut, positive handling, and superb discs-all-round braking system could manage a power hike of around 20 percent, perhaps even without recourse to the wider rubber of the close-ratio 'box HF model.

In most respects the 1600GT is thoughtfully equipped. External changes include a deeper front bumper/spoiler assembly and a discreet rear lip roof spoiler which are between them said to improve the drag coefficient by 10 percent. Most noticeable of the internal changes is a typically Italian (and that means High Style) wool seat fabric, designed exclusively by one Ermenegildo Zegna (just thought you'd like to know that).

The rest of this Chivasso-engineered car is as you would expect: complete. The only options are a steel sliding sunroof (excellent and quite draught-free at ordinary UK cruising speeds), and metallic paint. This means that the Delta 1600GT is one of the best equipped small/medium hatchbacks on the market.

A comfortable interior features such discreet niceties as individual reading lamps, spring-loaded grab handles, glovebox-mounted passenger's vanity mirror in addition to the expected full instrumentation, individually split flip-down rear seat, rear wash/wipe, tinted glass (laminated screen of course), electrically operated front window lifts, etc.

The driver and his or her passengers are indeed pampered, but Lancia's name is synonymous with the image of producing drivers' cars. Here the 1600GT delivers its promise, with no small thanks to the fitment of a Fiat/Lancia Digiplex ignition system which serves to smooth-out the twin cam to a remarkable degree, without any of the hesitancy and major flat-spot traumas previously associated with carburetted versions of this venerable engine.

To this end, it feels as though a higher final drive could be employed, for though the car will achieve a creditable 110mph plus top speed, the snarl of this comparatively high revving twin cam does become obtrusive.

But the most rewarding single aspect of this sports saloon is the quality of the gearchange. I do not know whether the linkage has been re-engineered, or if it is thanks to the new UK importer's insistence on better quality control, both in Italy and when UK cars are inspected here, but the gearchange of the *CCC* test car was radically better; not far short of the standard achieved by VAG and Ford in their sporting small saloons.

This one single improvement serves to transform the car and increase driving pleasure tenfold. No more worries with 'will-it-or-won't-it' stirring of the stick; the driver is given peace-of-mind to explore the Delta's superb front wheel drive handling (tyres are 165/65 14 inch diameter on alloy rims) and beautifully progressive and performance-orientated brakes.

Needless to say there is a shade of swings and roundabouts here, for the Fiat engine has never been renowned for its abstinence as a fuel tippler when the going is hard. A lesser transformation is the Delta's

ROADWORDS

steering: sharper and with less 'rubber' in the system than we remember.

Our verdict: a good, well-meaning motor car, which has finally been given the attention to detail of build quality and mechanical refinement that it always deserved. At a shade under £6000 it is also remarkable value by UK standards. Perhaps the GT nomenclature is for Gentleman's Transport, for a 2-litre engine would complete the picture; but with Italian tax laws the way they are, engines are rarely larger than they really need to be.

Ian Sadler

A series of original Giugiaro body design concepts for the Delta with all the major production elements in evidence. Note three-door version (top) that never was.

73

LANCIA'S SECOND WIND

After a period in the doldrums, Lancia blow back and follow their mechanically supercharged VX models with a turbocharged Delta HF due to come to these shores in the spring. Ian Wearing went to Lancia's Chivasso factory in Italy to see the HF being built and try one on the road

Dashboard is similar to Delta GT but with vacuum fluorescent bar gauges

Interior is very Italian with plenty of black plastic. Recaro seats are optional in Italy

THAT Lancias rust has become something of a cliche. Old ladies in the street who couldn't tell a Cortina from a Cavalier will talk knowledgeably about the corrosion problems of the Beta range — Esther Ranzen saw to that. Such is the power of the popular press that after the Beta 'expose', Lancia might well have packed up and gone back to Italy for all the cars they were going to sell over the following months.

But that was in the past. Since then, coincident with a complete reorganisation of Lancia's Chivasso factory near Turin, rust proofing has demonstrably improved to a level that the factory say is up with the very best in the world.

I saw for myself the lengths they go to in order to hold the dreaded brown stuff at bay with a high-thickness cataphoresic dip being the crucial process. Only time will tell just how successful this has been because the full treatment has only been in operation since 1982 and the six year corrosion warranty now offered by Lancia has nearly four years left to prove itself.

Certainly it would seem logical and possible that a shock-horror public outcry should guarantee an over-reaction to the rust problem that can only be in the consumer's interest.

But the reason I was in Turin (with due respect to Dr Pianta's clean and efficient looking factory) was to get my hands on what will almost certainly be the next Lancia to be released in the UK, the turbocharged Delta HF.

Such was the tight schedule that my drive was more cursory than detailed, but I drove far enough to get the feeling whether it was (and here's another cliche) a hot hatchback in the GTi/XR3i mould.

Starting with a Delta GT as a base, the HF has the ubiquitous Garrett T3 turbocharger mounted behind the 1585cc engine collecting the exhaust via very short manifold pipes. The driven side of the turbine goes via an air to air intercooler and thence to the Weber 32DAT progressive twin-choke carb. The carb is housed in a sealed box and thus uses the 'blown' method favoured for fast response. Despite a compression ratio drop from 9:1 on the GT to 8:1, a Marelli Microplex black box ignition system incorporates a knock sensor with a selection of 16 different advance curves available over the complete rev range.

By sensing the approach of knocking and computing the engine speed, a suitable curve is selected until conditions change again (perhaps less throttle is being applied) and more advance can be fed back. Maximum power comes at 5600rpm and is rated at 130bhp, Max torque is at 3,700rpm.

A ZF five speed gearbox is used with top gear giving just 21mph/1000rpm, so an economy gear it is not! In the true tradition of Italian machinery, gearing is low throughout giving the consequent hard working thrum that is music to Latin ears.

Internally, UK specification is yet to be decided but is likely to be Recaro seats, electric windows and central locking as standard. A high-tech dash with bar-chart vacuum fluorescent 'moving' bars is unique to the HF model being used for oil pressure, water temp, oil temp, volts, fuel level and turbo boost.

Exterior changes are few. Most obvious are the subtle black 'Miniskirts' along the

Delta HF Turbo

sills carrying a subtle 'Turbo' motif, the same deep bumper/spoiler used on the GT at the front and a roof spoiler at the rear. Air intakes are dropped into the bonnet and, for the observant, TRX tyres are fitted on bastard size (around 13½") rims. Throw in a bit more matt black than the GT version and that's about it.

I drove the left hand drive HF mainly on dual carriageways and the occasional town road — including a bit of Turin rush hour traffic. The most immediate trait to its character was its docility — in every respect. Light controls, fuss-free engine, easy gearchange and, until you started to push the car over 3500rpm, no hint that 130bhp lie under the bonnet. With only a fleeting chance to try a standard car some time before, it was obvious that the HF was considerably quieter — without the characteristic Fiat/Lancia exhaust crackle — and smoother to boot. A turbo unit does much to act as an additional silencer.

Over some pretty diabolical Turin *pave* roads, the suspension (softer springs, harder shockers) was quiet and compliant with only the odd crash over the most extreme pot-holes. Once past the magic 3500rpm and with the right foot hard down, the turbine whine built up and a smooth flow of power came in, not suddenly but in a progressive surge. The little twin-cam thrummed away happily past 6000rpm never sounding strained and always eager. Throttle lag was minimal and (often a turbo shortcoming) low speed response was exemplary with none of the fluffiness so often found even in more upmarket installations.

Lancia reckon the HF is good for 122mph and 0-60mph in around 8.0 seconds. The VW GTi is their target and, whilst these two figures look highly competitive, the HF does not have that irrepressible urge in the mid range when the GTi's naturally aspirated 1800cc engine makes itself felt.

But comparisons with the GTi are unnecessary because the HF has its own character. And it's comfortable.

Those Recaro seats are not of the superhard variety and the adjustable thigh support (sounds like surgical underwear) helps suit all drivers. What I didn't like about the interior was the typically Italian black plastic look around the roof pillars — much too claustrophobic.

Back on the road again, with a sustained burst on a main dual carriageway 200kph came up on the clock which translates to 125mph — so the quoted 122mph looks good. At this speed the HF showed excellent straight-line stability and had a sure-footed and forgiving feel around the curves. As I said, there was no opportunity to take liberties with the handling, that will have to come later.

As we went to press, no firm decision has been made about bringing the HF to Britain. It will come, probably around May, but the specification is up for discussion.

Most likely will be a high-spec with a price around £7,500, making it a viable alternative to an XR3 or GTi. One thing is sure, in a market where upwards of 20,000 units are sold each year, Lancar's aspirations of about 400 HF sales in the UK would seem to be attainable.

Turbocharger is tucked away at the back of the engine and feeds an air to air intercooler at the side

Above: *From above there's nothing to tell you that this is turbocharged apart from the intercooler*
Left: *Subtle changes to body, such as small black side skirts and black roof spoiler, distinguish the HF from the ordinary non-turbocharged 1600 Delta GT*

ROAD TEST

The handling of the new Lancia Delta Turbo is in the fun category with steering and gearing characteristics encouraging aggressive driving.

Lancia Delta HF Turbo: power packed hatch

MIKE McCARTHY examines a lively new Lancia turbocharged sports hatchback.

Twin cams *and* a turbocharger is a combination usually only found on out-and-out racing cars: yet that is part of the specification offered by Lancia on their Delta HF Turbo model, the top of the Delta range. It goes part of the way to explain Lancia's claim that the model is 'The fastest 1600 5 door hatchback' in the UK. Does it live up to the claim?

The Turbo is based on the successful GT1600, itself a distant off-shoot of the Fiat Strada. The chunky, razor-edged styling of the body is by Giugiaro, and our test car was a special edition version, featuring white paint work with highly distinctive 'Martini' red and blue stripes down the side. This is a no-cost option on the white models, and, if it is a little boy-racerish for your tastes, you don't *have* to have it, while it is agreeable to see sponsors getting their money's worth elsewhere than on the track or in advertisements.

The engine is a 1585cc four, featuring the afore-mentioned belt-driven twin overhead camshafts. Modifications to cope with the extra power and heat generated by turbocharging include the use of an oil cooler, an increase in engine oil capacity, pistons with lower crowns to reduce the compression ratio from 9.3:1 to 8.0:1, sodium-cooled valves, molybdenum-coated rings, a special head gasket, and part of the exhaust system made out of stainless steel.

The turbocharger is a Garrett T3 unit which, at full blast, is pressurising the intake system at 1.7 bar (about 24psi), which is quite high. There is an air-to-air intercooler between compressor and carburettor, which is a Weber 32 DAT twin-choke device.

With this set-up, Lancia claim 130bhp (DIN) at 5600rpm, and 142lb ft (DIN) torque at 3700rpm. Coincidentally, this is almost exactly the same as Fiat claim for the 2-litre twin-cam engine in the Strada Abarth 130TC (130bhp (DIN) at 5900rpm and 130lb ft (DIN) torque at 3600rpm), so this is a fair guide as to how much *in practice* can be gained by turbocharging. On the track, of course, the sky's the limit . . .

As is common nowadays, the engine and transmission are set across the nose of the car and drive the front wheels. Clutch diameter has been increased to cater for the higher torque, while a ZF five-speed close ratio gearbox is fitted. Fifth gives about 21mph/1000rpm, so is not the usual, very high, 'economy' ratio but a useable one.

Above: Prominent HF grille insignia. Left: space age cockpit with plenty of push button instrumentation. Below: Transverse 1586cc engine, belt driven twin overhead camshafts and Garrett T3 turbocharger. Lancia claim a maximum speed of 121mph.

Rerated springs

Suspension is by MacPherson struts all round, with anti-roll bars at each end. For the turbo, the springs have been rerated: up from 65mm/kg to 75mm/kg at the front, but *down* from 65mm/kg to 60mm/kg at the rear, while damper calibrations have been reset as well. More positive response from the rack and pinion steering has been achieved by adjustment to the geometry. There are servo-operated disc brakes all round, and the light alloy wheels are fitted with Michelin TRX AS 170/65R340 tyres.

Inside, 'bar-graph' instrumentation is used for minor information (fuel level, oil pressure, oil temperature, water temperature, volt meter and turbo pressure), though analog units are retained for the speedometer and tachometer. There were special Recaro seats on the test car, part of the 'executive pack', which also includes a steel sunroof, central door locking, and a headlamp wash-wipe facility.

Lancia claim a top speed of 121mph, and a 0-100kph (62.2mph) time of 8.9s for their 'fastest 1600 five-door hatchback'. In fact, they're relying rather heavily on the '1600 five-door hatchback' part of the claim. *What Car?* recorded a maximum of 118mph, and took 9.5s to reach 60mph from a standstill, a little lower than those claimed, but with what was admittedly a fairly tight car. The same team recorded 119mph and 7.9s respectively for the Fiat Strada 130TC, 118mph and 8.4s for the Golf GTi, 118mph and 8.4s for the Vauxhall Astra GTE and 115mph and 9.1s for the Ford Escort XR3i. These all provide close competition in the same class as the Delta Turbo, and as can be seen are quicker — but only the Escort has a 1600cc engine, the others being bigger but not blown, and of course all are *three-door* hatchbacks . . .

Sluggish at low revs

In practice, we were disappointed with the power unit. When cold it was a reluctant starter, and tended to stall and hesitate for the first few minutes after driving off. More significantly, though, was the marked contrast between the on-boost and off-boost modes. At low revs and throttle openings it feels distinctly sluggish and, though it pulls cleanly, it doesn't do so strongly. On the other hand, when the turbo cuts in, there is a fairly rapid increase in power, and you find yourself at the next corner somewhat sooner than expected. It isn't quite the 'all or nothing at all' characteristic of very early turbocharger set-ups, but it isn't the smooth, transitional flow of power that we've come to expect from the latest arrangements. If you're in a boy-racer mood, it can be great fun: if you're not, it can be simply irritating. In this sense we feel that driveability has been sacrificed somewhat for sheer performance, and this is a pity. (To be honest, we tried a second car in which this characteristic was reduced, but not by much — it was also a rather difficult car to drive smoothly.)

Unpleasant gear change

To make flowing driving even more difficult, the gear change on our test car can only be described as nasty. First and second in particular were notchy and obstructive if you tried to change gear gently. Slam the lever through from one gear to the next, however, treating it roughly, and then the change became quick and slick — once again, pandering to the boy-racer instincts. If you are in a hurry, and do want to use all the performance, the closeness of the ratios

ROAD TEST

in the box is ideal, with no appreciable gaps. Fifth is just right, too — high enough to make fast cruising reasonably relaxed, but not so high that you have to drop down to fourth or even third for every slight obstruction on a motorway.

Considering that, on the whole, we tended to drive the car hard, preferring the characteristics when the turbo was in operation, the fuel consumption we measured, 30.8mpg, was not unreasonable, and was very close to the official government figure for a steady 75mph, 30.4mpg.

Yet another factor which rather spoiled town driving for us was the steering. At parking speeds it was impossibly heavy, and doesn't really lighten up throughout the speed range. It also feels as if there's an excess of rubber in the system at first, but as you become attuned to it you realise that in fact it is quite good, with plenty of feel and very little kick-back or torque reaction.

'Fun' handling

Like its rivals which we've mentioned before, the handling of the Delta Turbo is in the 'fun' category. Most of the time it is delightfully neutral, but not unexpectedly, it understeers more and more strongly as the limit is approached, which may sound boring but in this particular car isn't. Nor is it deflected by mid-corner bumps, and lifting-off after a dose of over-enthusiasm into a corner produced mild, predictable tuck-in. The grip from the Michelin TRXs is excellent in the dry, but we were not too impressed with their wet-weather adhesion, though this could be as much a factor of the engine characteristics as those of the tyres.

And, like many Italian cars, we found the brakes over-servoed at low speeds, so that coming to a gently stop required practice. At the other end of the scale, though, they work remarkably well, being more progressive and fade-free. The ride is comfortable, taut, but not rock-hard, while there is a fair amount of roll in tight corners in spite of the anti-roll bars at each end.

Where the Delta Turbo will appeal to many who might not generally consider a sporting car is the fact that it has four doors, thus eliminating the battle past the front seat back to enter the rear seat.

A no cost option on the (white) Lancia Delta Turbo is Martini race image striping as worn by the Lancia Group C endurance challenger.

Having said that entry and exit are easier, though, we don't reckon there's all that much more space than the others inside. Those Recaro seats in the front, though superbly comfortable, do take up quite a lot of room. Thus, when the front passenger's seat is pushed forward to give legroom in the back, the driver finds his left hand knocking the knee of his passenger when selecting first and second. Another common complaint from taller drivers was lack of headroom.

Space-age cockpit

The front cockpit of the car may look rather space-age, but to us has some peculiar flaws. Ergonomically it is very strange: there's a row of identical push buttons in the middle of the top of the facia, poorly identified; the 'bar-graph' instruments in the instrument pod are hidden by the steering wheel and left hand, and those down on the console almost invisible as well; other minor controls are fiddly; those who are anti-social enough to smoke find that the gear lever gets in the way of the ashtray when it is in first, third or fifth; and the switches for the electric windows disappear beneath the gear lever gaiter . . .

In the hot weather we enjoyed while we had the Delta Turbo we made good use of the sliding roof to give additional air-flow to that provided by the plethora of vents on the facia. We didn't try the heating . . .

On the whole, the Delta Turbo is not a particularly refined car. The engine has that typical, throaty, crisp roar of a twin-cam Italian machine, though the turbocharger mutes it somewhat, while wind and road noise are about average. Fit and finish, too, is respectable if not outstanding — certainly not as good as that of the Golf GTi.

Summing up, the Delta Turbo is like the little girl who, when she was good, was very, very good, but when she was bad was horrid. On the plus side is the performance if the engine is used hard, the excellent road manners, the economy, and the fact that it has four doors. Against it is the poor low-speed performance, the gearchange, the steering when parking, and the strange interior and control layout. The sort of car that becomes more fun the harder you drive but no fun in traffic or town conditions. ■

Pronounced Giugiaro body styling on Lancia's five door 1600cc hatchback gives a little extra space inside, but insufficient headroom.

LANCIA DELTA HF TURBO £7990 (with Executive Pack)

Specification

Cylinders/capacity	Four in line, 1585cc
Bore × stroke	84 × 71.5mm
Valve gear	Twin ohc, belt driven
Fuel system	Turbocharger, intercooler, Weber 32 DAT carburettor
Power/rpm	130bhp (DIN) at 5600rpm
Torque/rpm	142 lb ft (DIN) at 3700rpm
Gear ratios	3.583, 2.235, 1.542, 0.903:1
Final drive	3.4:1
Steering	Rack and pinion
Brakes	Servo-assisted discs all round
Wheels	Light alloy, 135 TR340 FHB
Tyres	Michelin TRX AS 170/65 R340
Suspension (F)	Independent, MacPherson strut, anti-roll bar
Suspension (R)	Independent, MacPherson strut, anti-roll bar

Dimensions

Wheelbase	97.3ins
Track (F/R)	55ins F/R
Length	153.1ins
Width	63.7ins
Weight	19.6cwt

Performance
(*What Car?* figures)

Maximum	118mph
0-60mph	9.5s
50-70mph (4th/5th)	5.2/9.6s
Urban/56/75mph	26.2/41.5/30.4mpg
Overall test consumption	30.8mpg

LANCIA

ROAD TEST

"The fastest 1600cc, five door hatch back on the road," is how Lancia bill their Delta HF Turbo, which sells for nearly £8,000 as tested. It certainly is fast with a turbocharged twin cam engine, but how well is it made?

The milky white Lancia Delta we tested definitely provided "A Lotta Bottle" in terms of stripes and no less than six screaming Turbo logos. Although you could not mistake the turbocharging message, many were puzzled by the HF appelation instead of the usual GTi/XR/GTE performance initials. What's it all about?

In the sixties and seventies (Lancia was purchased entirely by Fiat in 1969) the Turin-based company competed in races and rallies with increasing success that revived pride in past achievements too. For Vincenzo Lancia had founded the marque with road racing success and considerable engineering ingenuity. The fifties saw Lancia in sports cars racing and providing a GP car that was eventually handed over to Ferrari. Cesare Fiorio begun a career that continues as Fiat-Lancia motorsport supremo at Abarth's premises, by running the fledgling Lancia competitions effort of the sixties, as well as driving for the team. Fiorio told us, with a distinct twinkle, that HF simply stood for "High Fidelity," which was regarded as a very chic sounding club tag in the Italy of the sixties.

The initials stuck — as did the distinctive emblem of a galloping red elephant — for many seasons. They have come to denote the quickest in Lancias for over 20 years.

The Delta design owes mechanical parentage to Fiat's Ritmo/Strada front drive hatchbacks and made its debut in September 1979. The Giogetto Giugiaro five door lines were distinguished enough for it to win a Car of the Year Award, but in Britain the Lancia name became synonymous with rust and importers (then Fiat owned, now part of the Heron Corporation who also control UK Suzuki sales) hardly benefitted from that prestige title. In November 1982 the Delta was restyled slightly, still around a Golf II-comparable 97.4 in wheelbase, and a 1600 Fiat Twin Cam GT variant introduced. In June 1983, about a year before Britain, the LHD markets started to receive the first 130bhp turbo HF models, still with the five door feature that stands out in the ranks of three door hot hatchbacks.

As ever, Lancia offer a lot of engineering features, even considering that UK

Although liked by the ladies, the Martini striping of the test car was apparently provocative enough to evoke a degree of hostility from other road users we have not previously encountered

DELTA HF TURBO

ROAD TEST LANCIA DELTA HF TURBO

tax-paid prices start at £7,250 and the model we tried retails for £7,990. The double overhead camshaft engine provides 130bhp, like the recently introduced Fiat 130 TC in the same price bracket. Instead of two litres from Fiat's contender, this 1585cc uses a Garrett T3 turbocharger and air heat exchanger to provide its carburettor horsepower. The gearbox is the same, save one ratio, for Fiat and Lancia — with excellent close ratios from ZF.

Effective ride and handling have long been a Lancia front drive bonus characteristic. Lancia also provide four wheel disc brakes to tame a pace that we measured slightly beyond 120mph.

For £740 over the lowest HF price in UK, we tried an Executive Pack that comprised a rapidly efficient manual sunroof, front Recaro seats, central locking and headlamp wash/wipe. As the car was in white it gained the Martini striping and hostility from other road users to a degree we have not previously experienced.

Interior and controls

Whereas the white exterior panels were adorned with (functional) twin bonnet vents, stripes, badges and the apparently mandatory menu for hot hatchbacks of additional aerodynamic aids, the interior was soberly and effectively trimmed "Exclusivo per Lancia" by Ermene Gildo. Even the efficiency of front Recaro seats is perfectly matched to a standard of trim fit and finish that is competitive with the best Japanese/Germanic efforts, but with the bonus of imaginative flair: score Lancia a maximum on this front, including the usefulness of five centrally locked doors.

Look forward from that superb seating and Lancia's resident interior instrumentation demon, scores on high tech style, with bar-graph illuminated subsidiary dials. But that designer should be awarded the Design Council's Supreme Inefficiency Trophy for the siting and legibility of a supplementary six-gauge layout. The chances of reading the central boost gauge, or battery volts, without running up the back of another road occupant are remote; for you have to look down and across to decipher these central console scales. As for the similar illuminated trickery recording oil pressure and temperature, water heat, and an erratic readout of remaining fuel within the 45 litre/9.9 gallon tank, the initial fascination turns to irritation. That is long before the 200 or so miles fuel range, indicated by the fuel gauge, has expired.

Conventional round dials cover a 140mph speedo and an 8,000rpm rev-counter, with a yellow warning line from 6,000rpm to a 6,300rpm redline. As ever Lancia have supplied a lot, but the presentation, and therefore usefulness suffers. Typical of their approach is a heating and ventilation system with the adequate outlet grilles controlled by two rotary and one direction knob for each outlet group. Or the daunting seven button line that has to be mastered to produce a single desultory sweep of the rear wiper with attendant spray.

The driving position deserves no Italian ape jibes, a simple and elegant black three spoke steering wheel adjusting to provide excellent man and machine interface with the complimentary seating comfort. Visibility is good straight ahead, and immediately behind. Yet Delta's shape has dated and numerous pillars — especially those hindquarter supports for the largest TURBO billboarding — which can produce worrying moments at road junctions. There are vision restrictions simply when joining a motorway, when the manually adjustable mirrors tend to start shaking from 60-65mph.

The test car came complete with a £77.80 Panasonic LW/MW/FM radio and stereo cassette player, with only one of the Uher speakers connected! We fixed that in five minutes, but we were never able to make the light switch produce more than sidelights or one headlamp and no taillights; neither did the interior light function on any switch position. During a busy day's testing we also repaired the throttle linkage at the carburettor end, when it simply came unscrewed and flopped apart... At this point we begun to wonder if Lancia ought to be pricing in the Golf GTi league?

Driving the HF

"Watch it when it's cold" commanded a colleague as we took over the HF and the advice was luckily heeded, for this turbo liked to stall on its auto choke. It would start promptly enough, but then would die within seconds. At about the third attempt it would drive normally, water temperature coming on the scale at 60°C within a Summer mile, with any sign of oil temperature delayed for several miles.

At the opposite end of the water temperature scale, the HF turbo developed the equally annoying habit of stalling repeatedly when queued for more than ten minutes at a time in an ambient temperature of 25°C, hardly more than a mild summer day to a Southern Italian.

Driving memories are dominated by a superb chassis and a distinctive engine note. The twin camshafts are slightly muffled by the turbocharger, which provides a seductive low whistle of pleasure whenever the accelerator is depressed between 2,000 and 4,000rpm, after which the toothed belt driven overhead camshafts and the increasing rpm beneath that alloy head predominate on the noise level front.

Generally the upright Delta is dominated by engine noise that will be enjoyed by most enthusiastic owners until 90-95mph, when the wind starts to make its presence felt around the front door edges.

Although our brake testing figures did not produce exceptionally curtailed distances from quadruple discs, and there was a degree of braking swerve that echoed the lighter and stubbier Honda CRX, the feeling on the road was of repeatable stability. At first the vacuum servo and unique (in our experience) 175/65 Michelins, on TRX 13.4 in diameter derivatives of the familiar Lancia eight spoke alloy wheels, tended to lock-up wheels rather too easily. Generally the braking system is a fitting counterpart to that vigorous 130 horsepower thrust.

Although the gearbox is internally the same, (bar an 0.903 fifth gear in place of an 0.967), as that of Fiat 130 TC, there was an initial problem selecting second. This occurred whenever the front passenger seat was pulled forward to accommodate shorter adults. Otherwise the shift is not quiet so precise as that of the Abarth, and not quite so pleasurable to operate as the best in the class, but has absolute speed on its side when hurrying.

Judged purely as a driving machine the Lancia provides pleasurable speed with outstanding poise under duress.

Handling and roadholding

There is a case for stating that Lancia know more about effective and enjoyable front drive suspension layout than anyone in the

Left, exteriors: Original Delta lines were by Giugiaro and distinguished enough to help it win the car of the year award in 1979. There was a slight revision in December 1982
Left, dash: Steering wheel is adjustable producing a driving position better than the average for an Italian car, but some of the instrumentation leaves much to be desired
Top right: A close up of the instruments shows perfectly acceptable dials for rev counter and speedometer, bar chart 'high tech' instrumentation can be seen on the left

ROAD TEST LANCIA DELTA HF TURBO

mass production business. The seventies Beta coupes and even the ungainly Gamma flat fours balanced Lotus-like cornering speed with well-controlled ride that made any cross country trip a worthwhile experience. The HF Delta does not dull the gloss on that Lancia legend.

Building around an all-independent strut system, with front and rear roll bars to contain cornering lean, the HF romps over British B-roads with a confidence. A reassurance that is surely felt through rack and pinion steering, which errs only marginally toward pedestrian parking manoeuvres in its 3.8 turns lock-to-lock.

Pointed hard at a succession of downhill test track corners the Delta simply flew downhill with an accuracy that was truly sporting, yet any mid-corner change in line or the wicked placement of mid-curve bumps, were absorbed with only the steering reporting that the car had changed line obediently. This rather than the full blooded slides that some of the opposition would have been tempted toward.

The 170 section TRX is not a generous width by current standards, and it is the limited amount of rubber on the road that provides a gently achieved final limit. The Lancia will then drift slowly off the intended line, led by the nose. However the comparatively unfashionable rubber provides better wet weather manners (particularly over standing water) to complement steering that has less front drive tug than you would expect of a 130 horsepower machine of this performance potential.

Traction under full power, or hard braking, has the same rubberwear limitations as the roadholding. However, since the best acceleration to speeds beneath 60mph was achieved with the clutch dropped upon an engine crankshaft speed on 4,800rpm, that cannot be regarded as a defect. For the resulting wheelspin was quickly contained, even with the turbo on full boost.

Coupled with the sheer brute speed of the HF, the cornering capabilities and absorbent ride provided pleasures that outweighed the niggling quality problems we had with this press demonstrator.

Engine and performance

The twin cam Fiat engine debuted in the sixties and has done service in an enormous number of production and competition roles during the best part of 20 years. Although its race and rally activities were generally topped by a 16-valve cylinder head, rather than the two-valve-per-cylinder layout shared by Lancia and Fiat production lines. Normally it is made in sizes from a really short stroke 1-3 litres to a long stroke 2 litre, our 1585cc test model biased to a shorter stroke at 84 x 71.5mm.

Top left: The engine bay of the Delta turbo, neat if a trifle crammed
Left: Wheel stowage is neat providing nice flat floor space (unlike Golf GTi for example) and without resorting to side stowage which wastes space
Far right: Front Recaros give excellent support without the resorting to the unnecessary extremes (however keen you are) of those in the Strada 130 TC. Rear seats are made easily accessible through the adoption of a five door specification while space is adequate for adults. The level of trim throughout is good with upholstered door panels following what now seems to be a popular trend.

ROAD TEST LANCIA DELTA HF TURBO

As is the current turbocharged fashion, a much higher compression ratio (8:1) is offered than used to be thought mechanically sympathetic. Lancia restrain any head leak potential by an Astadur gasket with stainless steel rimming and reinforcement plates to withstand a maximum 10psi boost.

The T3 Garrett is a widely used turbo and incorporates a wastegate system beyond maximum boost with an air-to-air heat exchanger radiator placed in front of the engine water coolant. The turbocharger boosts via the sealed and pressure-resistant Weber 32 DAT compound carburettor. The Weber sits forward of the transverse engine with the exhaust system — stainless steel for the top manifolding, studs and nuts — closest to the passenger compartment, shrouding an extremely heated turbine casing. Those bonnet scoops, particularly the one in front of the RHD wheelman, yield a blast of rising heat that could be helpfull to a balloonist!

Additional engine work included sodium-filled exhaust valves, an electric fuel pump to replace GT's mechanical unit; an oil capacity increase to six litres with an engine oil cooler; molybdenum coating to replace chrome upon the piston rings, and the low crown pistons that allow 8:1 in place of 1600 GT's 9.3:1.

Aside from that well advertised 130bhp at 5,600rpm, this turbo has been installed with torque in mind, some boost served up from 2,000rpm in fifth, and maximum boost indicated by 2,750rpm. From 1.6 litres Lancia offer 141 lb.ft at 3,700rpm pulling power, whilst the best the long stroke (84 x 90mm) two litre TC within the Fiat Abarth can manage is 130 lb.ft at 3,600rpm.

In practice you can prove the smaller TC's mid-range clout by comparing it against 130 TC in the fourth and fifth gear timing sheets. In fourth gear between 30 and 90mph each 20mph increment shows Lancia quicker than Fiat, pronounced at 50-70mph in fourth, where the Lancia recorded 5.66 seconds and the Fiat 7.0 seconds.

In fifth the smaller capacity turbo struggled a bit versus the bigger twin cam, but from 60mph to 100mph it proved superior after a fractional disadvantage between 30-50mph and 40-60mph. However, if you want the ultimate middle-of-the-power band pull, the supercharged two litre VX Lancia coupes and HPEs are the recipe you require. See Peformance Car, January 1984, but be prepared for surprises as the Delta HF Turbo outdoes its two litre supercharged cousins in many of the test results.

In sheer straightline acceleration the Lancia HF is one of the most impressive machines we have measured in the hatchback hothouse, dipping fractionally beneath 8 seconds for 0-60mph (Abarth 130 TC; 8.0 seconds for 0-60mph) and stomping up to 100mph in a best time nearly two seconds better than the Fiat 130 TC.

Even pitched against the latest Golf GTi, the Lancia emerges as a strong performer with a slight edge in top speed (120.4mph was our best quarter mile) and an overall turn of speed that is unlikely to be beaten by another 1600 except perhaps,

ROAD TEST LANCIA DELTA HF TURBO

by the Mitsubishi Colt Turbo. By any standards the Lancia is fast, but not frugal.

Our fuel consumption average over 750 miles was boosted to 24.24mph by initial restraint allowing 34.1mpg, the kind of figure normally aspirated hotshots such as the Golf GTi and Escort XR3i can provide. However the more typical road consumption return was 22-23mpg, slightly superior to the similarly powerful Fiat Abarth, but confirming our suspicions of official figures for turbo models from any manufacturer. In Lancia's case the official consumption returns range from just over 26mpg urban to beyond 40mpg at a constant 56mph.

Lancia have been tremendously successful in extracting considerable power with mid-range flair from the venerable twincam, plus fuel consumption that is competitive with turbo rivals. The cold and hot stalling we experienced is we think not likely to be typical, but we can only report — as for the throttle linkage — as we find. A press test car is hardly likely to presented in worse condition than a car for public sale.

Summary

This is a fashionable area of the car market and we suspect that many potential customers may feel the five door Delta dated in the showroom styling stakes. Remember that rare five door configuration and performance as counterpoints, plus an amazing chassis that shows you are not forced to put up with a ride like that offered by XR3i/Peugeot 205 GTI and Honda CRX to discover the thrills of cross country speed with safety.

Inevitably the £7,000 plus bracket brings the Golf into the equation, or the competition-inspired Fiat Abarth 130 TC tested by us in August. The Lancia, particularly the heavily-equipped model tested, is the most expensive at £7,990 versus £7,867 for Golf II GTi and £7,800 for Fiat's 130 TC. Without those Recaro front seats, central locking and the efficient single-handed sunroof the £7,250 Lancia, especially in white, is quite likely to appeal to the £6,999 Vauxhall Astra GTE customer. However, we all know there is a new Kadett/Astra on the way, so Vauxhall must remain off our shopping list until the replacement can be assessed.

Closest rival on paper — 1.6 hatchback turbo, albeit with slightly less claimed power, no rear disc brakes and three doors — is the Mitsubishi 1600 Turbo for £7,749. We have yet to test this model, but over a similar mileage, it impressed as extremely quick, better built than the Lancia, but with nothing like the suspension and braking sophistication supplied by Turin's HF.

The Lancia Delta HF Turbo is likely to appeal to the driver who appreciates fast fun miles, served with supple grace and an individuality that grown-up Alfasud owners would particularly savour. Sadly we must place a question mark against the build quality in view of disappointing minor failures, but our enthusiasm and respect for a fine team of engineers remains the outstanding memory of a fine car, one spoilt between factory gate and *Performance Car's* portals . . .
Jeremy Walton

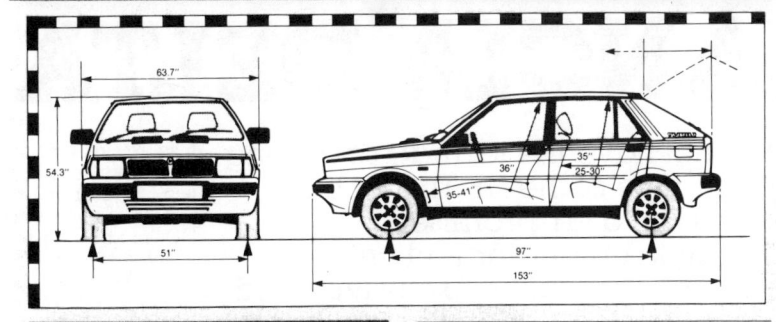

LANCIA DELTA HF TURBO TEST DATA

PERFORMANCE TEST RESULTS

All tests with a crew of two and a full tank of fuel.

Through the gears:
0-30mph	2.5secs	0-80mph	13.7secs
0-40mph	4.0secs	0-90mph	17.8secs
0-50mph	5.5secs	0-100mph	23.4secs
0-60mph	8.0secs	0-110mph	33.8secs
0-70mph	11.5secs		

STANDING ¼ MILE	16secs
TERMINAL SPEED	85.2mph
AVE TOP SPEED BANKED CIRCUIT	119.9mph
FASTEST ¼ MILE BANKED CIRCUIT	121.6mph

ACCELERATION IN 4th/5th:
30-50mph	6.5/10.8secs	60-80mph	6.1/8.6secs
40-60mph	5.5/9.2secs	70-90mph	7.2/9.7secs
50-70mph	5.7/8.3secs	80-100mph	—/11.7secs

MINIMUM SPEED COMFORTABLE PULL AWAY:
4th	17mph	5th	20mph

MAX SPEED IN GEARS @ 6,300rpm:
FIRST	30.9mph	FOURTH	96.0mph
SECOND	50.5mph	FIFTH	120mph/5750rpm
THIRD	72.4mph		

OVERALL FUEL CONSUMPTION	24.24mpg
GOVT FIGURES	41.5mpg @ 56mph; 30.3mpg @ 75mph; 26.2mpg urban driving
PROVING GROUND FUEL CONSUMPTION	17.38mpg

BRAKE TESTS

10 STOPS FROM 100mph, MAX BRAKING EFFORT:
AVERAGE	456ft	WORST	482ft (first)
BEST	426ft (fourth)		

10 STOPS FROM 100mph, MAX BRAKING EFFORT:
AVERAGE	418ft	WORST	435ft (stop 1)
BEST	401ft (stop 4)		

Testing carried out by Performance Car staff at Motor Industry Research Association Proving Grounds, Lindley, Warwickshire and Millbrook Proving Grounds, Bedfordshire. Sun rolling road chassis dynamometer facility provided by Auto Technique, Unit C, Kingsway Industrial Estate, Kingsway, Luton. Tel: 0582 414000. Radio equipment provided by Citizens Systems Ltd, 56 Lee High Rd, London. Tel: 01-852 4607

TRACK CONDITION	Dry
TEMPERATURE	22°C
WIND SPEED	15-20mph

SPECIFICATION

ENGINE TYPE:	Water cooled in-line four, dohc
DISPLACEMENT	1585cc
BORE	84mm
STROKE	71.5mm
COMPRESSION RATIO	8:1
MAX QUOTED POWER (DIN)	130bhp @ 5,600rpm
MAX QUOTED TORQUE (DIN)	141lbs/ft @ 3,700rpm
BHP PER LITRE	82.0
POWER TO WEIGHT RATIO (UNLADEN WEIGHT)	132.6bhp/ton
POWER TO WEIGHT RATIO (TEST WEIGHT)	108.3bhp/ton
FUEL SYSTEM	Weber 32 DAT 18/250 carburettor, turbo pressurised
CYLINDERS	Cast-iron block
CYLINDER HEAD	Aluminium, toothed belt, dohc drive
GEARBOX	5-speed ZF, front drive transaxle

GEAR RATIOS:
TOP	0.903	2nd	2.235
4th	1.154	1st	3.583
3rd	1.542	REVERSE	3.667

CLUTCH	Single dry plate of 8.46in, diaphragm spring
FINAL DRIVE RATIO	3.4:1
FRONT SUSPENSION	Independent MacPherson strut, lower link and anti-roll bar, hydraulic dampers
REAR SUSPENSION	Trailing arms, struts, transverse links and anti-roll bar, hydraulic dampers
BRAKES	Solid discs 10.1in front and 8.94in rear. Vacuum servo
WHEELS AND TYRES	Alloy 5.3in x 13.38; Michelin TRX AS 170/65 R340
UNLADEN WEIGHT	2200 lbs
TEST WEIGHT, CREW AND EQUIPMENT	2716 lbs
WHEELBASE	97.4in
TURNING CIRCLE	34.77ft/10.6 metres
FUEL TANK CAPACITY	9.99galls/45 litres
BASIC PRICE (INC TAX)	£7,250
OPTIONAL EXTRAS FITTED TO TEST CAR	Executive pack: Recaro seats; manual locking sunroof; headlamp washers; central locking. Total, as tested, £7,990

STAR RATINGS

ENGINE ★★★
Excellent intercooled turbo installation means a wide spread of torque and the horsepower of a sports 2-litre with competitive economy by turbo standards. Could be smoother and car supplied stalled frequently.

GEAR CHANGE ★★★
Very quick change for ultimate performance, but rubbery linkage can be more obstructive in town use.

GEAR RATIOS ★★
Biased toward mid-range turbo performance. Frequently feel that more than 50mph in second would be useful, but third covers a wide range of country road situations. With a 60mph second gear would cover 0-60mph in less than 7.5secs

CLUTCH ★★★★★
Smooth engagement and manfully resisted a full day's testing at two circuits.

BRAKES ★★★
Plenty of repeatable road retardation available, but under test conditions a lot of weaving before ultimate lock-up. Little detectable fade.

HANDLING ★★★★★
Perhaps the best ride/handling compromise in the class with outstanding balance for a machine with 61.5% of its weight over the front driven wheels.

RIDE ★★★★★
Of course its not a Jaguar, but by the class standards it might as well be a limousine! Sporting is bias, but controlled so well that B-roads are a pleasure rather than a bum-numbing torture.

ROADHOLDING ★★★★
Modest Michelin 170 section tyres are the limiting factor for sheer adhesion, but their 65% profile compromise traction and ride with grippy expertise.

COMFORT ★★★★
Above average: ride bonus and superb seats are welcome in daily use, along with leather rim wheel. Adequate ventilation with complicated controls. Excellent driving position: Italian Apeman is extinct!

INTERIOR ★★★
Super Recaros and individual cloth trim with flair is centered by confusing dials and masses of switchgear, some of it superflous.

ENJOYMENT ★★★★
Sheer driving pleasure, unrewarded by five stars only because of small failures on test car. Engineered to excite . . .

The more stars the better. Max. 5 stars.

RoadTest

LANCIA DELTA 1600 GT

It may not have the tarmac-scorching performance of the HF Turbo, but Lancia's Delta GT promises Italian-style entertainment with five-door practicality. Exclusivity and a bargain price are on its side, but the little Lancia is not without its shortcomings

AT LONG last, things are looking up for Lancia in the UK. The Beta and its sister the Trevi have gone, taking at least the seed of the rust stigma with them, the Delta-derived Prisma is selling steadily, and next year the executive-class Thema will arrive to top the range.

But it was the Giugiaro-styled Delta, launched in 1980, that marked the start of Lancia's crucial model-led recovery. Sales had taken a dive in the early part of the decade and the parent Fiat Group responded by dropping the UK Lancia concession. But since Lancar Ltd took up the reins, those lost sales are slowly being regained.

One of the first things Lancar did was to reshape the Delta range. The 1300 became an "economy" model and the 1500 (later deleted) was joined by the subject of this test, the twin-cam 1600 GT. And later came the rapid Delta HF Turbo.

The HF Turbo may be strong on image and performance, but the GT is likely to appeal to the wider market — 50 per cent of all Delta sales, according to Lancar. Especially as, at £6250, it's the cheapest sporty hatchback of its size. Two close rivals also come from Italy, namely Alfa Romeo's Alfa 33 1.5 (£6395) and Fiat's Strada 105 TC (£6345). The Fiat and Lancia share essentially the same twin overhead camshaft 1585 cc engine and a five-speed front-wheel drive transmission (though there are minor installation differences). But the Fiat has only three doors instead of five. Other rivals include Renault's 11 TXE which offers five doors and 1.7 litres at a good-value £6252, and if you're not bothered about sporting pretensions there's the Volkswagen Golf 1.6 GL (£6826) and the Ford Escort 1.6 GL (£6239), both again with five doors.

In true Italian tradition, the Delta GT's transversely-mounted engine has an alloy head with twin overhead camshafts. Breathing through a twin-choke carburetter it develops 105 bhp (DIN) at 5800 rpm and 100 lb ft of torque at 3300 rpm — figures which, incidentally, are within one bhp or lb ft of the outputs of the injected Ford Escort XR3i and Peugeot 205 GTI, also 1600s. But weighing, at 18.8 cwt, a hundredweight more than the XR3i and 2 cwt more than the Peugeot, the Delta cannot match the performance of these two overtly sporting hot hatches. Still, a time of 10.3 sec from rest to 60 mph is more than respectable, beaten only by the 9.7 sec of the Fiat among the five rivals selected here, though the R11 (10.6 sec) and Escort (10.7 sec) aren't far behind.

The Delta GT leads the pack on maximum speed: in windy conditions we couldn't match Lancia's claimed 112 mph, but the 108.6 mph achieved around Millbrook's high speed bowl is a respectable figure, representing just over 200 rpm beyond the power peak in fifth gear. Fourth-gear flexibility is up to scratch, too, with the Delta covering the 30-50 mph increment in 8.7 sec. But despite the short overall gearing, fifth gear acceleration is nothing special: 50-70 mph takes 15.0 sec against the similarly-geared and only slightly lighter Strada's 11.8 sec. And, even more significantly, although the 1600 GT has a higher top speed and better outright acceleration than the old Delta 1500, its acceleration in fourth and fifth gears is for the most part inferior, even though its higher peak torque is developed at a lower engine speed. But it's only fair to point out that windy test conditions have a disproportionately adverse effect on fourth and fifth gear acceleration figures.

Nevertheless, the engine does feel rather flat unless you work it hard. Certainly it's a smooth and eager revver — pulling well into the red with ease — but unless you make frequent use of the second part of the progressive-choke carburetter's throttle travel, the engine can feel lethargic under load. Drive the Lancia hard, and it will deliver the goods; drive it gently and you'll be irritated by the effort needed to maintain reasonable progress.

The low gearing, while adding to the joys of B-road driving, does nothing for the Delta's economy. The overall figure of 27.2 mpg is worse than that returned by any of the selected rivals apart from the very thirsty (25.5 mpg) Fiat, while the biggest-engined car, the R11, achieves the *best* figure (33.0 mpg). Nor is the estimated touring consumption of 33.6 mpg particularly impressive, although the 9.9 gallon tank will allow a typical range of over 300 miles between fill-ups.

Unlike the HF Turbo which has an obstructive ZF gearbox, the GT's gearbox is built by the Fiat Group. The change quality is better than the Turbo's, with a positive if rather stiff action provided you are definite in your movements. But if you are not, a missed gear can be the result. Selection of first gear can often baulk at rest, and there's an annoying rattle in neutral.

The Delta GT's alloy wheels are similar to the Turbo's, though they carry slightly narrower-section 165/65 × 14 tyres. And while the damper settings for the front and rear MacPherson struts are shared with the Turbo, the springs are softer and the anti-roll bars are thinner. These differences are significant, for while the Turbo exhibits exceptional poise, stability and accuracy combined with an absorbent ride, at low to medium speeds the GT feels a shade underdamped, its low-geared steering uninspiringly vague around the straight-ahead position.

Crack up the pace, though, and the Delta GT becomes more accurate and responsive, the steering gaining weight and feel without tugging in the bends under power. It's basically an understeerer, but when the Lancia's limit is reached it's the tail that will lose its grip first. Lift off mid-bend and the Delta will tighten its line more abruptly than many, but used with care this can be a useful trait. And although the shuddering over transverse ridges never disappears, at higher speeds the ride is otherwise reasonably smooth and well-controlled.

With disc brakes all round you would expect the Lancia to have good stopping power, and with a firm, reassuring pedal and good bite it doesn't disappoint. The only quibble is that there's a mite too much servo assistance.

As a less overtly sporting hatchback than the likes of a Golf GTi or an Escort XR3i, the Delta's five doors are by no means a drawback. Moreover the Delta has excellent packaging: it's very roomy for a car less than 13 feet long, although the boot's capacity (7.1 cu ft) is compromised by the suspension turrets. Generous oddments space includes a full-width facia tray, and the rear seat can be folded in two separate halves.

The front seat cushions are too short, and there's a prominent bar just where there should be more thigh support. For tall people the seats are too small and unsupportive, and even small people might find them over-soft, although lumbar and lateral support are acceptable. The actual driving position is good, with all major controls within easy reach. The height-adjustable steering wheel is larger than it needs to be, but to fit a smaller one would be a mistake as it already obscures some of the generous complement of finely-calibrated minor gauges. The minor controls are too far away in a row in the centre of the facia and the dip-switch is illogical in that down gives main beam and up dip. The indicator stalk is on the right.

Saab helped to develop the heating and ventilation system (the Delta is sold as a Saab 600 in Sweden) so the Delta's copious heat output and rapid demisting comes as no surprise.

The Rivals

Other possible rivals include the Ford Escort XR3i (£7035), Nissan Cherry Turbo (£6698), Peugeot 205 GTi (£6645), Rover 213 SE (£6422), Talbot Horizon 1.5 GL (£5895) and Volvo 340 GL 5-door (£5971)

LANCIA DELTA 1600 GT — £6250

Power, bhp/rpm	105/5800
Torque, lb ft/rpm	100/3300
Tyres	165/65 SR 14
Weight, cwt	18.8
Max speed, mph	108.6
0-60 mph, sec	10.3
30-50 mph in 4th, sec	8.7
Overall mpg	27.2
Touring mpg (computed)	33.6
Fuel grade, stars	4
Boot capacity, cu ft	7.1
Test Date	December 22, 1984

In 1600 GT form, Lancia's Delta has lively performance but loses out on refinement and economy. Surprisingly woolly handling at low speeds, but sharpens up considerably when driven hard. Inconsistent gearchange, choppy low speed ride, curious ergonomics and a plasticky facia are other shortcomings, but in the Delta's favour are exclusivity, good external finish and elegant looks. Well equipped, but disappointing compared with the original 1500.

ALFA ROMEO 33 1.5 — £6395

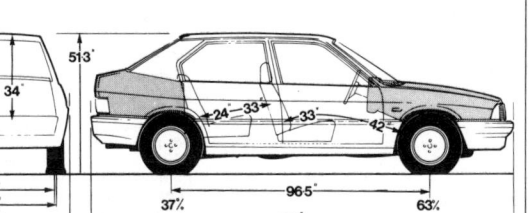

Power, bhp/rpm	85/5800
Torque, lb ft/rpm	89/3500
Tyres	165/70 SR 13
Weight, cwt	17.2
Max speed, mph	103.1
0-60 mph, sec	11.6
30-50 mph in 4th, sec	9.5
Overall mpg	28.8
Touring mpg	36.9
Fuel grade, stars	4
Boot capacity, cu ft	12.2
Test Date	June 11, 1983

Alfa Romeo's Alfasud derivative is aimed at a slightly higher market sector than the 'Sud and shares much of its mechanical componentry. Good performance and economy are matched to excellent accommodation and poised and balanced handling. Mechanical refinement, brakes and transmission are other good features of the car. Ride comfort is only mediocre, however, and the rather plasticky interior trim doesn't match its upmarket aspirations.

FIAT STRADA 105 TC — £6345

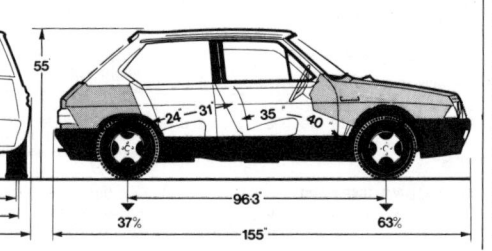

Power, bhp/rpm	105/6100
Torque, lb ft/rpm	98/4000
Tyres	185/60 HR 14
Weight, cwt	18.3
Max speed, mph	104.0
0-60 mph, sec	9.7
30-50 mph in 4th, sec	8.4
Overall mpg	25.5
Touring mpg	31.2
Fuel grade, stars	4
Boot capacity, cu ft	9.9
Test Date	March 6, 1982

High equipment levels and good accommodation at a keen price make the Strada very good value, but in a number of important areas — economy, gearchange, comfort, refinement and heating/ventilation — it falls somewhat short of class standards. Good smooth road grip, but handling lacks finesse, brakes are over-sensitive and ride comfort is poor. Reasonable performance but overall a rather disappointing car. Face-lifted version due soon.

FORD ESCORT 1.6 GL 5-DOOR — £6239

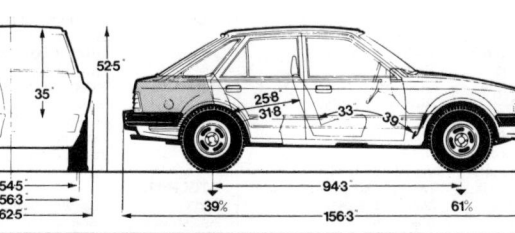

Power, bhp/rpm	79/5900
Torque, lb ft/rpm	92/3000
Tyres	165/SR 13
Weight, cwt	17.0
Max speed, mph	103.0
0-60 mph, sec	10.7
30-50 mph in 4th, sec	8.9
Overall mpg	32.2
Touring mpg	33.6
Fuel grade, stars	4
Boot capacity, cu ft	10.3
Test Date	September 27, 1980

In 1600 cc form, Ford's new Escort has very good performance, without any sacrifice in fuel consumption. It is also one of the roomiest cars in its class, has a comfortable driving position, sweet gearchange, versatile heating and ventilation, and a high standard of finish. Excellent roadholding and handling on smooth roads, powerful brakes, but ride and bumpy-road handling still mediocre, as also is refinement.

RENAULT 11 TXE 5-DOOR — £6252

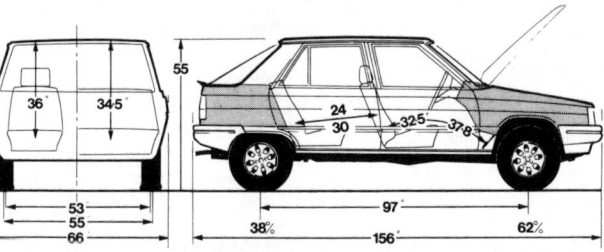

Power, bhp/rpm	80/5000
Torque, lb ft/rpm	100.5/3250
Tyres	175/70 SR 13
Weight, cwt	16.9
Max speed, mph	102.2
0-60 mph, sec	10.6
30-50 mph in 4th, sec	8.7
Overall mpg	33.0
Touring mpg	39.0
Fuel grade, stars	4
Boot capacity, cu ft	12.1
Test Date	November 19, 1983

New high-torque Renault 11 derivative isn't the GTi-basher its engine size (1700) suggests. In combination with high gearing gives a fine blend of performance and economy, with good refinement except when revved hard. Light, easy gearchange and good ride/handling compromise are other virtues. Roomy and versatile for luggage, but passenger accommodation is poor, with cramped driving position for tall drivers. Good heating and ventilation, well priced.

VOLKSWAGEN GOLF 1.6 GL 5-DOOR — £6826

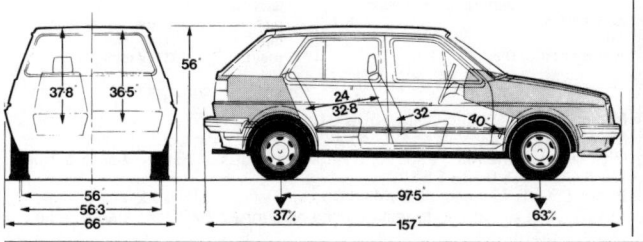

Power, bhp/rpm	75/5000
Torque, lb ft/rpm	92/2500
Tyres	175/70 SR 13
Weight, cwt	18.1
Max speed, mph	100.5
0-60 mph, sec	11.6
30-50 mph in 4th, sec	9.9
Overall mpg	32.7
Touring mpg	39.5
Fuel grade, stars	2
Boot capacity, cu ft	10.8
Test Date	February 18, 1984

In 1.6 GL form Volkswagen's new Golf has excellent fuel economy, though performance is only fair for the class. Very good handling, and other good points include accommodation (apart from some irritating details), instruments, driving position, build quality and most aspects of refinement. Ride comfort fair, but visibility could be better, and lack of heater-independent ventilation is a serious shortcoming. Rather expensive and averagely well equipped.

Steering wheel obscures minor gauges, and facia looks more functional than it is. Prominent parcel shelf can look untidy when full (above). Good-looking twin-cam engine is lively but unrefined (left)

Unusual pleated wool trim is continued on doors and headlining. Seat cushions are too short (below, left). Plenty of room in the back (below)

But all the vents are heater-linked, which can lead to stuffiness, and the dark green labelling of the heater controls against a black background is an ergonomic disaster.

Compared with the tranquillity of the original Delta's cabin, the 1600 GT is disappointingly noisy, the more so as it has to be worked so hard. The engine is raucous when extended, with ever-present induction gruffness and an insistent drone when cruising. The sporty exhaust note is quite pleasant, but road roar isn't well suppressed and wind noise is even worse. On a long journey the Lancia can be somewhat wearing. Effortless it is not.

But it is well-equipped: it boasts electric front windows, tinted glass, a multi-function digital clock (distractingly located above the interior mirror) and alloy wheels. The sliding steel sunroof is an option, and Lancia leaves the choice of in-car entertainment to the buyer.

Finish is good, too, with well-fitting colour-keyed bumpers and spoilers. Paint is deep and lustrous — and the help that Saab has given Lancia in corrosion protection technology should reassure customers as to its lasting qualities. Inside, the woollen trim material is attractive if unusual, but there's too much black plastic and the facia does look as though function has been sacrificed for style.

It's a car with a dual personality, the Delta GT. For long distance cruising or leisurely driving with minimum effort on the part of the driver, there are better bets. But if you're in the mood it can be a rewarding car. It has the advantages of elegant good looks and comparative rarity — and, at the price, it's well-equipped. But overall, the Delta GT doesn't live up to its promise.

MOTOR ROAD TEST
LANCIA DELTA 1600 GT

PERFORMANCE

WEATHER CONDITIONS
Wind	15-30 mph
Temperature	46 deg F/7.7 deg C
Barometer	28.54 in Hg
Surface	Dry tarmacadam

MAXIMUM SPEEDS
	mph	kph
Banked Circuit	108.6	174.7
Best ¼ mile	112.0	180.2
Terminal speeds:		
at ¼ mile	77.3	124.4
at kilometre	91.1	146.6
Speeds in gears (at 6500 rpm):		
1st	31	50
2nd	50	80
3rd	73	117
4th	97	156

ACCELERATION FROM REST
mph	sec	kph	sec
0-30	3.3	0-40	2.6
0-40	5.0	0-60	4.6
0-50	7.2	0-80	7.1
0-60	10.3	0-100	11.1
0-70	14.0	0-120	15.9
0-80	18.9	0-140	26.8
0-90	31.5		
Stand'g ¼	17.6	Stand'g km	32.9

ACCELERATION IN TOP
mph	sec	kph	sec
20-40	12.7	40-60	7.9
30-50	12.0	60-80	7.2
40-60	13.1	80-100	8.8
50-70	15.0	100-120	10.7
60-80	17.7		

ACCELERATION IN 4TH
mph	sec	kph	sec
20-40	9.3	40-60	5.6
30-50	8.7	60-80	5.4
40-60	8.6	80-100	5.6
50-70	9.5	100-120	6.8
60-80	12.2	120-140	11.1

FUEL CONSUMPTION
Overall	27.2 mpg
	10.4 litres/100 km
Govt tests	27.7 mpg (urban)
	44.8 mpg (56 mph)
	34.0 mpg (75 mph)
Fuel grade	97 octane
	4 star rating
Tank capacity	9.9 galls
	45 litres
Max range*	333 miles
	535 km
Test distance	952 miles
	1532 km

*At estimated 33.6 mpg touring consumption

STEERING
Turning circle between kerbs
	ft	m
left	35.4	10.8
right	34.3	10.5
Lock to lock	3.8 turns	

NOISE
	dBA	Motor rating*
30 mph	68	14
50 mph	74	21
70 mph	78	28
Maximum†	83	38

*A rating where 1 = 30 dBA and 100 = 96 dBA, and where double the number means double the loudness
†Peak noise level under full-throttle acceleration in 2nd

SPEEDOMETER (mph)
True mph 30 40 50 60 70 80 90 100
Speedo 32 43 53 64 76 87 98 109
Distance recorder: 2 per cent fast

WEIGHT
	cwt	kg
Unladen weight*	18.8	955
Weight as tested	23.0	1168

*with fuel for approx 50 miles

Performance tests carried out by *Motor*'s staff at the Motor Industry Research Association proving ground, Lindley.

Test Data: World Copyright reserved. No reproduction in whole or part without written permission.

GENERAL SPECIFICATION

ENGINE
Cylinders	4 in-line
Capacity	1585 cc
Bore/stroke	84.0/71.5 mm
Cooling	Water
Block	Cast iron
Head	Aluminium alloy
Valves	Dohc
Cam drive	Toothed belt
Compression	9.3:1
Carburetter	Weber 34 DAT twin-choke downdraught
Ignition	Marelli Digiplex electronic
Bearings	5 main
Max power	105 bhp (DIN) 77 KW at 5800 rpm
Max torque	100 lb ft (DIN) 135 Nm at 3300 rpm

TRANSMISSION
Type	5-speed manual
Clutch dia	200 mm
Actuation	Cable
Internal ratios and mph/1000 rpm	
Top	0.959/18.0
4th	1.163/14.9
3rd	1.550/11.2
2nd	2.235/7.7
1st	3.583/4.8
Rev	3.714
Final drive	3.59:1

BODY/CHASSIS
Rust warranty	6 years against perforation corrosion
Aerodynamic drag coefficient (Cd)	0.37

SUSPENSION
Front	Independent by coil springs; MacPherson struts; lower wishbones; anti-roll bar
Rear	Independent by coil springs; MacPherson struts; transverse and longitudinal links; anti-roll bar

STEERING
Type	Rack and pinion
Assistance	None

BRAKES
Front	Discs, 257 mm dia
Rear	Discs, 257 mm dia
Park	On rear
Servo	Yes
Circuit	Diagonally split
Rear valve	Yes
Adjustment	Automatic

WHEELS/TYRES
Type	Alloy, 5½J × 14
Tyres	165/65 SR 14
Pressures	29/29 psi F/R (normal)
	32/32 psi F/R (full load/high speed)

ELECTRICAL
Battery	12V, 40 Ah
Earth	Negative
Generator	Alternator, 55 A
Fuses	14
Headlights type	Halogen
dip	110 W total
main	120 W total

Make: Lancia **Model:** Delta 1600 GT
Maker: Lancia, Via Vincenzo Lancia 27, 10141 Torino, 33311 Italy
UK Concessionaires: Lancar Ltd, Henwood Industrial Estate, Ashford, Kent TN24 8DN. Tel: 0233 25722
Price: £5016.72 basic plus £418.06 Car Tax plus £815.22 VAT equals £6250.00. Extras fitted to test car: sliding sunroof £260.00, stereo radio/cassette player to buyer's choice.

LANCIA DELTA HF TURBO IE

A HIGHER PROFILE

Don't be fooled by the minor body alterations — the Delta HF Turbo ie is all-new under the skin. We test a leading contender in the hot hatch stakes all too often overlooked by potential buyers

FOR:
PERFORMANCE
REFINEMENT
RIDE
AGAINST:
REAR HEADROOM

Go on — name three hot hatchbacks... VW Golf GTI, Escort XR3i and Peugeot 205GTi? Yes, but what about the Lancia Delta HF Turbo? It seems to be the one that everyone forgets and yet it deserves better than this, especially in its 1986 specification.

Technical and cosmetic modifications to the Delta appear throughout the range and include two completely new models, the HF 4WD and the turbo ds. Only the former will be available in the UK, however, and even this will not be until some time early next year. The UK range consists of just four models, two carburettored 1301cc low and high specification options, a 1585cc GTie fuel-injected model and, at the top of the range, the car tested here, the HF turbo ie. This of course ignores the 'real' flagship of the Delta range, the road-going S4 homologation special.

The most obvious external changes are to the front of the HF Turbo with a new bumper and accentuated deep spoiler incorporating blanks for fog lights. The light clusters are also new. At the rear the most noticeable change is to the roof profile which has enabled the vestigial roof spoiler of the previous model to be removed. These changes are all fairly minor but serve to clean up cosmetically what was essentially a good design to begin with. The Delta HF Turbo ie remains an understated hot-hatchback with no striping (optional on the white versions of the old car to emphasise the Martini rallying connection) and only subtle badging.

Improvements to the interior are substantial. They include new seats, a completely redesigned facia incorporating clearer instrumentation, a new heating and ventilation layout and a whole range of new fabrics and colours for the carpets and trim.

The most significant changes, however, take place under the skin. The engine has been modified to make it more compact, then rotated through 180deg and inclined forward by 18deg. By doing this, the exhaust manifold is now at the front of the engine bay and exposed to direct cold air-flow for better cooling and a reduction in engine bay temperatures. The engine's centre of gravity has effectively been lowered by 1.8ins and it is mounted to the body by four (one more than previously) flexible blocks. Alterations have also been made to the cylinder head, piston crown profile and camshafts to make combustion within the cylinders more efficient.

Weber Marelli IAW fuel-injection replaces the single Weber carburettor fitted previously and, combined with the other improvements including an intercooler and Garrett AiResearch T2 turbocharger, has raised the maximum power by 10bhp to 140bhp at 5500rpm with the peak torque remaining at 141lb ft but developed at 3500rpm, 200rpm lower than in its predecessor.

Suspension remains essentially the same, but incorporates a few changes. These include flexible rubber links to the bodyshell, ball-type thrust bearings in the steering rack to reduce steering effort at low speeds, progressive rebound bump-stops, different damper ratings, and changes to the relative angle between springs and dampers and front/rear toe-in.

As can be seen from these modifications, Lancia has done a comprehensive revamp of what was already a very competent car. But have all the changes been worthwhile?

In pure performance terms, the 1986 Delta HF turbo ie remains very similar to its predecessor with a mean maximum speed in fifth gear of 119mph at 5690rpm and a wind-assisted best of 122mph corresponding to 5825rpm. These figures are lower than the last HF we tested (*Autocar*, 14 July 1984) by 2mph but this is probably due to the differing test conditions. This test was undertaken with a 10mph stronger mean wind speed.

The test car was fitted with a rev-limiter set at 6800rpm — the rev-counter is red-lined at 6500rpm — and taking the HF up to the limiter under full acceleration in each gear resulted in maxima of 36, 58, 84, and 111mph for first, second, third and fourth gears respectively.

On the day, the mean 0-60mph time was 8.5secs which, although 0.3secs down on the previous test car, is still very respectable and allows the Lancia to remain well in contention with the obvious competitors. The quarter-mile post was reached in 16.1secs at a terminal speed of 84mph with the kilometre post coming up in 30.1secs at 105mph.

Where this latest example of the Delta HF turbo really shows that Lancia has got its sums right is in the incremental gear times, and especially in the mid-range of each gear — as the graph clearly shows. It is an improvement in this area that really matters in the course of everyday driving as nobody does full-blooded standing-starts from the traffic lights or drives at maximum speed on the motorway. The quickest 50-70mph incremental time occurs in third gear and is an impressive 4.6secs — easily quick enough for a clean, safe overtaking manoeuvre.

For a car with this level of performance on tap, Lancia has managed to keep the fuel consumption at a sensible figure. During the 1338 miles the HF spent with us, it returned an overall consumption figure of 24.5mpg, with a worst interim figure of 20.4mpg following the test session. A best of 28.6mpg was recorded after a leisurely motorway journey. It should be reasonable to expect a figure closer to 27mpg during the course of normal driving and, combined with the enlarged 12.6 gallon fuel tank, will give the HF a possible cruising range of 300 miles.

Throw the HF turbo hard into a corner and it is reassuring. The driver can feel the balance of typical front-engine, front-wheel-drive understeer along with the power delivery characteristics of a turbo. Lift off mid-corner and there is a certain amount of front-end tuck-in with a resulting tightening of line. Roll oversteer can be provoked, as in most front-wheel-drive cars, but it is controllable due to the smooth input from the turbo. It is almost possible to corner the Lancia on the accelerator by balancing the understeer and tuck-in.

Ride quality in this revised Delta is as good as, if not better than, the last example we tested. There is a certain amount of bump-thump evident but it is generally well-damped, unless one of the infamous urban-potholes is negotiated at speed; but few cars take this sort of obstacle without complaint. The ride can be described as firm but supple and, in most cases, ▶

Fuel-injected, *flat-four turbo gives top speed of 119mph*

Three-spoke, *leather-bound wheel adds to sporting feel*

TEST UPDATE

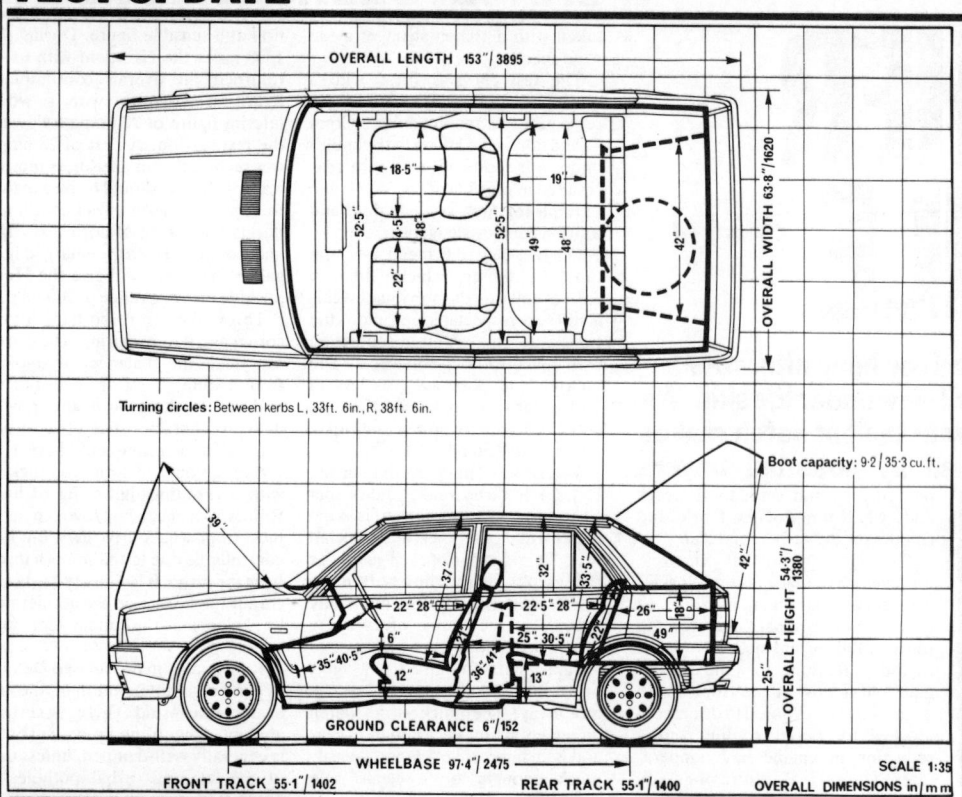

MODEL

LANCIA DELTA HF TURBO IE
PRODUCED BY:
Lancia,
Via Lancia 27,
Turin 10141, Italy

SOLD IN THE UK BY:
Lancar Limited,
46-62 Gatwick Road, Crawley,
West Sussex RH10 2XF

SPECIFICATION

ENGINE
Transverse front, front-wheel drive. Head/block al. alloy/cast iron. 4 cylinders in line, bored block, 5 main bearings. Water cooled, electric fan.
Bore 84.0mm (3.31in), **stroke** 71.5mm (2.81in), **capacity** 1585cc (96.7 cu in).
Valve gear 2 ohc, 2 valves per cylinder, toothed belt camshaft drive.
Compression ratio 8 to 1. Marelli electronic ignition with mapped advance control and knock-sensor, electronic IAW Weber fuel injection. Garrett AiResearch T2 turbocharger with intercooler, max boost pressure 12 psi (0.85 bar).
Max power 140bhp (PS-DIN) (104kW ISO) at 5500rpm. **Max torque** 141lb ft at 3500rpm.

TRANSMISSION
5-speed manual. Single dry plate clutch, 8.5in dia.

Gear	Ratio	mph/1000rpm
Top	0.903	20.93
4th	1.154	16.38
3rd	1.524	12.40
2nd	2.235	8.46
1st	3.583	5.27

Final drive: helical spur, ratio 3.40.

SUSPENSION
Front, independent, MacPherson struts, lower wishbones, double-acting telescopic dampers, anti-roll bar.
Rear, independent, MacPherson struts, transverse links, trailing arms, telescopic dampers, anti-roll bar.

STEERING
Rack and pinion. Steering wheel dia. 14.5in, 3.75 turns lock to lock.

BRAKES
Dual circuits, split diagonally. **Front** 10.1in (257mm) dia self-ventilating discs. **Rear** 8.9in (227mm) dia discs. Vacuum servo. Handbrake, centre lever acting on rear discs.

WHEELS
Alloy, 5.5in rims. Radial tubeless tyres (Michelin MXV on test car), size 165/65R14, pressures F28 R28 psi.

EQUIPMENT
Battery 12V, 40Ah. Alternator 65A. Headlamps 110/120W. Reversing lamp standard. 14 electric fuses. 2-speed, plus intermittent screen wipers. Electric screen washer. Air blending /water valve interior heater.

PERFORMANCE

MAXIMUM SPEEDS

Gear	mph	kph	rpm
Top (Mean)	119	192	5690
(Best)	122	196	5825
4th	111	179	6800
3rd	87	135	6800
2nd	58	93	6800
1st	36	58	6800

ACCELERATION FROM REST

True mph	Time (sec)	Speedo mph
30	2.9	32
40	4.4	42
50	6.0	53
60	8.5	64
70	11.1	75
80	14.4	85
90	19.5	96
100	25.6	106

Standing ¼-mile: 16.1sec, 84mph
Standing km: 30.1sec, 105mph

IN EACH GEAR

mph	Top	4th	3rd	2nd
10-30	—	9.0	6.2	3.6
20-40	10.8	7.3	4.7	2.8
30-50	9.6	5.9	3.8	2.9
40-60	8.5	5.1	4.0	—
50-70	7.7	5.6	4.6	—
60-80	8.3	6.6	5.5	—
70-90	10.0	7.6	—	—
12.4	12.4	10.4	—	—

FUEL CONSUMPTION
Overall mpg: 24.5 (11.5litres/100km) 5.4mpl
Autocar constant speed fuel consumption measuring equipment incompatible with fuel injection.
Autocar formula: Hard 22.1mpg
Driving Average 27.0mpg
and conditions Gentle 31.9mpg
Grade of fuel: Premium, 4-star (98 RM)
Fuel tank: 12.6 Imp galls (57 litres)
Mileage recorder: 1.7 per cent long
Oil: (SAE 15W/40) — synthetic recommended — negligible

BRAKING
Fade (from 84mph in neutral)
Pedal load for 0.5g stops in lb

	start/end		start/end
1	20-25	6	40-140
2	20-35	7	45-120
3	25-50	8	45-120
4	30-120	9	45-80
5	40-120	10	45-60

Response (from 30mph in neutral)

Load	g	Distance
10lb	0.23	131ft
20lb	0.46	65ft
30lb	0.62	30ft
40lb	0.76	40ft
50lb	0.95	32ft
55lb	1.00	30ft
Handbrake	0.32	94ft

Max gradient: 1 in 3
CLUTCH Pedal 26lb; Travel 4.5in

WEIGHT
Kerb 21.0cwt/2352lb/1064kg
(Distribution F/R, 61.3/38.7)
Test 24.5cwt/2747lb/1243kg
Max payload 995lb/450kg
Max towing weight 2431lb/1100kg

COSTS

Prices

Basic	£7055.52
Special Car Tax	£587.96
VAT	£1146.52
Total (in GB)	**£8790.00**
Licence	£100.00
Delivery charge (London)	£200.00
Number plates	£20.00
Total on the Road	**£9110.00**
(excluding insurance)	
Insurance group	6/7

EXTRAS (fitted to test car)
Blaupunkt Cambridge radio/cassette player	£224.25
Speakers (2)	£67.85
Total as tested on the road	**£9402.10**

SERVICE & PARTS

Change	Interval 3000	6000	12,000
Engine oil	Yes	Yes	Yes
Oil filter	Yes	Yes	Yes
Gearbox oil	No	No	Yes
Spark plugs	No	No	Yes
Air cleaner	No	Yes	Yes
Total cost	£22.33	£44.80	£101.32

(Assuming labour at £18.40 an hour inc VAT)

PARTS COST (inc VAT)

Brake pads (4 wheels)	£38.27
Exhaust complete	£282.21
Tyre — each (typical)	£66.76
Windscreen	£126.49
Headlamp unit	£76.68
Front wing	£76.81
Rear bumper	£176.51

WARRANTY
36 months/unlimited mileage, 6-year anti-corrosion

EQUIPMENT

Automatic	N/A
Self-levelling suspension	N/A
Trip computer	N/A
Front headrests	●
Heated seats	N/A
Height adjustment	N/A
Lumbar adjustment	N/A
Rear seat belts	●
Seat back recline	●
Split rear seats	●
Door mirror remote control	●/●
Electric front windows	●
Heated rear window	●
Interior adjustable headlamps	N/A
Manual sunroof	●
Tinted glass	●
Tailgate wash/wipe	●
Central locking	●
Child proof locks	●
Fog lamps	N/A
Internal boot release	N/A
Luggage cover	●
Metallic paint	£115.00
Radio	DO
Radio/cassette	DO
Aerial	DO
Speakers	DO

● Standard N/A Not applicable DO Dealer option

TEST CONDITIONS
Wind: 10-18mph
Temperature: 18deg C (64deg F)
Barometer: 29.8in Hg (1009mbar)
Humidity: 31per cent
Surface: dry asphalt and concrete
Test distance: 1338miles

Figures taken at 3040 miles by our own staff at the Motor Industry Research Association proving ground at Nuneaton and at the General Motors proving ground at Millbrook.

All *Autocar* test results are subject to world copyright and may not be reproduced in whole or in part without the Editor's written permission

LANCIA DELTA HF TURBO IE

Marelli *fuel injection and T2 turbo combine to give 140bhp*

Rear seat passengers *will find their heads hitting the roof-lining*

Body roll *is limited by the fitment of front/rear anti-roll bars*

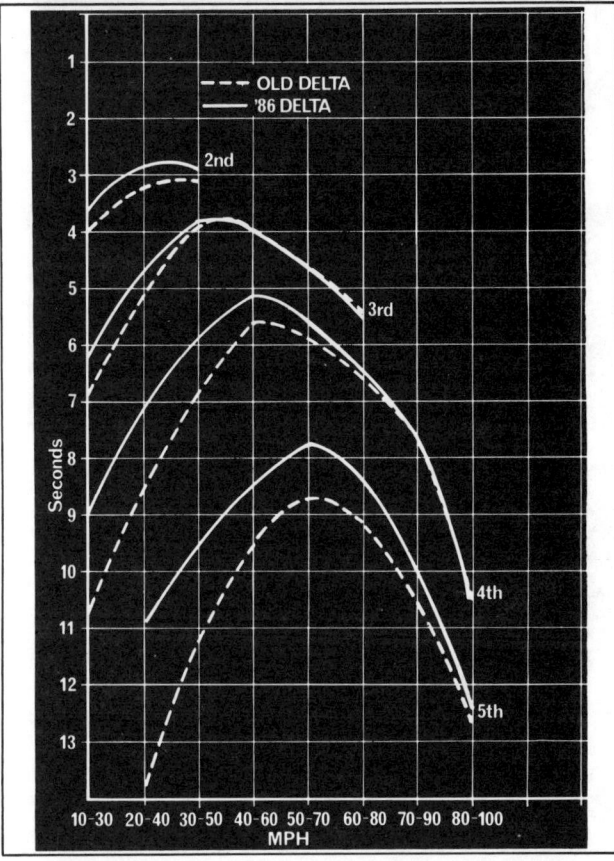

All-important *incremental in-gear times have improved substantially over the old model — especially in the mid-range of each gear*

appropriately damped while body roll is kept in check with the fitment of front-rear anti-roll bars.

One of the problems of transferring 140bhp to the road via the front wheels is the almost certain onslaught of torque steer. The HF turbo does exhibit this to a certain extent, but once again it is well controlled. Putting the power on mid-corner does not result in the driver having to fight the steering wheels.

For an Italian car with obvious sporting aspirations, the refinement levels in the HF are commendable. There is a slight whistle from the turbo as it begins to operate at about 1500rpm and builds up to full boost at 2800rpm, but as speed rises it becomes mixed in with engine, wind and road noise. It does not rise to an annoying level, however, and it is always possible to hold a conversation without the need to raise voices. As we said, commendable for a car of this nature.

Sit behind the wheel of the HF turbo and the improvements are noticeable immediately. The instrumentation is comprehensive, to say the least. All the dials can be easily read at a glance, including the ancillary readouts for oil pressure and temperature which are positioned to the left of the steering column.

Ergonomics are one of the strong points of this latest Delta with a good seating position for average-height drivers — although taller testers did comment that the seat slide could do with an extra inch of rearward travel — and a comfortable, leather-rimmed, three-spoke steering wheel. It is a shame that more manufacturers do not fit such simple and purposeful steering wheels, instead of designing new ones for the sake of being different.

The seats themselves are comfortable and generally supportive, although they could do with improved lateral support for the squab. The HF 4x4 we drove in Sardinia recently was fitted with Recaro-type seats with just this sort of support and it would have been nice to see these fitted to the HF Turbo ie. The clutch and brake pedals are well spaced but the distance between the brake and accelerator pedals is such that it makes heel-and-toeing difficult; it *is* possible, but requires practice and a flexible right ankle.

On first acquaintance, the gearchange feels a little rubbery and imprecise, but after a short time with the car gearchanges become quick and smooth, if a little long-throw for a car of this performance and character.

For a small, five-door hatchback, the Delta has a spacious interior — at least for the front seat passengers. The new facia design slopes away from the occupants to give an impression of roominess and, combined with the adequate elbow and headroom, makes for good interior comfort. The re-designed rear roofline, however, seems to have affected the rear headroom to the extent that even an average-sized person has to be careful not to make contact with the roof-lining by the top of the tailgate.

One notable feature of the Delta is the excellent manual sunroof which was fitted to the test car. A lever is pulled down, the roof slides back and can be locked in any position by pushing the lever up. Who needs the additional weight and expense of an electric sunroof when this type is so simple and smooth in operation?

Luggage space is adequate but does suffer from a certain amount of suspension intrusion. The tailgate opens from just above bumper level so that a high sill does not have to be negotiated with heavy items.

Lancia seems to have done its homework with the revamped Delta range. Not content to rest on its laurels and settle for a cosmetic re-skin, Lancia has done a major re-engineering job, to bring the range bang up to date. This will include the adoption of a four-wheel drive system developed with the experience gained from rallying.

It may be at the top end of the price range in this very competitive area of the market but the performance, refinement and practicality mean that the Lancia Delta HF Turbo ie deserves to be considered seriously by anyone looking to buy a family hot-hatchback.■

LANCIA DELTA HF TURBO ie
Vs
MITSUBISHI COLT TURBO

THE MARKETING men call it a USP. It's a way of telling Stork from butter. A blue whitener. Six appeal. It refreshes the parts other products can't reach.

It's the Unique Selling Point. The feature that makes your product stand out. Everyone's got one – somewhere. That's how marketing men earn their coronaries and ad agencies pay for those flashy office suites.

Don't go thinking it doesn't apply to cars. It does. And the more crowded the competition, the more desperate everyone is for a USP.

Take a look at the hot hatch market. Some sell thanks to their chic looks, others, seemingly, thanks to their reassuringly commonplace background. There's a fastest, a best handling and quickest 0-60mph, too, for those in the know.

But what about those others that can't claim such definitive USPs? We've been looking at two such hatches. Hot ones, true, but lacking anything incendiary enough to fan the fire in most enthusiasts' hearts.

Not that the Lancia Delta HF Turbo ie, to give it its full tongue-twisting title, is without possible USPs. There's the charisma of the Lancia marque and the magic of the famous HF logo – though both may be lost on younger enthusiasts. Then there's the power output of the twin-cam turbo – newly upped to a class-leading 140bhp. But that doesn't seem to be reflected in tarmac-tearing performance.

The Mitsubishi Colt Turbo is even shorter on uniquely marketable qualities. There's that Japanese reliability – as legendary as ever. High technology, of course, but no higher than anyone else's, and an unspoken implication that Mitsubishi ought to be considered as the Japanese BMW. Quite why, though, remains a mystery.

Are their virtues simply more subtle – or just contrived by desperate marketeers? In other words, they may be different from the 205s, Golfs and XR3s of the street scene but is either of them different enough to be distinctive?

Certainly the pair follow a fairly familiar formula – but then that's the name of the hot hatch game. Take a family box and turn it into a fast box with the minimum of disruption. Lancia's family box happens only to come with four passenger doors. Hence the HF, like the MG Maestro, will double as a granny carrier.

Both it and the Mitsubishi are transverse engined, front-drive and all independently sprung. They are also turbocharged but that's about as far as the real similarities go.

The Lancia is powered by the same capacity 1,585cc short stroke twin-cam as ever, though it has been extravagantly re-engineered to wring out an extra 10 horsepower. To begin with, it has been turned through 180 degrees to expose the exhaust system to the cooling airflow, and internal mod's have been made to the head, pistons and belt-driven cams. However the big change is the use of Weber IAW electronic fuel injection in place of carburation, along with a mapped ignition-cum-engine management system and an intercooler for the turbo – now a Garrett T2 rather than the former, bigger and slower-responding T3. The result is 140bhp developed marginally lower at 5,500rpm and no change to the 141lb ft of torque, though this is developed marginally lower, at 3,500rpm. Worth all the effort? We'll see.

The 1,597cc single-cam Colt engine uses a broadly similar system of Mitsubishi's own devising – electronic management of ignition and injection plus a turbocharger. There's no intercooler, though, which perhaps explains away some of its power deficit to the Italian – 123.3bhp at 5,500rpm.

Capacity-tied torque is much nearer – 137.2lb ft at 3,500rpm.

Both are five-speeders, naturally. The Colt has a new 'box with juggled ratios that make little difference to its overall gearing: that's a couple of points higher all through than the Lancia's set of Latin lows.

Lancia suspension is by struts, coils and lower links all round, with antiroll bars front and back. It's been revised in detail to reduce steering effort among other things. The Mitsubishi has struts and rear trailing arms, with a pair of anti-sway bars, too. Again, it has benefited from attention to noise suppression and steering response.

Both cars use vented front discs but only the Delta has rear discs as well. The difference in tyre sizes is startling. The Lancia has the sort of puny 165/65s you see on everyday Escorts while the Colt uses those par for the hot hatch course, 185/60s. Fat or thin, both come fitted to 5.5in alloy wheels.

And the pair both have had the usual exterior titivations to see them through another showroom year. The Colt has a new grille, bonnet, bumpers and light clusters: the Delta has had a nose job, too, and a detail change to the rear roofline so it could lose the rear roof spoiler. But the Lancia has also had more substantial cosmetic surgery to give it a new facia and new seats.

Comparative dimensions still point up the older Lancia as unfashionably shorter, narrower and taller than the Colt, though it runs on a usefully longer wheelbase and wider track. The five-door Delta is heavier, too, by nearly two hundredweight, and probably less slippery though neither ranks as much of a wind cheater.

QUITE A catalogue of changes, then, make the '87 HF a higher flyer than its predecessor. But don't expect anyone to notice. Only the most knowledgeable eye could tell this year's model from the last and it's still the most subtle of its ilk. Trim, boxy and standing on those slender slips of rubber – who'd have thought it packed a 140 horsepower punch?

You might even be inclined to doubt its potency on first driving the Delta. Where's the wheel scrabble, the steering tug, the sudden manic surge of turbo-boosted acceleration?

Not there, is the answer. None of them. The Delta Turbo is a cultured, discreet driver just like its outward appearance. Yes it can torque-steer – but not much – and yes it does have that turbocharged punch – but it's not a clubbing left hook, rather an immaculately timed sequence of blows. And here the HF ie is considerably improved over its predecessor which was most certainly a member of the haymaking fraternity, wildly letting fly above 3,000 revs. The new car, with its lighter turbo and intercooler, has a smoother, much more progressive response to the throttle.

There's still some of the inevitable turbo-motor flatness at the bottom end but the mid range feels much crisper and the translation between no power and full power is much cleaner. There never was much turbo lag, as such – just in or out response. Now the pick-up is as quick and clean, and the response more profiled.

In terms of outright performance, there is precious little difference between the new car and the old. Indeed, as the newcomer gets up to racing pace so unfussedly, it seems at first to lack its predecessor's legs. But then just start stretching the super little twin-cam out and feel it go. It revs all the way to an eager 6,800rpm and the low, close, intermediate ratios let it keep on singing just as long as you keep the throttle floored.

A gearshift from the low second just before 60mph takes the edge off its 0-60mph sprint time. An impressive 7.6 sec nonetheless, this is still close to the old car's best. So, too, is the top speed of 124.1 mph when the HF is just creeping over its power peak in fifth gear. None of those fuel conscious overdrive top gears for this racer – a fact that is perhaps reflected in its 25.4 mpg overall economy.

If these performance figures show little change for all Lancia's efforts, the in-gear times do confirm the road impressions of greater flexibility and smoother response. Sizeable chunks have been chopped off both the fourth and fifth gear figures – especially down at the bottom end.

Behind the engine, the HF is still very much the same machine to drive. Enjoyable but enigmatic. It dosen't have the razor sharp character of some rivals: the steering is a shade heavy (still), the gearshift a touch clumsy, the pedals awkwardly placed for heel-and-toe shifts. There are a few degrees more body roll, too, and on those narrow tyres, there's inevitably less grip.

And yet it is the *lack* of kart-like qualities that appeal. It's a composed, unflustered performer. The steering may be heavy but it is taut and accurate. Grip, in the end, may be less than the best but the Delta's handling balance is first-rate. And the supple suspension provides a much better than class-average ride as well as ensuring that it doesn't need to be on a billiard table to handle properly.

In our test of the previous model we concluded by saying: "What the Delta could really do with is an all-new facia that would improve the quality of materials used, the control layout and the heating and ventilation." We're not claiming anything but . . . the new HF ie *has* an all-new facia and the quality of materials is better. The control layout is improved and the heating and ventilation system heats and ventilates, which it didn't before. The all-analogue dash is a great improvement, in fact, even if the Italians still haven't quite grasped the notion that elegance can best be subtle. The Delta has new seats, too, for the comfortable but bulky Recaros have been dropped. The new ones have a snappier trim but not quite the Recaro-style support. They do provide a little more space for those in the back, however. The Delta may have five doors but it's a five-seater only at a pinch: the rear seat is for the small or the sociable.

At nearly £9,000, the HF comes almost completely kitted, as you would expect. Electric front windows, central locking, split rear seat and steel sunroof are all standard. Only an ICE system is conspicuously missing.

MITSUBISHI COLT TURBO

SUBTLE IT ain't. Bespattered with spoilers and with wheels like waffle makers, the Colt is out to cut a dash. Quite where, though, is another question.

And subtle its performance ain't either. What the Delta does with style, the Colt does with savagery. Haymaking is its speciality. Floor the throttle and after the first low hesitant revs as the turbo charges up to speed, the Colt streaks off like the proverbial scalded cat. The front wheels scrabble for grip, the steering wheel writhes from side to side and as the revs burst up towards the 6,000rpm ceiling, the longish stroke power plant becomes decidedly raucous. The figures will tell you that the Colt has only about par performance among the increasingly hair-raising hot-hatches. Eight and a bit seconds to 60mph is quick but breaks no records. The 122mph top speed is better but can still be beaten.

The figures won't tell you about the style of its performance. It feels damned fast but, in fact, it's the beginnings of panic that stimulate the senses. Like a teenage tearaway, the Colt doesn't much like being told what to do.

The turbo motor doesn't suffer serious lag problems—even if it lacks the Delta's intercooler. There's a momentary pause in response: just enough time to glance down at the boost gauge and see it climb after flooring the throttle. All the same, it is still very much in the hit-em-between-the-eyes school of turbo technology. Boost is evident, though its effects are still well disguised at a couple of thousand revs. After 3,000, however, comes the punch to wobble the knees and then the engine flies round to its noisy maximum.

Exciting stuff and good fun sometimes, save for the wheel scrabble and steering fight which range from simply tiresome to downright tricky, depending on the conditions.

It doesn't do a lot for fuel consumption, either. Even with a long-striding fifth, the Colt only averages 26mpg and all too easily drops below this when you are stuck in the intermediates. The high top does make it a quieter than average cruiser, however.

Shift quality of the new gearbox is an improvement. The change is still vaguer than one might have hoped for but quicker and easier. A slightly higher fourth speed also leaves less of a hole to the high fifth. The handling, unsurprisingly, can be dominated by the power delivery. Take it smoothly and the Colt has the familiar predictability of a front-drive hatch: too much power too soon and the ragged edges show. The steering has been lightened, though it is still quite heavy and it lacks the feel and precision of the Lancia. Grip is good and understeer builds gently if you don't bash that throttle too hard. On the other hand, shutting the throttle can tighten the line and nip the tail end out.

The ride has a sporty tautness that is generally acceptable in this class. That doesn't mean it's comfortable because it isn't very, reacting sharply over larger jolts.

Like so many Japanese cars, the Colt has retained an "easy-to-drive" quality despite its high performance. Your granny could drive it. The pedals are all well placed and light to use, the driving position is straightforward, all round visibility is good and the switchgear has a simple functionalism. Rally-style front seats may be a little thin and hard for some but give very good support to the shoulders and hips in cornering. In the back, the 60/40 split fold rear seat is rather more of a squeeze for adult passengers, but again about par for the course.

As usual, though, the interior is lacking in class or style. (Not that this is any more a criticism of the Colt than of most Nissans, Mazdas etc.) The plasticky facia and clumsy looking switchgear speak volumes about volume production, while the garishly striped interior is a suitable match for the exterior.

Instead of European elegance, the Turbo offers an impressive array of equipment. Height adjustable driver's seat, radio/cassette, headlamp washers, an electric sliding sunroof, electric front windows and door mirrors.

VERDICT

SO WHAT about those USPs? The Delta's are fairly easy to pinpoint. The four doors are a useful clue. It has a certain mature appeal, does the Delta. The ride is a little more comfy than average; the technology is a little more sophisticated; the looks are a little more discreet; the performance is a little more subtle. And those rear doors give a little more versatility to drivers with a little more than themselves to consider.

In short, it could appeal to the more mature enthusiast. To the family man who still likes his fun or to anyone looking for a car that is good looking, refined and comfortable as well as simply quick and fun to drive. The Colt is harder to pin down. It's brash and garish. Unsubtle to look at and to drive. It offers a certain raw driving excitement and says so in capital letters.

Of the two, the Delta not only has the more 'marketable' personality but it is also quite the nicer car all round. Where you'll soon tire of the Colt's quirks, the Lancia's more sophisticated balance of performance and comfort looks like a recipe for long-term happiness.

Unfortunately the Delta HF's amalgam of virtues slap it straight up against powerful opposition in the shape of the Golf GTI – a car of very similar appeal.

Lancia's past problems may be behind them (the HF showed good build quality) but VW remain the masters of the marketplace. And, despite the Delta's undeniable qualities, until Lancia can come up with a winning USP it looks like staying that way.

TEST MATCH

	LANCIA DELTA HF TURBO IE	MITSUBISHI COLT TURBO
Price	£8,790	£8,869

ENGINE

Cylinders	4 in-line	4 in-line
Capacity, cc	1,585	1,597
Bore/stroke, mm	84.0×71.5	76.9×86.0
Valves	dohc	sohc
Compression ratio	8.0:1	7.6:1
Fuel system	Weber IAW fuel injection, Garrett T2 turbocharger, intercooler, mapped ignition	Mitsubishi electronic injection and turbocharger, mapped ignition
Max power, bhp/rpm	140/5,500	123.3/5,500
Max torque, lb ft/rpm	141/3,500	137.2/3,500

TRANSMISSION

Type	5-speed manual	5-speed manual
Internal ratios and mph/1000 rpm:		
Fifth	0.903:1/20.9	0.731:1/23.5
Fourth	1.154:1/16.4	0.896:1/18.9
Third	1.524:1/12.4	1.240:1/15.4
Second	2.235:1/8.5	1.833:1/9.7
First	3.583:1/5.3	3.166:1/7.1
Final drive	3.40:1	4.067:1

SUSPENSION, STEERING, BRAKES

Front	Independent, MacPherson struts, anti-roll bar	Independent, MacPherson struts, anti-roll bar
Rear	Independent, MacPherson struts, transverse links, trailing arms, anti-roll bar	Independent, trailing arms, coil springs, anti-roll bar
Steering	Rack-and-pinion	Rack-and-pinion
Brakes, front/rear	Vented disc/disc	Vented disc/drum

WHEELS, TYRES

Wheels	Alloy, 5.5×14	Alloy, 5.5×14
Tyres	165/65R14	185/60HR14

DIMENSIONS

Length, in	153.3	156.1
Width, in	63.8	64.4
Height, in	54.3	53.5
Wheelbase, in	97.4	93.7
Track front/rear, in	55.2/55.1	54.7/52.8
Fuel tank, gall	12.5	9.9
Kerb weight, cwt	20.0	18.3

PERFORMANCE

Maximum speed, mph	124.1	120.6
Speeds in gears	(at 6,800rpm)	(at 6,000rpm)
First	36	43
Second	58	58
Third	84	92
Fourth	112	113
Acceleration through gears, sec:		
0-40mph	4.3	4.1
0-60mph	7.6	8.2
0-90mph	16.3	18.9
Acceleration in fourth, sec:		
30-50mph	6.5	6.9
40-60mph	5.3	6.1
60-80mph	6.2	6.7
Acceleration in fifth, sec:		
30-50mph	11.2	12.4
40-60mph	9.6	11.8
60-80mph	8.4	7.9

FUEL CONSUMPTION

Overall test average, mpg	25.4	26.0
Government test figures, mpg:		
Urban cycle	28.2	28.0
Steady 56mph	43.5	45.6
Steady 75mph	33.6	35.3
Manufacturer/importer	Lancar Ltd, Crawley, W. Sussex	Colt Car Company, Cirencester, Glous

A HARD DAY'S NIGHT

Lancia lent NZ Car a Delta GTie to remind us what a good car it is. Ross Horsburgh wasn't totally convinced...

Placed blindfolded behind the wheel of a Delta GTie you'd be in no doubt you were in an Italian car.

Firstly there would be the smell. What ingredient of Fiat/Alfa/Lancia upholstery gives it that distinctive smell? Whatever, the smell says Italian, and promises excitement.

Then there would be the driver's seat. Supportive in all the right places, despite only fore/aft and backrest-tilt adjustments being available. Who needs multiple adjustment if the seat's design is okay to start with?

Next there'd be the driving position. The straight arms and tight squeeze down in the footwell. The pedals, close together and perfectly placed for heel and toeing. The long, door-mounted elbow rest encouraging a quarter-to-three hold of the steering wheel.

Remove the blindfold though, and some home truths might hit hard. The Delta has been around for a long time now, and it shows. The interior still smells exotic, and still looks it. But late 1970s exotic, not 1980s. There's lots of that heavy ill-fitting black plastic trim the Italians have moved away from with more recent designs like the Uno and Tipo.

Switchgear has the 1970s Italian "random-scatter" look. You will of course eventually find the power window switches jammed down between the front seats and the handbrake – but not as readily as usual in a modern car. Heater controls are quaint though – three large round knobs looking like escapees off an old mantle radio control all ventilation functions. Different, and great conversation pieces which work surprisingly well.

A flimsy plastic lid covers a poky glovebox, which at least has an interior light. While there are plenty of storage bins, because they are all made of that hard shiny plastic, objects skate around annoyingly in them. Small rear vision mirror size is yet another reminder of days gone by.

Overall then, the cabin looks dated. Moreover, there is minimal headroom for taller drivers, and predominant black

Squarish lines for the 5-door Lancia Delta GT, a low profile Italian in New Zealand.

plastic makes the interior a little claustrophobic.

Similarly with the body styling. What was once a styling lesson to other manufacturers on how to combine 5-door practicallity with distinctive, sporty lines is now a period piece. Sure, the lines are still elegant, the layout practical, and the whole package distinctive. But the styling is too angular, the window area too small, the roof gutters too prominent, and the detail trim too cluttered for anyone to be fooled into thinking this is a recent design.

Opening the bonnet reveals the familiar 1600cm^3 twin cam motor, its glorious alloy "Lancia" emblazoned rocker covers sure to please any car enthusiast. Mounted east-west in the car, there is surprisingly little room left in the engine bay. Power has increased to 80kW (108 bhp) since fuel injection was recently added to the motor.

The Delta reaches 100 km/h in 10.4 seconds, runs from 50 to 80 km/h in 3rd gear in 5.3 seconds and has a top speed of 185 km/h. Gearing is a low 30.1 km/h per 1000 rpm in 5th.

Driving the Delta provided more of the good and bad picture. Powerful from low revs yet willing to spin out to 7,000 rpm, the motor seems the epitome of a hot-blooded, sporty Italian powerplant. Who cares that is is much noisier than an oriental motor when it is such a willing accessory to driving pleasure?

Front-wheel drive cars of the Delta's vintage all had some degree of torque steer. However, in the Delta's case it is not excessive, and front wheel grip overall is good. Unfortunately with the steering there is noticeable vagueness in the straight-ahead position, and the steering is very heavy at low speed.

Also heavy is the clutch action. That by itself might be acceptable, except that its action is annoying because of the sudden and often unpredictable take-up point. Similarly, the throttle action did not have the sensitivity expected in an Italian car. Of all the controls though, the most disappointing was the brake pedal – long travel, spongy and not particularly powerful, the Delta's brakes did not inspire confidence.

Clutch, throttle action, and the heavy steering all conspired to make the Delta a chore to drive in town – I never thought I would catch myself in an Italian car hoping the traffic lights would stay green so I didn't have to go through the gears again!

That's the bad news about driving the Delta. The good news is that it still is a great handler. Push it harder and harder, and it feels better and better. Caution through. If you are used to relying on the predominant understeer of a Japanese car to keep you out of trouble, the Lancia's neutral handling may catch you out. Fortunately oversteer appears predictably and is easily controlled. It certainly enhances the car's entertainment factor.

One other point worthy of comment is the great sunroof. This is steel and has such a simple lever-down, slide-back action that one wonders why others bother with electric motors or cumbersome manual winders.

The Delta GTie is a mix of good and bad. My wife, normally rather one-eyed in her love of any Italian car, summed it

Delta shows its age on the inside with a somewhat claustrophobic interior.

Twin cam power for the compact Lancia, a motor that produces 135 Nm of torque at 3500 rpm.

up best when she said she "liked it in principle".

Exactly! It's Italian. It's got character. It's fun to drive quickly. It wears a famous maker's name. And it's got status.

But it's also a bit long in the tooth in many areas. For its $33,000 pricetag you look at an Alfa 33. The Alfa is also hard work to drive, but at least its controls have more feel to them. Then of course there is the Peugeot 405GR. It also possesses European status, but without the stigma of old age.

In the end it comes down to personal taste. More than most cars, the Delta's mix of good and bad really needs sampling for the individual to decide for him or herself which is the dominant influence.

Tried

LANCIA DELTA HF TURBO

A car strictly for enthusiasts, feels Mark Hales

THIS IS depressing. It costs the same as a Golf GTI, is faster than an Escort RS Turbo, and looks like a world rally champion. Why is it not a best seller? Before some off-the-shelf cynic points out that it's a Lancia and what did you expect, let's look at the good bits.

First and foremost, the engine. It looks great; twin cams nestling beneath sculptured aluminium castings, it's extremely powerful (140bhp/141lb ft of torque) with minimal turbo lag (*and* big brother integrale's 30 second extra power overboost facility). It revs to oblivion without noise, vibration or fuss, and hell, it's quick enough to zig zag in and out of any traffic queue you care to name. And the chassis; there's almost no understeer worth the mention, no tug at the hands when you give it some good boost in second or third, and a degree of obliging, rather than challenging, tuck-in when you lift off in a corner. Power squats the tail and keeps it wide, more like rear-drive than front, and the overall balance is really good once you learn that the squat-and-sway feel of the rear never develops into anything more serious than that. And the ride is excellent; with no jarring or jiggling, it floats over imperfections like a big car and yet displays the body control of a sports saloon.

Sadly, objectivity demands that we look at the less good bits as well. The information from an excellent chassis reaches the driver through steering which needs muscles of Garth to operate, even when on the move. Unfathomably, power assistance is not an option. Equally unfathomable, given the on the move integrity of the chassis, is its behaviour away from rest. Apply full power in first, and wait for a little lag. Then *pow*. The front wheels light up, and the car dives all over the road, accompanied by crashing and banging from the front suspension. Once you know this happens, you avoid it, and select second early. Then, having gathered speed, the brakes are effective enough, but the pedal moves way too far before anything happens and it has that sticky overservoed feel which makes a smooth stop difficult.

Ergonomically, Lancias are vastly better than they were, and the Delta's traditional arm stretch to the (rake adjustable) steering wheel seems less at least, and the pedals are better placed, but why does the wheel still obscure the top half of the speedo and rev counter, and why is the dip and indicator on the left of the steering wheel? Why, after only 10,000 recorded miles did the dash creak and squeak like a Hammer Horror portal, and the trim rattle and thump? Perhaps the press are to blame.

This car *should* be better than Ford's most expensive, and lest we forget – flawed – Escort. The Lancia has a vastly more refined engine which propels it faster, it has a better basic chassis, better ride and costs less. The sad truth is that the Lancia can't match the Ford's build quality or its interior packaging, or its basic user friendliness. Which is another way of saying that the Ford does nothing supremely well, or disastrously badly.

Lancia, thank goodness, still make cars for enthusiasts rather than the general public. But for how much longer?

Price: £10,195. **Performance:** 122.4mph/0-60 7.6sec/30-50 in 4th 6.5sec. Superb engine, class-topping performance. **Handling:** able chassis, minimal understeer, balanced against awful steering. **Fun factor:** considerable but qualified. **Verdict:** A super basis for a class victory spoiled by poor ergonomics, dubious build quality, heavy steering.

98

GROUP TEST: HONDA INTEGRA EX 16, LANCIA DELTA GTie, PEUGEOT 309 SRi, VW GOLF GTi

REVALUING THE·MARQUE

The Golf GTi has long been considered the hot hatch benchmark, but the soaring Deutschmark is making an already expensive purchase price prohibitive. Are cheaper rivals – including Peugeot's hotshoe 309 – undermining the Golf's superiority? We test the upstarts

As well as being the car that started off the Hot Hatchback craze more than a decade ago, the Golf GTi is still widely regarded as *the* car in its class to beat. For years now, motoring critics everywhere have heaped praise upon the Volkswagen, and though others are now quicker and have sharper handling, it's still hard to beat as an all-round package.

This all-round competence has given the car an enviably wide appeal: boy-racers, sloane-rangers, junior executives all fall under the spell of the GTi, and even senior managers whose budgets would finance a far more more lavish car than the Golf have been known to forsake their luxurious carriages in preference for the GTi.

However, despite the apparent inability of rival firms to come up with a really first-rate GTi copy, it seems a force beyond the control of even the most high-tech manufacturer may soon be exerting a big influence on the market. That force is the exchange rate, and its influence is the ever-strengthening Deutschmark which is now forcing the price of German cars through the roof.

Along with BMW and Porsche, Audi/Volkswagen have been forced to up their prices (they rose twice in just over a month at the beginning of the year), and the Golf GTi – which was never a bargain in any case – is beginning to look very expensive indeed. The five-door we test here is a heavyweight £10,449, over £1500 more than it was one year ago.

So, this might be the chance for others of the hot hatch fraternity to steal an advantage; with the Golf now carrying a five-figure price tag, it could be that it has become unacceptably expensive even to committed Golf fans, and that they will now look elsewhere when buying. For this test we have gathered three likely five-door candidates.

The cheapest of these Golf alternatives is the Lancia Delta GT i.e., priced at a very attractive £7990; next up on the price scale is the Peugeot 309 SRi at £8795; the Honda Integra EX 16 costs £8980.

All three do, of course copy the Golf's basic drivetrain layout: an engine placed transversely over the front driving wheels and transmitting power through a five-speed gearbox. Their overall dimensions aren't dissimilar either (though the Honda does have some extra inches in overall length), and all sit on a wheelbase just under 100 inches.

However, in the engine department it's Honda who offer the most innovation. Following on from the 1.5-litre Integra launched in March last year, a quicker version has been introduced, and it's set to turn the innocuous hatch into a hot performer. The engine is the 1590 twin cam, also found in the diminutive CRX, and four-valves-per-cylinder along with Honda's own PGM Fi computerised fuel-injection results in an output of 125bhp.

As might be expected from a 16-valve engine peak power occurs at 6500rpm, and the maximum 103lb ft torque isn't produced until the engine is spinning at 5500rpm.

Modifications have been made to the Integra's chassis to complement the extra power from the screamer of an engine, though the changes aren't sweeping ones. Disc brakes are added to the rear, and there's power-steering, but the spring rates of the front torsion bars and rear trailing-linked dead axle suspension are not stiffened.

In terms of equipment and appearance, the EX 16 is also surprisingly undifferentiated from the 'cooking' version; perhaps this is why the price premium over the 1.5-litre isn't a vast one. An electric sunroof is added, but there's no sign of electric windows or central locking.

Peugeot will of course soon be launching their ultra-hot 1.9-litre 309 GTi, but until it arrives in the UK the hottest 309 is the SRi. This uses the same Bosch-injected 1580cc single cam engine used in the delightful 205 GTi, and which produces 115bhp at a heady 6250rpm. This is backed up by 98lb ft torque at a high-ish 4000rpm. Gearing is exactly the same as for the 205 GTi, with fifth set for a sporty 18.7mph/1000.

Clearly, Peugeot are saving the big visual hit for the red-hot 309 GTi, because the SRi looks much like any of the cheaper models in the range. There's a bootlid spoiler and 205-style alloy wheels running low-profile 60-series tyres, but the whole effect is one of understatement. It doesn't fare badly on equipment, however: included is full central locking, electric front windows, and a radio/

GROUP TEST: HOT HATCHES

HONDA INTEGRA EX 16

The Honda has sleek, low body styling but its good interior space packaging is not on the same par as the other hatches. Though the interior is stylish and comfortable with light and easy-to-use controls – enhanced by effortless power steering – rear space is cramped and the boot is small. Engine is more innovative, a smooth-revving 16-valve design which combines performance with good economy

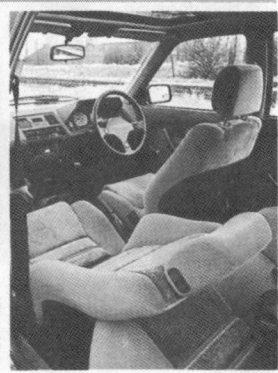

cassette player for example.

The chassis is standard 309, with front struts and transverse torsion bars at the rear, though the springing is stiffened for less travel. Power-steering isn't included (it's a £310 option), and brakes are disc/drums.

The Delta is something of an old stager now, the same bodyshape having been on sale for seven years, and indeed its replacement will soon be reaching the scoop-photo stage. However, its impending demise didn't stop Lancia giving it a thorough going-over last year, with substantially re-tuned engines.

As such, the Lancia seems something of a bargain in this group. With a fuel-injected twin-cam engine, central locking, a sunroof, and electric windows as well, the Delta GT i.e. is very well priced indeed.

Its 1585cc engine has shed the double carburettors it used originally, and in their place is the Weber IAW combined electronic injection and ignition; output is 108bhp, and torque is 100lb ft at 3500rpm.

The Strada-based chassis uses a simple suspension recipe, with independent struts all round, though there is at least a touch of Lancia's sporting heritage to be found in that braking is by discs all round. Steering is rack and pinion; power-assistance is neither fitted nor available.

And finally to the Golf itself. On paper, there's nothing sensational about it; simply a well tried formula.

Though the Golf's engine has a 200cc capacity advantage over the other three, Volkswagen have chosen to extract maximum torque, rather than outright power. The result is 112 bhp at 5500rpm from the Bosch-injected unit – less than the smaller-engined Peugeot – but the dividend comes with an impressive 115 lb ft torque at a lazy 3100rpm. The transmission is five-speed, with closely spaced ratios.

The chassis of the current Golf is much the same as it has always been, with struts at the front, and a torsion beam axle at the rear. The introduction of the Golf II, however, saw the adoption of much-needed rear disc brakes, and at the same time the springing was made a little more compliant.

Gone are the days when Volkswagens were very poorly equipped: in five-door form the GTi comes with a new design of alloy wheel shod with 60-series tyres, and there's also a sunroof. But central locking, tinted glass, and a split-folding rear seat, for instance, all bump up the price.

PERFORMANCE	
GOLF	●●●●●
PEUGEOT	●●●●
INTEGRA	●●●●
LANCIA	●●●●

Awarding a full five points isn't something we do often, but the Golf's engine is one of the few four-cylinder units around which really is hard to fault, so we feel it to be worth top marks. Some other manufacturers have now matched its turbine-like refinement, but what none have managed to do is equal the delight-

100

HONDA INTEGRA EX 16, LANCIA DELTA GTie, PEUGEOT 309 SRi, VOLKSWAGEN GOLF GTi

LANCIA DELTA GTie

Lancia's hatchback styling is dated, and a replacement is rumoured to be on the way. However, it's a good value package since it's competitively priced and packed with such equipment as sunroof, electric windows and central locking. Facia is a bit untidy with its scattering of brightly-coloured dials, but it does look sporty. Seats offer adequate support, but could be firmer. Fuel-injected twin cam engine has sweet Italian raciness, but it's not as fast as the Golf

fully even power-spread which begins virtually at tickover and continues until peak revs.

The engine's ample torque means that instant acceleration is available regardless of what gear is being used, and it's a delight to keep the GTi unit turning at 1500 to 3500rpm. This lazy pulling power is ideally suited to stop-start town driving.

This time round we achieved a zero to 60mph time of 9.6 secs (the three-door might go a fraction faster because of marginally less weight) and a maximum of 114mph.

The gearchange is precise, allowing the driver to flash through the ratios and thus make the most of the engine, and the close-ratio 'box enhances the 1.8-litre's willingness.

The 309 may not have the super-chic looks of the 205, but it does have the same revvy engine, and the same instant response which is a constant encouragement to exploit the power. Not surprisingly, in view of the marginal extra weight, the 309 doesn't have quite the pace of its smaller brother, but it's still rapid.

It will reach 60mph in 9.9secs, and despite the capacity disadvantage over the Golf, it still matches the German car on acceleration all the way up to 100 mph. On top speed, it's pretty close, too, reaching 113.

As well as delivering sparkling acceleration, the SRi's engine is also very flexible: a good spread of low-down torque coupled with low overall gearing means ready acceleration in any gear. Refinement, too, is worthy of praise: somehow Peugeot have made this motor a lot sweeter in injected guise.

It's a pity, then, that the 309's gearchange has a long, inexact gate plus a lifeless feeling lever. Changes have to be made carefully to avoid missing the slot, and this contrasts rather markedly with the sporting nature of the engine.

The Honda's engine is very smooth indeed; despite the extra cam gear there is no undue harshness or vibration anywhere in the rev band. But we must admit to an initial disappointment, for the considerable extra horsepower did not translate into the neck-snapping acceleration we had expected. Certainly, a 0-60mph time of 9.6secs is hardly slow, yet below 5000 rpm the Integra feels ordinary; flat, even.

But in true 16-valve fashion, the car comes alive towards the top of the rev range: after 90mph, the added breathing efficiency of the engine comes into its own, and the car begins to leave all the others a long way behind.

Of course, just how relevant this type of performance is to average road conditions is another question, particularly as the Honda is hardly dressed to kill. But the engine is flexible enough when operating at normal revs – even if it lacks the iron-glove punch of the Golf – and there is no sign of temperament when chugging along at low revs.

As is to be expected from Honda, driveline refinement is beyond reproach. The gearchange is well-sprung and precise, and the clutch

101

GROUP TEST: HOT HATCHES

PEUGEOT 309 SRi

GTi power is hidden under a conservative interior in this Peugeot, which goes better than it looks – it matches the Golf to 100mph in hard acceleration. Interior also lacks sports atmosphere – and more importantly lacks good side support in seats, needed for a car with such handling characteristics. Six-dial facia is busy, but still manages to appear bland. It seems good value though, since equipment is good, including remote central locking, electric window and mirror adjustment, tinted glass and alloy wheels

engages smoothly and lightly.

The Lancia has definitely benefitted from the addition of fuel injection: it feels sweeter and has the edge of refinement over the previous carburettored unit.

The engine offers a wide spread of power and torque, and many will feel it has the all-important soul that the Honda and Volkswagen lack. Performance isn't notably quick, but it is respectable. The Delta can be wound up to a maximum of 112mph (the slowest of the four, by a small margin), and the 30-80mph 3rd/4th/5th gear times are suitably short.

Something that doesn't help the Delta's cause at all is the gearchange. As well as feeling very dead the lever works through a notchy, and often obstructive gate.

HANDLING AND RIDE

GOLF	●●●●
PEUGEOT	●●●●
LANCIA	●●●
INTEGRA	●●

As we've already mentioned, there are plenty of cars which will easily out-grip the Golf, but not so many incorporate an acceptably compliant ride, or are so forgiving when the driver strays too near the limits of the chassis. By the standards of some current hot hatches which feature board-like springing the Golf might seem soft, but it only takes a short spell on a tightly-cornered handling course to appreciate that standards of tyre adhesion are still very high.

You couldn't by any means describe the ride as soft, but there is a well-damped firmness that many drivers will prefer in any case, and only over the very worst road bumps does the suspension feel unforgiving. But the Golf does have its Achilles heel, and that's the steering: though well-geared (3.8 turns), it's simply too heavy; parking calls for a sweat-inducing effort.

The Peugeot has a very chuckable feel to it, but its reflexes are nowhere near as quick as those of the 205 GTi. It lacks the razor-sharp turn-in of the little car, showing a substantially greater degree of understeer.

You can still have a lot of fun with the Peugeot, and many will actually prefer its more reserved chassis tuning to the almost outrageously firm go-kart like set-up of the 205.

It also beats the Volkswagen hands down on ride comfort: though well-damped and virtually free of body roll, there still remains the traditionally excellent ride which one expects from Peugeot.

Our car didn't come with power-steering fitted so we can't comment on its quality. But it isn't particularly expensive, so all we can say is that it surely must be better than the manual system, which is low geared and heavy, and with some vagueness about the centre position.

In our last test of the Integra, we bemoaned its vague handling; again, unfortunately, we have to repeat our criticism, but with added emphasis seeing as the EX 16 is meant to be a sporting hatchback.

HONDA INTEGRA EX 16, LANCIA DELTA GTie, PEUGEOT 309 SRi, VOLKSWAGEN GOLF GTi

VOLKSWAGEN GOLF GTi

Golf's styling has classic charm but striped, boy-racer seat trim might not appeal to all tastes. Seats themselves are very supportive, though, encouraging a commanding driving position. Controls are well placed, but steering is heavy for parking. Rear seat space is reasonable; space-saver spare means flat boot floor, but split-folding rear seat is a costly extra – as are central locking, electric windows and electric mirror adjustment. Engine is still exciting, though, particularly in its punchy flexibility

Honda chose not to substantially beef up the spring rates on the car, and that's a pity, for at a very early stage the Integra begins to lose its composure: thrust it even moderately hard through a bend and a good deal of understeer is to be felt.

There's lots of body roll, and it takes little further provocation to see the Integra get into an undignified wallowing rhythm which, compared to the tautness of the Golf, is most unimpressive. This general chassis softness suggests that at least the ride will be good, but not so: bumps and crashes are felt, and there's some bounciness.

No complaints about the power-steering, though: it's pleasant to use (even if it lacks feel), and the weighting is good. It complements the car's light controls, and makes the Integra an effort-free car to handle.

Lancia have struck a good balance between handling and ride. The Delta has a good level of roadholding, and also transmits a useful amount of information back to the driver. The all-disc brakes feel crisp and powerful, and in complete contrast to the Integra, the driver knows he's in a sporty hatchback.

This hasn't compromised the ride, though, which has good bump and noise absorption qualities, without relying on a long spring travel. But while the Lancia's handling is undoubtedly good, it is beginning to show its age in a couple of respects. To get the comfortable ride, Lancia have had relied on rubbery, rather than soft, suspension settings, and at times the Delta can feel vague.

Power-steering can't be had on this Delta; that's a pity. The system is lifeless, and parking calls for tiresome arm-twirling. What would really improve handling would be the well-weighted assistance found on the four-wheel-drive HF.

ACCOMMODATION

GOLF	●●●●
PEUGEOT	●●●●
INTEGRA	●●●
LANCIA	●●●

Not everyone will like the striped, boy-racer seat trim of the GTi, but the two front seats are near to perfection. They have excellent thigh support and plenty of lateral location, and the driving position is high-up, providing commanding forward vision. Quite recently, Volkswagen added a seat-height adjuster to the GTi as standard equipment.

The five-door bodyshell is worthwhile paying for if rear passengers are to be carried often, for access to the rear seat in the three-door is anything but impressive. Rear seat space is quite generous, but by no means the best in the class.

Boot space is large once the rear seat is folded flat, and now that Volkswagen have fitted a slim space-saver spare wheel, the floor is virtually flat. Oddly, though, you have to pay £213 extra for a split-folding rear seat.

The Honda is the biggest car in the group, measuring a foot more than the next longest, the Peugeot. But

GROUP TEST: HONDA INTEGRA EX 16, LANCIA DELTA GTie, PEUGEOT 309 SRi, VW GOLF GTi

from the inside, you wouldn't be aware of this, for there's little extra space. Indeed, the front cabin even feels a little cramped.

However, the BX 16 has superior seating to the ordinary Integra: instead of the shapeless, small seats of the cheaper version, here we have well padded frames with added side-cushioning, and the result is far better comfort.

The driving position is low-set, a necessity due to the Integra's low roofline, but lots of glass prevent any feeling of being hemmed in, and also gives good visibility. Steering wheel, gear lever and pedals are all well placed, and there's a clutch footrest.

Fold the rear seat down, and there's a long platform, but with the seat up (which split-folds) the boot is surprisingly small. It sorely lacks height, and the limited space is further compromised by a high load sill. There are a couple of design points which irritate, too: the parcel shelf which is hinged in the middle is unnecessary, and it's a pity that only the top portion of the rear seat comes down.

In contrast to the poor space-efficiency of the Integra, the 309 makes very good use of every available inch. Head space in the front is good (though of course the lack of a

> **"Peugeot's gutsy 309 SRi goes well but it's just too bland to be a real GTi rival"**

sunroof means no valuable inches are taken up), and rear passengers shouldn't feel cramped unless they're above-average size.

The main reason the Peugeot falls second to the Golf is because of the front seats. While they're comfortable enough, and feature a lumbar support adjustment, we feel that in this sporty version some extra side-location would be welcome. Also, the ergonomics are less satisfactory than in the Golf: some may find the steering wheel positioned a shade too low down, and the gear lever a little too far away. These points mentioned, though, the Peugeot's good ride enhances comfort.

The Peugeot's load sill is disappointingly high, but once the rear seats, which split-fold, are flat, the load bay is a good size and generally free of obstructions, thanks to the compact suspension design.

The Lancia has some good points such as a bumper-level load sill and good rear cabin space, but generally it's a design on which time has caught up. While most Italian cars have thankfully lost the long arms/ short legs driving position, you'll still find a trace of it in this Delta, and this does nothing for comfort. The seats themselves are reasonably supportive, but nothing special bearing in mind Lancia re-designed them last year. An adjustable steering column is, however useful.

The Delta's generous rear seat space is at the expense of some boot space, and a pair of large suspension struts take up a good deal of space. However, once the split-folding rear seats are folded flat, the car is turned into a good load carrier. Because of the size of the tailgate, it's not particularly light to lift up.

LIVING WITH THE CARS
INTEGRA	●●●●
GOLF	●●●
PEUGEOT	●●●
LANCIA	●●

Instant starting and light controls make the Integra effort-free to drive, and as we've already mentioned, the 16-valve engine doesn't become difficult and temperamental when working at low revs in town traffic. It's a quiet unit, too, and noise is only appreciable once 5000rpm has been exceeded. The only complaint in everyday driving is that the rear body pillars create big blind spots making city driving sometimes difficult.

The dash is attractive to look at, and also very clear. The four dials are easy to assimilate quickly, mainly because they're large enough and not overcrowded with confusing calibrations. All switches work with a high-quality feel, and in general the Honda feels the best constructed of the group.

The heater system is powerful, and makes particularly short work of demisting the front screen. However, ventilation is less impressive. As ever, the Honda scores on thoughtful little details such as the remote boot/fuel filler release.

If the Honda is a relaxing car in town, the Golf is anything but. Heavy steering dominates low-speed driving, and the low gearing makes the engine noisy and resonant at pretty well all revs.

All the controls are within easy reach, and are sensibly grouped around the steering wheel. But this also means the number of blank switches becomes very apparent, emphasising how little equipment the car comes with. Little praise either for the instrumentation; five years ago we thought the design attractive, but it looks dated now, and the small dials can be hard to read at a glance. What the Golf does have, though, is a splendid trip computer. Simply pressing the tip of the right-hand indicator stalk produces an

> **"Five-figure price tag could put the Golf out of the running"**

instant read-out on the facia.

There is no independent ducting for the ventilation system, so no cool air for the cabin in winter. But the heater is powerful, and begins blowing out hot air a very short time after the engine has been started. Like the Integra, the Golf suffers from bad rear blind spots caused by the thick body pillars.

The Peugeot may be a great little sprinter, but it's less happy trickling through traffic. The accelerator pedal has no play in it, the result being that power comes in too suddenly. Also, the fuel cut-off causes a jerkiness when deaccelerating.

Peugeot's dash is ordinary-looking, though it certainly doesn't lack instruments. There are six dials in all, and they're all clear to read. Generally, the interior looks very plasticky, and one is left with the impression Peugeot thought little about style when designing the interior. For instance, the heater controls (which work efficiently, but still can't supply cool air along with heat) are placed on an unimaginative black panel.

However, quietness is on the 309's side: compared to the buzzy 205 GTi, the SRi is subdued at speed, despite the fact that engine revs are high when cruising at 70 mph. But what the 309 does share with its smaller brother is an only average build quality – the finish doesn't impress after the Golf or Integra.

There are two things which are immediately obvious about the Delta. The first is that it's the best equipped of the group: the sunroof, electric windows, and central locking give it a luxury air, and the fact that no radio is included isn't all that important because many customers want a unit of their own choice.

The second aspect is a less attractive one, this being that the car doesn't seem particularly well made; the shiny black plastic isn't what quality-seekers are after these days, and around the seams the trim doesn't all fit too neatly.

The dash is a busy one, with a multitude of brightly-coloured dials, and this adds to the sporty feel of the car. Heating is powerful (the system was recently redesigned, say Lancia), though some more thought might have been given to the almost silly-looking rotary controls which regulate heat and air distribution.

With the twin-cam engine well muted, the Delta is quiet at speed and makes a reasonably good cruiser. Around town it can't match the drivability of the Honda – the low-geared steering and and baulky gearchange see to that.

COSTS
LANCIA	●●●●
PEUGEOT	●●●
GOLF	●●●
INTEGRA	●●●

In this test the Integra turned out to give the best fuel consumption, with an average of 32.1 mpg – good for a high-power 16-valve engine. However, because these cars are built to be enjoyed, we don't want to place too much emphasis on fuel consumption as a significant cost element; economy will vary greatly according to how the cars are driven.

Our Golf gave us an average of 27.0 mpg (though we know from previous tests the car can easily better 30 mpg); the Peugeot 26.1 mpg, and the Lancia 27.5.

What is probably more important is value, and here the Lancia scores strongly with its high equipment specification and low price. It also comes with a three-year warranty, and this is a particularly attractive benefit for the private buyer. Were it

GROUP TEST: HONDA INTEGRA EX 16, LANCIA DELTA GTie, PEUGEOT 309 SRi, VW GOLF GTi

not for Lancia's traditionally high rates of depreciation, the Delta might even have scored more points.

The Peugeot doesn't look particularly good value, and it can't match the Italian car on fittings. But what it does do is slip into group five insurance as opposed to the group six which applies to the others.

As we've said, the Golf is now very expensive, but because it's a well-made, sought-after car, depreciation has always been low so it makes a lot of sense for drivers who have to finance their own motoring. We'll give Volkswagen the benefit of the doubt and say that despite the high price we expect first and second year depreciation to remain low.

The Integra is certainly a good buy considering its potent engine, but it has to be placed an honourable last because spare parts costs will most likely be the most expensive of the group; one service could easily wipe out any savings made by good fuel consumption.

On servicing the Golf would initially seem to have the advantage with its 20,000 mile schedule (or 10,000 miles after one year), though many drivers may not wish to cover this many miles without their car receiving attention, and so may opt

"The GTi wins because it is the best – and the best always has to be paid for"

for an additional check service. The other three cars rely on the more traditional pattern of an oil change at 6000 miles, and major attention at 12,000 miles.

While the Lancia has the longest mechanical warranty, the Golf comes with a useful three-year paintwork guarantee. Mechanical warranties for the Honda, Peugeot and Golf are 12 months/unlimited mileage, and all four cars are covered for six years against rust perforation. Peugeot and Volkswagen score on reasonably large dealer networks; Honda and Lancia owners will have to make do with a smaller choice of dealers.

VERDICT	
GOLF	●●●●
PEUGEOT	●●●●
INTEGRA	●●●
LANCIA	●●●

This is certainly no easy verdict for every car has its virtues, yet at the same time none emerges as the obvious winner.

If price did not come into it, we would go for the Golf: by our reckoning it's still the best all-round hot hatchback, offering secure handling, a lovely engine, good comfort, and a high build standard. Yet our enthusiasm has to be tempered just a little by the heavy steering and high cruising noise. But if price did come into it, which of course it does in the real world, we would have to think very carefully indeed about buying a GTi.

Whether it's worth paying such a lot for the Golf is obviously a personal decision, but of course it should be remembered that you can cut down at least some of the expense by ordering a three-door model on steel wheels. In this test, then, we'll award the GTi the top spot on the basis that it is the best, and the best always has to be paid for.

Some might feel the Peugeot is too bland to be a true Golf rival, and indeed we have to admit we weren't knocked out with the styling. But what the SRi does very successfully is to offer practically the same performance as 205 GTi, but in a less nervous chassis which gives far superior comfort. The 309 will run with the hot hatch pack, but at the same time is a civilised family car when the pace is slacker.

The Integra looks the part and has a good engine. It's also well built and reliability is virtually guaranteed. So it's a pity the car has very mediocre handling and a surprisingly poor amount of interior space. Considering the price premium on the Golf 16V, the Integra seems quite a reasonably priced package, but you have to remember that equipment is fairly basic.

Had Honda sorted out a decent, firmed-up suspension pack, then the EX 16 could have been a real GTi-challenger, and we could have turned a blind eye to the limited seat and luggage space.

The Lancia is an old favourite, and last year's revamp was an intelligent development which improved on some of the previous weak points, but which left the good aspects, such as the good-looking body, untouched. But the thorough update can't disguise the fact that the car feels old-fashioned now, and all eyes are on its replacement.

Its handling now feels dated, and the still-lovely twin cam engine isn't enough to make one turn a blind eye to the car's many weak points. And as other manufacturers seem to be gradually improving the quality of build on their cars, unfortunately we see no evidence of this in the Delta.

The one big saving grace is the price: it's undoubtedly the best buy of the four, and there's also that very comprehensive warranty. So if economics were the sole factor in this contest, we would find ourselves Lancia-choosers. But over a wider range of considerations, the Delta has to be placed fourth.

HOW THE CARS COMPARE

CAR	HONDA INTEGRA EX 16	LANCIA DELTA 1600 GT ie	PEUGEOT 309 SRi	VOLKSWAGEN GOLF GTi 5-dr
PRICE	£8980	£7990	£8795	£10,449
Other models	One	three	10	11
Price span	£7550-£8980	£5945-£8790	£5575-£8795	£8528-£11,949

PERFORMANCE

Max in 5th (mph)	116	112	113	114
Max in 4th (mph)	108	102	103	100
Max in 3rd (mph)	84	82	79	80
Max in 2nd (mph)	58	57	57	54
Max in 1st (mph)	35	36	33	33
0-30 (sec)	3.5	3.1	3.2	2.9
0-40 (sec)	5.4	4.8	5.2	4.9
0-50 (sec)	7.3	6.8	7.1	6.7
0-60 (sec)	9.6	9.7	9.9	9.6
0-70 (sec)	12.8	12.8	13.2	12.5
0-80 (sec)	16.3	17.5	17.4	17.2
0-90 (sec)	22.3	23.1	23.4	22.6
0-100 (sec)	29.5	32.9	33.7	33.5
0-400 metres (sec)	16.9	17.1	17.3	17.2
Terminal speed (mph)	83	80	81	80
30-50 in 3rd/4th/5th (sec)	5.4/7.5/4.8	5.4/8.4/12.6	5.5/7.6/10.7	4.9/6.4/9.2
40-60 in 3rd/4th/5th (sec)	5.5/7.6/9.5	5.6/7.9/12.1	5.6/7.7/10.9	4.8/6.6/8.4
50-70 in 3rd/4th/5th (sec)	5.8/8.4/10.4	5.8/8.7/12.5	5.9/8.2/11.8	5.3/6.9/9.4
60-80 in 3rd/4th/5th (sec)	6.5/9.2/12.0	6.8/9.3/14.0	6.9/9.1/6.7	7.0/7.5/10.5

SPECIFICATIONS

Cylinders/capacity (cc)	4/1590	4/1585	4/1580	4/1781
Bore x stroke (mm)	75 × 90	84 × 72	83 × 75	81 × 87
Valve gear	dohc	dohc	ohc	ohc
Compression ratio	9.3:1	9.7:1	9.8:1	10.0:1
Fuel Injection	PGM Fi	Weber IAW	Bosch L	Bosch K
Power/rpm (bhp)	125/6500	108/5900	115/6250	112/5500
Torque/rpm (lbs/ft)	103/5500	100/3500	98/4000	115/3100
Steering	PA/rack/pin	rack/pin	rack/pin	rack/pin
Turns lock to lock	3.8	3.8	3.8	3.8
Turning circle	34.1	34.7	34.5	34.4
Brakes	S/Di(v)/Di	S/Di/Di	S/Di/Di	S/Di(v)/Di
Suspension front	I/Tor/AR	I/McP/AR	I/Tor/AR	I/McP/AR
rear	DA/TA/C/AR	I/McP/TC/AR	I/Tor/C/AR	TCA/C/AR
Tyres	185/60 R14	165/65R14	175/65/HR14	185/60/HR14

COSTS

Test mpg	32.1	27.5	26.1	27.0
Govt mpg City/56/75	31.4/43.5/35.3	28.8/45.6/35.3	30.7/48.7/37.2	27.4/48.7/37.2
Tank galls (grade)	11.0(4)	12.5(4)	12.0(4)	12.0(4)
Major service miles (hrs)	12,000(2.8)	12,000(3.2)	12,000(1.5)	20,000(1.4)
Parts costs (fitting hours)				
Front wing	£73.16 (0.5)	£85.18 (—)	£42.00 (2.40)	£83.89 (1.9)
Front bumper	£65.01 (0.2)	£195.57 (—)	£70.00 (0.50)	£72.41 (0.2)
Headlamp unit	£49.52 (0.4)	£78.37 (—)	£38.00 (0.50)	£31.46 (0.4)
Rear lamp lens	£42.71 (0.6)	£7.37 (—)	£5.00 (0.10)	£42.13 (0.2)
Front brake pads	£25.77 (0.6)	£35.00 (—)	£22.50 (1.0)	£27.80 (0.7)
Shock absorber	£60.14 (0.7)	£44.54 (—)	£30.50 (2.0)	£44.90 (0.8)
Windscreen	£88.83 (4.5)	£120.20 (—)	£42.00 (1.50)	£59.80 (0.7)
Exhaust system	£111.73 (1.1)	£251.52 (—)	£N/A (1.0)	£210.75 (0.9)
Clutch unit	£71.82 (6.7)	£83.70 (—)	£73.90 (4.20)	£100.48 (2.90)
Alternator	£160.78 (0.5)	£111.93 (—)	£80.00 (0.50)	£102.23 (0.50)
Insurance group	6	6	5	6
Warranty	12/UL	36/UL	12/UL	12/UL
Anti-rust/Paint	6 yrs	6 yrs	6 yrs	6 yrs
Paint	none	none	none	3 yrs

EQUIPMENT

Power steering	yes	n/a	£310	n/a
Alloy wheels	no	yes	yes	yes
Seat height adjustment	no	no	no	yes
Seat lumbar adjustment	yes	no	yes	no
Adj. steering column	yes	yes	no	no
Split rear seats	yes	yes	yes	£213
Tinted glass	yes	yes	yes	£240
Central locking	no	yes	remote	£168
Electric windows	no	yes	yes	£477
Electric mirrors	no	no	yes	£174
Sunroof	glass	yes	n/a	yes
Sound system	rad/cass	none	rad/cass	rad/cass

DIMENSIONS

Front headroom (ins)	37	35	35	34-36
Front legroom (ins)	32-39	37-42	33-41	33-41
Steering-wheel-seat (ins)	15-22	15-22	12-21	12-20
Rear headroom (ins)	35	32	34	37
Rear kneeroom (ins)	26-36	26-32	25-33	25-33
Wheelbase (ins)	99	97	97	97
Boot load height (ins)	31	25	32	31
Kerb weight (cwt)	20.2	19.6	18.3	18.5
Boot capacity (cu ft)	10/22	9/35	11/22	14/50

KEY. Valve gear: ohc, overhead camshaft; dohc, double overhead camshaft; Steering: rack/pin, rack and pinion; PA, power assistance. Brakes: Di(v), ventilated discs; Di, discs; Dr, drums; S, servo assistance. Suspension: I, independent; C, coil springs; AR, anti-roll bar; McP, MacPherson struts; TA, trailing arms; TCA, torsion crank axle; TL, transverse links; DA, Dead beam axle; Tor, torsion bars.

WORLD CHAMP on the cheap?

Brother to rallying's all-conquering Lancia Delta Integrale, a two-year-old Delta GT 1600ie for half its price new sounds too good to be true. Or does it? David Sutherland reports

The outskirts of Falmouth are a prime hunting ground for buyers in southern Cornwall. Both prestigious and run-of-the-mill marques are represented, and there are plenty of non-franchised traders, too. And, like everywhere else, Cornwall's motor trade is struggling through a recession, so salesmen are keen to do good deals.

Of the vast choice of hardware on offer it was this two-year-old G-registered Lancia Delta GT 1600ie which caught our eye. Offered for £4895 by a non-franchised garage, Autotrend, this was just under half its new price charged back in August 1989. It seemed amazing value for money considering that to buy a Golf or Peugeot 205 GTi for the same money you'd have to settle for a circa 1987 D-reg car. Go for the Lancia and you get a five-door hot hatch equipped with alloy wheels, electric windows front and rear and central locking.

However, our excitement was curtailed when we discovered that the odometer read 51,073, and that there was no documentation, which of course meant no service history.

But at least Autotrend's proprietor, Mr Aslam, wasn't making any optimistic claims about the Lancia. He more or less said it was for sale 'as seen'. He wasn't prepared to verify the mileage, but we guessed it was genuine. After all, if anyone had intended to 'clock' this car they'd surely have wound the mileage much further back than to 51,000.

He had bought the Delta at an auction in August, and thought it had probably been owned by a leasing company which had disposed of it on its second birthday. He said: "One in three cars you buy from an auction don't come with a log book or a service book — sometimes they arrive in the post afterwards, sometimes they don't."

The Delta would be sold with a three-month parts and labour warranty, though if desired the customer could extend this at extra cost to 12 months. Aslam told us that had there been full documentation including a service history he would have priced the car £1000 higher. And no doubt if it had been any other colour apart from white with tan interior trim — one of the less desirable Delta colour combinations — the value would have been higher still.

This Delta seems to have been the classic company car, driven by someone who wasn't paying the bills and who didn't care much what state it got into. The wheels were all scuffed, particularly the offside front. Inside, the trim looked dirty, with some unsightly stains on the seats. The boot clearly hadn't been used much, so at least that was clean.

Three of the tyres were Michelin MXLs, while the offside was a Goodyear Grand Prix. All four were worn to near the limit, while the outside tread of the spare was bald.

The Delta fired up first time, the classic twin-cam Fiat/Lancia 1.6-litre engine sounding in good fettle. The Weber electronic injection/management regulated idling and throttle response as it should, to the accompaniment of a constant whine from what we assumed was the fuel pump.

But while the engine felt good for another

SECONDHAND *spotlight*

Lancia Delta GT 1600ie

Date Registered *3/8/89*

Asking price *£4895*

Mileage *51,073*

Colour *White*

Service history *None*

Warranty *3 months parts/labour*

Number of owners *Not known*

Dealer *Autotrend, North Parade, Falmouth, TR11 2TG.*
Telephone: Falmouth (0326) 311114

Unsightly stains and dirty trim...

1.6-litre twin cam started first time

What the trade says

A LANCIA DELTA WOULD NOT BE MY CHOICE AS A SENSIBLE USED CAR BUY AT ANY age or mileage, but the fine handling and Italian image will attract those looking for fashion on the cheap. I cannot think of any other reason to choose one in preference to other GTi-type cars from other European and Japanese manufacturers which are far superior in build quality, mechanical reliability and long term residual strength.

Five grand buys you a lot of hot hatch these days and the choice is wide. Private advertisements are full of bargains for those who are prepared to take time and care to find the right car. The private buyer can call on the RAC or AA for a full mechanical survey at reasonable cost, and no one needs to buy a dog in a market place awash with unwanted used hot hatches.

This Lancia is a typical example of the kind of car that no one needs to buy at a price that no one needs to pay. Obviously bought from auction and sourced from the finance company repossession department this tired looking Delta GTie in white on 1989 G-plate had no registration document or service documents. Needing tyres and with kerbed wheels and dirty seats, this Delta had obviously had a hard life and its condition would suggest a mileage of 71,000 rather than the 51,000 showing on the clock.

A transmission whine was also a bit disconcerting and added to the overall impression of an uncared-for machine. As for the price, it may appear to the inexperienced that £4895 for a G-plate car which now costs over £11,000 new is fair value, but depreciation on all Lancias is very severe and first year falls are in the region of 48 per cent with subsequent depreciation of nearly 20 per cent per annum. The retail market is only interested in immaculate low-mileage Lancias with full service history.

Under the hammer this car would make around £3000 on a good day to trade buyers and brave private punters may pay up to £3500; £4895 is not within striking distance of sensible and even a discount of £1000 would not tempt the wise.

John Coates

SPECIFICATION

LANCIA DELTA GT 1600ie
ENGINE
Transverse, front, front-wheel-drive
Capacity 1585cc, 4 cylinders in line
Bore 84mm **Stroke** 72mm
Compression ratio 9.7:1
Head/block al alloy/cast iron
Valve gear dohc, 2 valves per cylinder
Ignition and fuel IAW Weber electronic injection, mapped electronic ignition
Max power 108bhp (79kW) at 5900rpm
Max torque 100lb ft (135Nm) at 3500rpm
GEARBOX
5-speed manual
Ratios top 0.96, 4th 1.16, 3rd 1.56, 2nd 2.24, 1st 3.58 **Final drive ratio** 3.59:1
SUSPENSION
Front Independent MacPherson strut, coil springs, anti-roll bar
Rear Independent MacPherson strut, transverse links, anti-roll bar
STEERING
Rack and pinion
BRAKES
Front 10.1ins (257mm) dia disc **Rear** 8.9ins (227mm) dia disc
WHEELS
Al alloy, 5.5J x 14ins rims, 165/65 R14 tyres
DIMENSIONS
Length 153ins (3895mm)
Width 64ins (1620mm)
Height 54.3ins (1380mm)
Wheelbase 97ins (2475mm)
PERFORMANCE
0-62mph 10.0secs
Max speed 116mph
FUEL CONSUMPTION (claimed)
28.8mpg urban, 45.6mpg at 56mph, 35.3mpg at 75mph

A 1989 GTie at this price with alloy wheels, electric windows and central locking

> 'It seemed amazing value for money considering that to buy a Golf or a Peugeot 205 GTi for the same money you'd have to settle for a circa 1987 D-reg car'

GTie: sounds exciting, but be careful

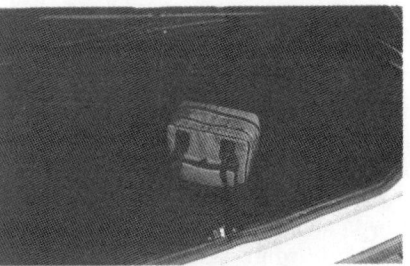

Boot was hardly used and clean

50,000 miles, the transmission didn't inspire confidence. The gearchange was as vague as any Delta, and so of little concern, but on the overrun the gearbox was worryingly noisy.

Our test drive reminded us that the Delta really is long in the tooth now. Its lively engine is still one of the nicer 1.6-litre units around, but the heavy, lifeless steering and bumpy ride don't impress.

Like most other Italian cars of the early '80s, driver ergonomics are slightly awkward, and the seats aren't a patch on the superb Recaros of the HF Turbo or Integrale.

Mindful of the Delta's below par build quality we listened carefully for rattles and squeaks, and looked out for trim that had come adrift and doors that didn't shut properly. But surprisingly, given the hard use to which it had clearly been subjected, the Lancia felt quite solid.

The instruments, electric windows and central locking all worked, as did the heated rear screen and rear wash/wipe. Given the dubious reliability of Italian electrics, it's essential to check these items out.

An Italian car kept by the spray-soaked Cornish coast — and in the absence of a log-book we could only assume it was a local car — is potentially bad news as far as rust is concerned. And we certainly noticed several brown pin spots on body seams and joints.

After detailed inspection our initial enthusiasm for this Lancia Delta all but evaporated. Half price, yes, but there were too many unknowns. Had there been just one owner? Was the service history really in the post? We reckoned this Delta had been driven hard for each of its 51,073 miles, and — notwithstanding the normal Lancia built quality — it had aged prematurely.

We would have given this car a miss and searched for a better preserved example, preferably not in white. But it wasn't a car that had to be avoided at all costs, and if Aslam had been prepared to make a drastic price cut we could have been interested. ■

Delta Four Wheel Drive

When the Delta was launched in 1979 four wheel drive was restricted almost entirely to practical off-road vehicles in the tradition of the Jeep and the Land Rover. However, experience with high-powered four-wheel drive cars in international rallies in the early eighties had shown that this configuration had advantages over two wheel drive both in putting power on the road and in improved roadholding. Lancia showed a prototype four wheel-drive Delta - which also used a turbocharged 1600 twin-cam engine - at the Turin motor show in 1982. There was at that time no further activity from the company in this direction as far as road cars were concerned, but a programme to develop a four-wheel drive rally car started in earnest in 1983. This car, unveiled in December 1984, was known as the Delta S4; the "S" indicating that it was supercharged and the "4" indicating four-wheel drive. Although it bore the Delta name and resembled the road car in its general shape, the S4 had nothing in common with the production model. Despite this, the knowledge gained in developing and running the Delta S4 in the World Rally Championship enabled Lancia to produce a very advanced and effective four-wheel drive road car.

The road-going four-wheel drive car, the Delta HF 4WD, was introduced in May 1986. It was powered by the turbocharged 1995cc twin-cam engine developed for the top-of-the-range Thema saloon. This was equipped with a water-cooled Garrett T3 turbocharger and electronic control of both fuel injection and ignition systems. Maximum power was 165bhp (EEC) at 5250rpm with normal maximum torque of 26mkg at 2500rpm. An overboost device enabled the maximum torque to rise to 29mkg at 2750rpm for short periods. As in the Thema, the Delta's engine utilised contra-rotating balance shafts to reduce second-order vibrations. The power was transmitted to the wheels through a five-speed gearbox and the most advanced permanent four-wheel drive system available at that time. The epicyclic central differential normally provided 56% of the torque to the front wheels and 44% to the rear, but incorporated a Ferguson viscous coupling which automatically varied the torque split to give the optimum drive depending on the amount of tyre grip front and rear. In addition, the rear differential was of the Torsen (torque sensing) type which further varied toque to each rear wheel in relation to the grip available.

The Delta HF 4WD was intended purely as a road car, but at the end of 1986 the Group B rally cars, such as the Delta S4, were banned from the World Rally Championship following a number of accidents in which participants and spectators were killed. From 1987 the Championship was to be contested by Group A cars, based on production vehicles. Lancia had just such a car in the HF 4WD and quickly developed a Group A version of this model which was highly successful, winning nine rounds of the 1987 Championship and easily taking the Championship title.

In order to maintain their leadership in

Delta HF Integrale with eight value engine

World Rallying Lancia needed to develop the cars further, but had to homologate any changes by building at least 5,000 cars in 12 months with these changes. this had considerable benefit to Lancia's customers as they were offered more and more advanced versions of the four-wheel drive Deltas in succeeding years. The next model was the Delta HF integrale, introduced in November 1987. "Integrale" is the term used in Italy to indicate four wheel drive whereas "4WD" is an English expression. The HF integrale had a more powerful version of the 1995cc twin-cam

Delta HF Integrale Evoluzione

with a larger turbocharger and bigger intercooler contributing to a power output of 185bhp. Maximum torque with overboost rose to 31mkg. Along with increased power the new car had uprated suspension and brakes, with extended wheel arches allowing wider wheels to be used.

A further upgrade was introduced in 1989 when a 16-valve version of the twin-cam engine was fitted. The car was now identified as the HF integrale 16V. The extra valves were not the only change; the fuel injection and ignition mapping of the ECU were also revised. Power increased to 200bhp at 5500rpm while peak torque remained at 31mkg. The brakes were also improved and an anti-lock braking system (ABS) was offered as an option. Externally the 16V was distinguished by its lower ride height and a prominent bonnet bulge. The final version of the Delta, known as the HF integrale Evoluzione, appeared in 1991. Power was increased to 210bhp but more important were the changes to the suspension. The wheel tracks front and rear were both increased and the suspension was modified to give greater strength while retaining the original layout. Further improvements were made to the brakes and wider wheels were fitted. The increased tracks and wider wheels meant that even bigger wheel arch extensions were needed, giving the car a very wide, aggressive look. When the Delta was replaced in April 1993 there was no high-performance four-wheel drive model in the new range so production of the HF integrale Evoluzione is planned to continue until late 1994.

Although the developments applied to the four-wheel drive, cars benefited Lancia's customers, their main purpose was to provide an enhanced base for the rally cars and in this they were extremely successful. The four-wheel drive Deltas set many records in World Championship Rallying and won the Makes Championship for Lancia six years in succession - a totally unprecedented achievement. Members of Lancia's works team also won the Rally Drivers Championship in four of those years.

The advanced configuration of these cars naturally attracted the attention of the specialist coach builders in Italy and several projects based on the four-wheel drive Deltas were put forward. The first was the Orca, proposed by Giugiaro of Ital Design, who had worked with Lancia on the prototype four-wheel drive Delta, in 1982. This was a low-drag, five-door hatchback with a CX of 0.245 using the prototype Delta's transmission but with a supercharged engine. The Orca was a practical, working car but never went into production. Another project, in this case based on the Delta HF integrale in eight-valve form, was the HIT (High Italian Technology) by Pininfarina, shown in 1988. As its name implies, the HIT used very advanced construction techniques and was based on a floorpan made of honeycomb Nomex, sandwiched between layers of carbon fibre reinforced resin. The three-door, four seater body also incorporated composite materials, together with polycarbonate sheet. Although longer and wider than the production car, the HIT weighed 220kg less. Like the Orca, the HIT did not reach production. One special version which did go into production was the Hyena, designed and built by Zagato. This was an even higher-performance car based on the HF integrale Evoluzione. A two-seater closed coupe, the Hyena was built of carbon fibre reinforced resin and aluminium to be 150kg lighter than the production car and could be supplied with a higher-powered version of the twin-cam engine. With such features as racing seats made to measure for the purchaser and four-point seat harnesses, the Hyena was intended for serious drivers. A limited production of 75 was built.

Models covered in this section: HF 4WD, HF integrale, HF integrale 16V and HF integrale Evo. Special versions: Orca, HIT, Hyena

Production Data

Model	In production from	to	No. built
2000 HF 4WD	1986	1987	7,665
2000 HF int	1988	1989	7,475
2000 HF int 16V	1989	1991	15,589
2000 HF int Evo	1991	1994	12,118 (at end 1993)

NEWCOMERS

LANCIA TURBO GETS NOVEL FOUR-WHEEL DRIVE SYSTEM

LANCIA'S DELTA

First of all it was Audi who redefined the supercar by bringing four-wheel-drive out of the farmyard and on to the tarmac. Then, rather suddenly, the original quattro concept seemed out of date when Ford came up with viscous-coupling differentials and asymmetric front/rear torque split.

Now it's Lancia's turn to claim the state-of-the-art technology, and the humble six-year-old Delta has suddenly been thrust into the limelight to show off the latest in 4WD hardware. It's called the Torsen ('torque-sensing') differential, and Lancia's engineers are saying it combines the best of all 4WD worlds.

The Torsen diff is able to monitor road conditions under the rear wheels, and consequently it feeds the correct amount of torque to each one – that is, the most torque possible without causing wheels to skid.

The basic difference between Lancia's innovation and existing systems is that because the diff never locks up completely, it is always able to read road surfaces and react quickly to changing circumstances. So at all times the rear wheel which has the better purchase is fed the more torque.

Lancia say the Torsen effect is particularly effective on fast, dry tarmac, but is equally good as the Ford and Audi transmissions on snow, ice, or mud. And the fact that the McLaren race team use such a system has to be a recommendation.

To avoid confusion, it should be noted that the 4WD Prisma doesn't use this system, instead having a normal manually lockable rear diff.

Apart from Torsen, the rest of the Delta HF 4WD is very interesting too. The centre differential is a Ferguson viscous-coupling unit, and the one at the front is a conventional free-floating type. Unusually, Lancia have split the front/rear torque 56/44 per cent – this is to retain some of the front-drive Delta's handling characteristics.

So that the Delta has the sort of performance which its sophisticated chassis demands, Lancia have squeezed in transversely the engine from the two-litre Thema turbo. This intercooled unit produces 165 bhp at 5250 rpm, and there's an aero engineering-derived overboost feature, which temporarily closes the turbo's wastage when maximum power is required.

But to avoid engine damage the Marelli/Weber injection and ignition prevents this from operating when the engine is either too hot or too cold. And for smooth revving, the twin cam engine is equipped with counter-rotating balancer shafts.

Completing the HF 4WD's mouth-watering specification is standard power steering, all-disc brakes, and alloy wheels shod with 60-series tyres. To look at, the HF is actually quite subtle: the main differences over other models are the side 'mini' skirts, double headlamps, and drilled wheels.

And is the car quick? Factory claims seem relatively modest, with 0–62 mph coming up in 7.8 secs along with a maximum speed of 130 mph. But third/fourth/fifth gear times are bound to be much more impressive than the manufacturer's somewhat misleading standing start acceleration figures.

With the 4WD version making its dashing debut, it's all too easy to forget that every other Delta has been given a useful revamp too. Giugiaro's original well balanced hatchback body has wisely been left alone, the changes being applied to the engines, and to the chassis which has been given improved levels of sound insulation.

On the 1.3 model the engine has a new carburettor equipped with an electronic cut-off system, and also fitted is second-generation breakerless ignition. The 1.5 also receives a new carburettor with cut-off, though this engine gets the additional benefit of Digiplex ignition.

The once-carburettored GT becomes the GT i.e., and in the process the engine has been subjected to a substantial redesign. The block is new, as are the pistons and head, and it's turned around so that the exhaust is facing forward, before being tilted forward 18 degrees.

Fuelling is by Weber Marelli injection, and this is the same single point system, used on the new Regata 100. Similar claims of extra fuel economy and reduced emissions are made, though the Delta is the more powerful of the two cars with an impressive 108 bhp at 5900 rpm.

The HF Turbo uses the same engine, except, of course that a turbo unit increases power to 140 bhp – that's an extra 10 bhp more than the original blown car.

You wouldn't associate the sporting name of Lancia with diesel engines, yet now for the first time there's an oil burner on offer. Sharing the KKK turbocharged 1.9 litre unit used in the new Regata, the Delta's 80 bhp at 4200 rpm is identical to the output of its Fiat cousin. And as with the Regata, it won't be seen in Britain for the time being.

That's the effective range for the Italian market, though Lancia do include the 250 bhp road-going version of the S4 rally car in their line-up, even if it is merely a nominal entry to satisfy the rally legislators.

Approaching the HF 4WD, one expects it to be a fierce, temperamental machine, intolerant of less than perfect driver control, and making heavy demands at all times. Yet in reality nothing could be further from the truth: the Delta's engine is superbly flexible, the handling light and beautifully well balanced.

The engine revs like a dream, and though its power characteristics are necessarily turbo-dominated, flexibility is strong, lag minimal, and boost nicely progressive. The turbo effect is first felt at around 2000 rpm, and from 3000–5000 rpm the Delta has effortless thrust instantly on tap. Not surprisingly, the twin cam spins easily past its power peak.

You don't have to be a racer to appreciate the 4WD's handling: cornering grip is of the highest order, yet at the same time the front-biased torque split allows a welcome degree of understeer. But throw the Delta into a bend at very high speeds and the

NEWCOMERS

REVISED ENGINES GIVE USEFUL EFFICIENCY BOOST

Subtle looks for punchy new 4WD Lancia – just the mini side skirts and drilled wheels give it away. The engine is a 165 bhp turbocharged two-litre twin-cam unit, transmission is to all four wheels via epicyclic centre diff and auto-locking rear Torsen diff. Redesigned facia is stylish and well instrumented

FORCE

wheels clamp down on the tarmac, biting with ferocity.

Power-assistance is a standard fit, and it's well weighted and quickly geared (2.8 turns), along with an attractive leather-bound wheel. The gearchange is also highly satisfying to use, having a far better movement than the 2WD cars. Praise too for the ride, which is a great deal less harsh than you'd expect in a car with this sort of chassis. Though taut, the springing soaks up bumps well.

As yet there's no anti-lock braking and that's a pity, for the brakes don't rate too highly. There's a lot of pedal travel, and despite the all-round discs, the set-up doesn't feel inspiring. however, a Lancia/Bosch ABS system is on the way, Lancia tell us.

It's actually the less powerful HF turbo that's the tearaway of the range. The turbo is much peakier, and with 400cc less you can't afford to be lazy with the gears if rapid progress is required. But it's still a lot of fun, even if there is a big difference between its off- and on-boil performance.

It is a shame though, that Lancia have not tried to improve the gearchange of the plainer Deltas. After the excellent 4WD gearbox, the 2WD car feels difficult and obstructive, and is poor by Golf/Escort class standards.

We also tried the turbodiesel, with mixed feelings. Whereas, for instance, the Citroen/Peugeot diesel engines are equally refined as the eqüivalent petrol units, the Fiat/Lancia unit isn't nearly as enjoyable to drive as its petrol counterpart.

So while by diesel standards the Turbo ds is acceptable – it pulls well and is smooth – buyers stand to lose a lot by missing out on petrol performance. Until the British market offers a true financial incentive for diesel motoring, the Delta would be unlikely to prove popular.

Across the range, Lancia have restyled the Delta's interior, and thankfully the treatment is kept as subtle as the changes made to the exterior.

The dash is new, and the minor switchgear has now been simplified; gone are the confusing switches lined up along the top of the centre console, and which were so difficult to use efficiently. Now they're contained in a neat panel below the radio, and are much easier to use.

Lancia have also paid attention to the original car's heating and ventilation system, but while the new set-up is improved, it's still not terribly impressive. The fan is capable of pushing out plenty of air (and it's reasonably quiet, too), but the ducting is not temperature-split so no way can hot air be fed to the feet and cool air to the face. Neither are the controls good; they consist of three large round knobs which work with a plasticky feel.

It's a pity this small, but immensely useful refinement (especially so in our climate) has not been added, especially as the Delta's close relative, the new Fiat Regata does have it.

Trim wise, the Delta's interior is changed subtly, and is still very Lancia. The seats are covered in a pleasant checked material, and that's matched by the door trimming.

Unchanged is the good rear seat, and of course the standard five-door bodyshell is something you don't have to pay extra for. The boot's big too and the rear seat split-folds.

We looked closely to see how well put together the Delta was, and came away with mixed feelings. While the car displayed a build quality superior to some Lancias of yesteryear, and had well fitting trim, it still couldn't be described as being in the BMW class.

Anxious to re-establish themselves in the British market which they once found so lucrative, Lancia will be bringing the cars to British showrooms as soon as July. Pricing will be keen: a five per cent increase will be asked for the range's improvements.

At the time of Delta's presentation, Italian bosses would say only that the 1.3, 1.6 and 1.6 Turbo would come to the UK. But it's a safe bet that Lancar executives are doing their level best to wring a few 4WDs out of Lancia – this is exactly the sort of prestigious model which would provide a much needed boost to an image which has suffered cruelly in the last few years.

MODEL:	LANCIA DELTA
RANGE:	5-door hatchbacks
DATE IN UK:	July 1986
ENGINES:	1301 cc, 78 bhp; 1585 cc, 108/140 bhp; 1995 cc, 165 bhp
PERFORMANCE:	1301 cc, 102 mph, 0–62 mph 14.3 secs; 1585 cc, 116 mph/126 mph, 0–62 mph 10.0/8.7 secs; 1995 cc, 130 mph, 0–62 mph 7.8 secs
MPG:	26.2 (HF 4WD) – 51.3 (1300)
PRICE GUIDE:	£6700–£11,500

What Car? August 1986

ROAD TEST
LANCIA DELTA HF 4WD

PRICE n/a, **TOP SPEED** 127mph, **0-60mph** 6.6secs, **MPG** 19.8
FOR Performance, Handling, Refinement. **AGAINST** Fuel consumption

EVERY CLOUD HAS A SILVER LINING

The passing of the Group B rally cars will be mourned by many, but it does at least mean the emergence of some potent Group A machines, like the Lancia Delta HF 4WD. We tested a left-hand-drive version and found it exhilarating, to say the least. The car is due on sale here next year

With the demise of Group B at the end of this year, and the uncertainty on FISA's part as to group classifications, it is certain that the Delta HF 4WD will be at the forefront of Lancia's Group A rally attack next season.

The Lancia Delta S4 is, perhaps, the most technically interesting of the Group B cars in that it is not only turbocharged, but supercharged as well. By using a supercharger at the bottom end of the rev range, Lancia has managed to negate most of the lag inherent in any turbo installation. When combined with a sophisticated four-wheel-drive system this gave Lancia a very potent rally weapon.

The tragic deaths of Henri Toivonen and Sergio Cresto in an S4 on the Tour De Corse this year was the straw that broke the camel's back.

The rallying world suffered a tremendous blow in terms of development as FISA decided to scrap plans for a proposed Group S as well as cancelling Group B. It was ruled that Group B cars were too fast and, as a consequence, too dangerous.

It turns out that Lancia is one of the more farsighted manufacturers since it already had the HF 4WD production car in the pipeline, using experience gained from the development of the S4 rally car. It is not supercharged, has a different four-wheel-drive arrangement and does not have the advantage of four valves per cylinder, but by utilising the existing turbocharged 2-litre engine from the Thema Turbo ie, Lancia has managed to create a potential Group A winner. Only time will tell.

Superseding the HF Turbo ie (*Autocar*, 27 Aug) as the flagship of the Delta range—S4 excepted—the HF 4WD has a lot to live up to. The HF Turbo ie is no slouch and its handling is praiseworthy for a front-wheel-drive car.

The Delta range was first introduced to the UK in 1980 and remained virtually unchanged until this year, when small changes were made to the body shape and the engines updated; plus, of course, there was the addition of a four-wheel-drive model. One of the features of the HF 4WD is the understatement of the body treatment. There is very little to distinguish it from the Turbo ie apart from the four-headlight system, fog lamps mounted in the front spoiler, discreet 4WD badging on the rear hatch and the small side skirts and two raised ▶

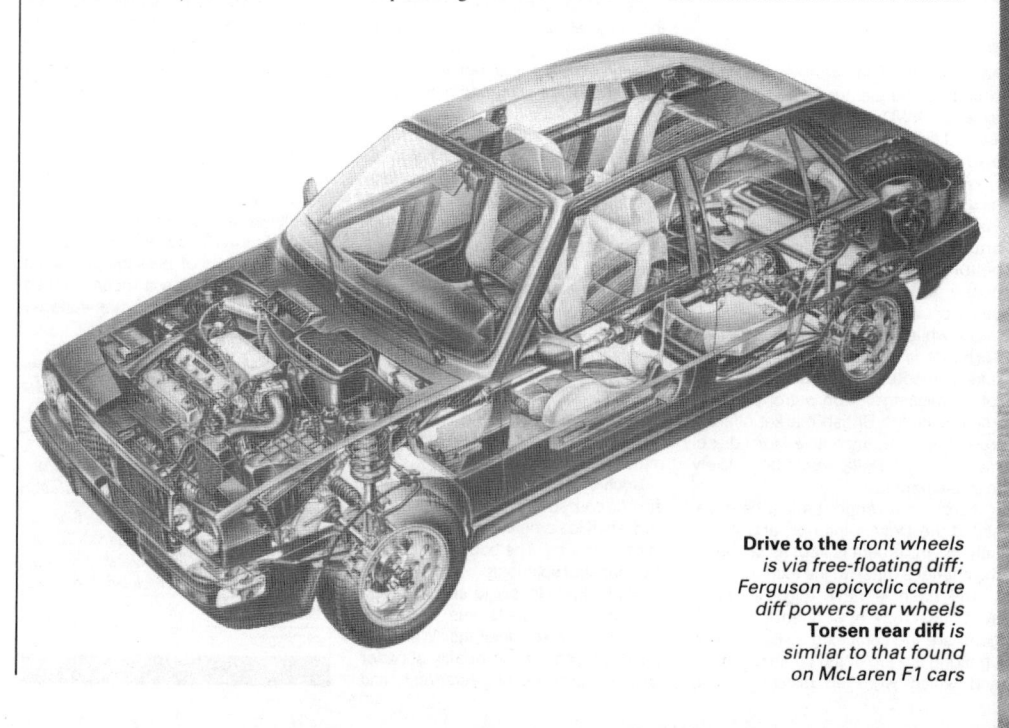

Drive to the *front wheels is via free-floating diff; Ferguson epicyclic centre diff powers rear wheels* **Torsen rear diff** *is similar to that found on McLaren F1 cars*

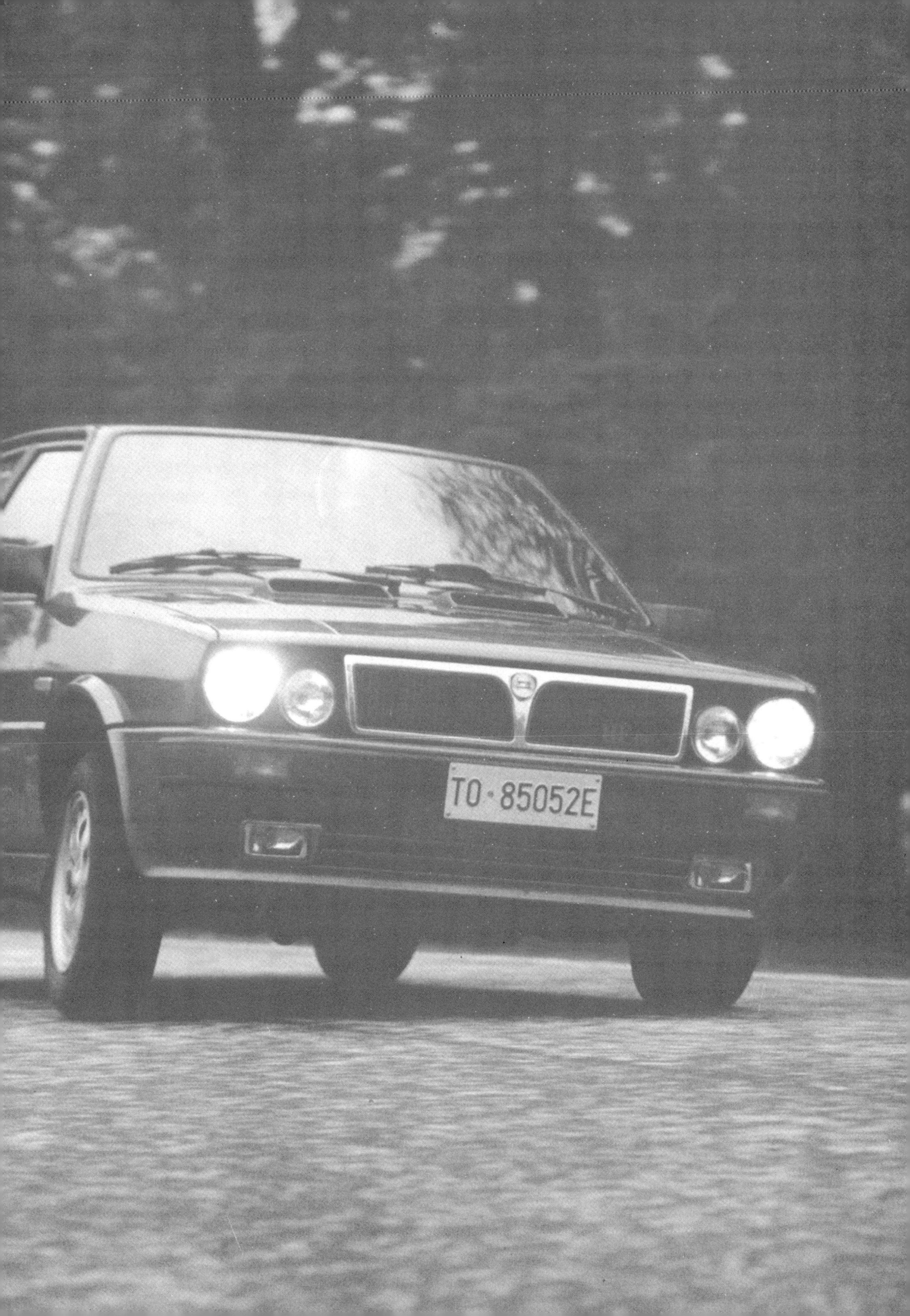
TO·85052E

ROAD TEST

TECHNICAL FOCUS

With 165bhp on tap the best way of transferring it to the road is via four, rather than two-wheel drive. In the Delta HF 4WD, Lancia was not content to go for a simple system but rather opted for one with an in-built torque-splitting action to ensure that the available power was going to the wheels with the most traction at any given time, thus ensuring the most efficient use of the available power and torque. Three differentials are at the heart of the system.

Drive to the front wheels is linked through a free-floating differential; drive to the rear wheels is transmitted via a 56/44 front/rear torque-splitting Ferguson viscous-coupling-controlled epicyclic central differential. The real innovation as far as production cars are concerned, however, lies between the rear wheels. The Torsen (torque sensing) rear differential is similar to that found on McLaren Formula 1 cars. The result of combining these differentials in this configuration is an automatic 'thinking' four-wheel-drive system which requires no manual input from the driver, yet ensures maximum potential traction at any given time.

The Torsen differential is a true 'intelligent' differential in the way it distributes torque. It divides the torque between the two wheels according to the available grip and it does it without ever locking fully; maximum lock-up is 70 per cent. Standard differentials are either free-floating or self-locking.

Free-floating systems are good at differentiating between wheel speeds on bends but always supply the same amount of torque to both wheels. In this situation, however, there is a risk that the wheel with the lighter load — on an incline, for example — or less grip will lose traction. To counteract this possibility, totally self-locking differentials ensure that both wheels rotate at the same speed but in doing this, prevent free differentiation in cornering, to the detriment of handling and stability.

The basic suspension layout of the HF 4WD remains the same as in the rest of the Delta range: MacPherson strut-type independent suspension with dual-rate dampers and helicoidal springs, with the struts and springs set slightly off-centre. There are a few more subtle changes, though, with the suspension mounting points to the bodyshell now better insulated by incorporating flexible rubber links to provide improved isolation. Progressive rebound bumpers have also been adopted, while the damper rates, front and rear toe-in and the relative angle between springs and dampers have all been altered.

The steering retains the rack and pinion mechanism of the rest of the Delta range and in this application it is power-assisted. Steering effort has been reduced further by fitting thrust bearings of the ball, rather than roller type. Additional steering sensitivity has also been obtained by adjusting the angle of incidence of the steering rack.

◀ air intakes on the bonnet. It is therefore virtually indistinguishable from the 1600cc HF Turbo ie.

The only requirement for driving the HF 4WD around London — our car was left-hand drive and on Turin number plates — is that it is essential to memorise one Italian phrase: *non parlate Inglesse* . . .

PERFORMANCE

Lancia has fitted a relatively large engine under the bonnet of the four-wheel-drive Delta, the same 1995cc unit as that fitted in the Thema Turbo, and it produces 165bhp at 5500rpm and 188lb ft torque at 2500rpm. There is an overboost facility on the Garrett T3 turbocharger taking torque up to 210lb ft at 2750rpm. Overboost comes in if the throttle is kept wide open on acceleration and acts on the wastegate valve temporarily increasing feed pressure and raising torque in the process.

The twin-cam also benefits from the fitment of the Marelli-IAW integrated electronic ignition and fuel injection system while the engine itself is claimed to be as smooth as a six-cylinder thanks to its twin counter-rotating balancer shafts running in roller bearings (rather than the earlier shell bearings) and that's said to give an extra 4bhp. The charge from the turbo is cooled by an air-to-air intercooler reducing the intake air temperature from 120deg C to 70-75deg C and with so much additional underbonnet plumbing necessary for this conversion Lancia has had to move the front section of the subframe forward.

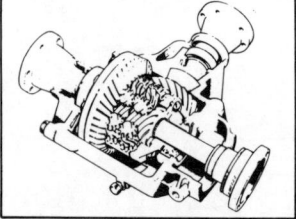

Torsen rear differential

When you drop the clutch at high revs in the four-wheel-drive Delta HF there is no dramatic wheelspin, in fact no apparent loss of traction whatsoever. Any excess of power is simply lost through the drivetrain so that acceleration from rest is very impressive. A 0-30mph figure of 2.2secs is exceptional for a road car and this is achieved in first gear, which incidentally pulls right up to 41mph and the tachometer redline of 6500rpm. The engine has little problem exceeding this, pulling right up to the rev cutout which is set at 7200rpm, although performance testing was limited to the manufacturer's safe maximum.

If a quick change into second gear is made through the light and well defined gate it is possible to hit 60mph in just 6.6secs and with such an eager and willing power unit it is easy to see that the Delta HF 4×4 is more than a match for just about any of its rivals. Third gear takes the little Delta to 95mph with two more gears to go. There is, in fact, very little difference in terms of top speed betwen those two gears. Fourth gives snappy acceleration right up to the mean maximum speed of 128mph (at maximum revs) while fifth allows a more

LANCIA DELTA HF 4WD

Garrett *turbocharged power unit develops 165bhp*

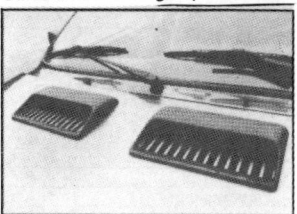

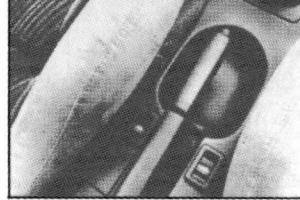

Twin bonnet *air intakes*

Windows *are electric*

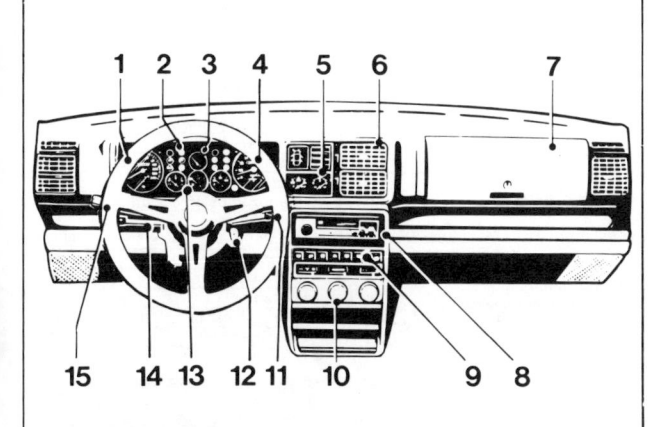

1 Speedometer, 2 Lighting indicator controls, 3 Turbo boost gauge, 4 Rev counter, 5 Oil pressure and temperature gauges, 6 Air vents, 7 Glovebox, 8 Radio/cassette player, 9 Auxilary switches, 10 Ventilation and heater controls, 11 Windscreen wiper stalk, 12 Ignition switch, 13 Voltmeter and fuel gauge, 14 Indicator stalk, 15 Headlamp stalk

Instrumentation *is plentiful but stowage space is lacking*

progressive surge of power up to 127mph at a slightly undergeared 5350rpm, but the car could be coaxed on to 130mph with a tail wind at a still leisurely 5450rpm.

Acceleration in the gears shows that Lancia has got its sums right in this area too. There is no noticeable turbo lag and the low speed acceleration figures on a wide throttle opening are not far removed from those of a normally aspirated car. The most impressive figures are those over the 50-70mph span (4.4secs) and the 70-90mph time of 6.6secs in third gear. Acceleration in the other gears — although less impressive on paper — is certainly more than adequate to provide quick, overtaking manoeuvres without the need to change down. In contrast, in the Mazda 323 Turbo 4x4 tested recently the more natural course of action would be to change down a gear.

ECONOMY

An overall fuel consumption figure of 19.8mpg may not appear to be too impressive but neither is it surprising. The problem with cars like the HF 4WD is that there is a tendency to utilise the available power and level of adhesion to their full extent. This, of course, is bound to have a detrimental effect on fuel consumption.

Intermediate fuel consumption figures worked out over the 1218 mile test period produced a worst of 17.7mpg, for the strenuous testing session at Millbrook, and a best of 24.1mpg which was returned during a restrained leg of motorway driving in the 60 miles of heavy traffic from Dover to West London.

By applying the *Autocar* formula, this gives a calculated average fuel consumption figure of 21.8mpg which is likely to be fairly realistic for normal day-to-day motoring. Combined with the 12.6 gallon fuel tank, this should enable the HF 4WD to cruise for about 270 miles on a full tank of petrol: not a massive distance, but probably sufficient for most people's needs.

A lockable, flush-fitting fuel filler flap is located in the nearside rear-three quarter panel and is unlocked with the same key that activates the central locking system.

Fuel consumption may be a little disappointing but when taken in context and weighed up against performance and the advantages of a full-time four-wheel-drive system it is not quiet as bad as it first appears.

REFINEMENT

One of the most impressive features of this small-sized, high-performance hatchback is its high level of refinement. Italian cars are not usually noted for low noise levels, but in the case of the HF 4WD, Lancia has managed to isolate and insulate the passenger compartment to a commendable level, especially bearing in mind the 'competition' nature of the car.

Sitting at the traffic lights with the 4WD ticking over, the most noticeable source of noise is the engine. It is not induction or exhaust related, however, but a faint electronic 'humming' from the high-pressure fuel pump. Normally these tend to buzz, but the Lancia's has a pleasing hum. Once on the move and working up through the gears, the sound of the engine becomes more evident — not in the usual induction/exhaust Italian fashion but in a distant, subdued way, accompanied by a faint turbo whistle as the boost builds up.

At cruising speeds on the motorway, the 4WD really comes into its own. Not only is the engine responsive at these sorts of speeds, but also engine, wind and road noise are kept to very acceptable levels, all of which go to make this little 'racer' an accomplished motorway cruiser — something which must be the envy of a lot of other hatchback manufacturers.

Cruising even longer distances proved to be a relaxing experience, even with the absence of a radio/cassette player to cover up refinement faults; the only real sources of noise, even at high speeds, are the A-pillars and door mirrors.

Road noise is well contained with only a small amount of tyre roar audible from the low-profile Michelin MXV 60s, which points to good window and door sealing.

Engine noise is worst at high revs, but only ever becomes a pleasing growl — it is lacking that Italian rasp so frequently encountered in Fiats, Alfas and some other Lancias. Exhaust boom is particularly well subdued.

ROAD BEHAVIOUR

With 165bhp on tap a simple front-wheel-drive system would have been totally unsuitable for this car, but the thinking four-wheel-drive system ensures that every last bit of power is fed to all four wheels in a very useable manner.

Suspension includes MacPherson struts and anti-roll bars all round with front lower wishbones and transverse links and longitudinal reaction rods at the rear.

But this is only half the story. The 56/44 front/rear torque split ensures well balanced handling which is apparent after only a short spell behind the wheel. The unusual addition of a third Torsen differential at the rear also allows torque between the wheels to be varied by up to 70 per cent to maintain good grip on varying surfaces.

In normal driving the Lancia Delta 4×4 produces gentle understeer and this stays, until much higher cornering forces are generated when the attitude becomes decidedly neutral. This characteristic is apparent on both tarmac and loose surfaces — on the latter this aspect allows a driver to set the car up very precisely.

Simply driving the car on tarmac reveals tremendous composure, little body roll and really outstanding traction. The Delta is deceptive in ▶

115

ROAD TEST

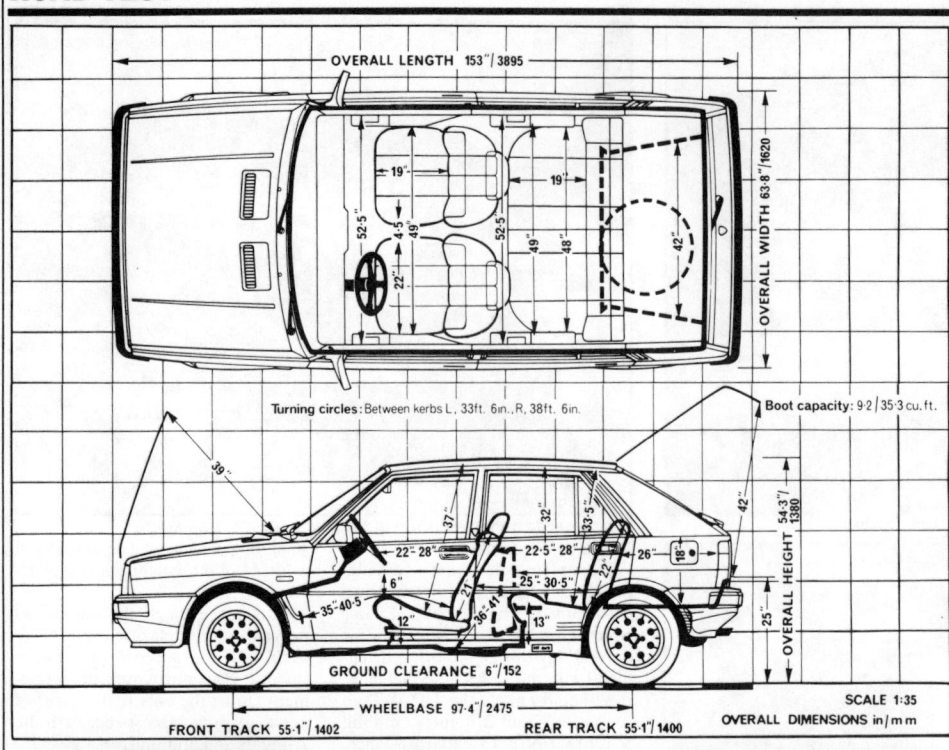

MODEL

LANCIA DELTA HF 4WD

PRODUCED BY:
Lancia
Via Lancia 27
Turin 10141
Italy

SOLD IN THE UK BY:
Lancar Limited
46-62 Gatwick Road
Crawley
West Sussex RH10 2XF

SPECIFICATION

ENGINE
Transverse, front, four-wheel drive. Head/block light alloy/cast iron with counter-rotating balancer shafts, 4 cylinders in line, bored block, 5 main bearings. Water cooled, electric fan.

Bore 84mm (3.3in), **stroke** 90.0mm (3.5in), **capacity** 1995cc (122 cu in).

Valve gear 2 ohc, 2 valves per cylinder, toothed belt camshaft drive.

Compression ratio 8.0 to 1. Breakerless electronic ignition, IAW Weber electronic fuel injection. Water-cooled Garrett T3 turbocharger with air-to-air intercooler, boost pressure 12.8psi (0.9 bar).

Max power 165bhp (PS-DIN) (122kW ISO) at 5250rpm. **Max torque** 188lb ft at 2500rpm, 210lb at 2750rpm with airboost.

TRANSMISSION
5-speed manual, single dry plate clutch, 9.1in dia. Torque converter front/rear 56/44 per cent.

Gear	Ratio	mph/1000rpm
Top	0.928	23.95
4th	1.132	19.63
3rd	1.518	14.64
2nd	2.235	9.94
1st	3.500	6.35

Final drive: helical spur, ratio 2.944.

SUSPENSION
Front, independent, MacPherson strut-type, lower wishbones, double-acting telescopic dampers, anti-roll bar.
Rear, independent, MacPherson strut, transverse links, longitudinal reaction rods, telescopic dampers, anti-roll bar.

STEERING
Rack and pinion, hydraulic power assistance. Steering wheel diameter 14.75in, 2.8 turns lock to lock.

BRAKES
Dual circuits, split diagonally. **Front** 10.1in (257mm) dia ventilated discs. **Rear** 8.9in (227mm) dia discs. Vacuum servo. Handbrake, centre lever acting on drums within rear discs.

WHEELS
Light alloy, 5.5in rims. Tyres (Michelin MXV on test car), size 185/60SR14, pressures F28 R28 psi.

EQUIPMENT
Battery 12V, 45Ah. Alternator 65A. Headlamps 55/60W. Reversing lamp standard. 14 electric fuses. 2-speed plus intermittent screen wipers.

PERFORMANCE

MAXIMUM SPEEDS

Gear	mph	km/h	rpm
ODTop (Mean)	127	205	5350
(Best)	130	210	5450
4th	128	206	6500
3rd	95	153	6500
2nd	65	105	6500
1st	41	66	6500

ACCELERATION FROM REST

True mph	Time (sec)	Speedo mph
30	2.2	31
40	3.3	43
50	4.9	54
60	6.6	64
70	9.4	75
80	11.8	87
90	15.7	98
100	21.4	109
110	27.8	121
120		131

Standing ¼-mile: 15.2sec, 91mph
Standing km: 29.0sec, 110mph

ACCELERATION IN EACH GEAR

mph	Top	4th	3rd	2nd
10-30	—	11.2	7.0	4.1
20-40	12.5	8.7	5.2	3.0
30-50	10.2	7.0	4.2	2.7
40-60	8.6	6.0	4.1	3.2
50-70	8.2	5.9	4.4	
60-80	8.7	6.7	5.1	
70-90	10.0	8.1	6.6	
80-100	12.5	9.9	—	
90-110	16.6	13.4		

CONSUMPTION

FUEL
Overall mpg: 19.8 (14.3 litres/100km) 4.4mpl

Constant speed
Autocar constant speed fuel consumption measuring equipment incompatible with fuel injection
Autocar formula: Hard 17.8mpg
Driving Average 21.8mpg
and conditions Gentle 25.7mpg
Grade of fuel: Premium, 4-star (98 RM)
Fuel tank: 12.5 Imp galls (57 litres)
Mileage recorder: 4.1 per cent long

BRAKING
Fade (from 91mph in neutral)
Pedal load for 0.5g stops in lb

	start/end		start/end
1	20-25	6	25-35
2	20-25	7	20-30
3	20-27	8	20-30
4	20-30	9	25-30
5	20-30	10	25-30

Response (from 30mph in neutral)

Load	g	Distance
20lb	0.50	60ft
30lb	0.70	43ft
40lb	0.90	33ft
55lb	1.20	25ft
Handbrake	0.20	151ft

Max gradient: 1 in 3
CLUTCH Pedal 26lb; Travel 6in

WEIGHT
Kerb 24.5cwt/2740lb/1240kg (Distribution F/R, 61.3/38.7)
Test 28.2cwt/3159lb/1429kg
Max payload 995lb/450kg
Max towing weight 2652lb/1200kg

COSTS

Not available at time of going to press.

SERVICE & PARTS

		Interval	
Change	6000	12,000	24,000
Engine oil	Yes	Yes	Yes
Oil filter	Yes	Yes	Yes
Gearbox oil	No	Yes	Yes
Spark plugs	No	Yes	Yes
Air cleaner	No	Yes	Yes

WARRANTY
12 months/unlimited mileage, 6-year anti-corrosion

EQUIPMENT

Ammeter/Voltmeter	●
Automatic	N/A
Cruise control	N/A
Economy gauge	N/A
Limited slip differential	N/A
Power steering	●
Rev counter	●
Self-levelling suspension	N/A
Steering wheel rake adjustment	●
Trip computer	N/A
Headrests front/rear	●
Height adjustment	N/A
Lumbar adjustment	N/A
Seat back recline	●
Split rear seats	●
Door mirror remote control	●
Electric windows	●
Interior adjustable headlamps	N/A
Tinted glass	●
Headlamp wash/wipe	N/A
Tailgate wash/wipe	●
Central locking	●
Fog lamps	●
Locking fuel cap	●
Luggage cover	●

● Standard N/A Not applicable

TEST CONDITIONS
Wind: 7-12mph
Temperature: 12deg C (54deg F)
Barometer: 30.1in Hg (1019mbar)
Humidity: 90per cent
Surface: dry asphalt and concrete
Test distance: 1218 miles
Figures taken at 4239 miles by our own staff at the General Motors proving ground at Millbrook.
All *Autocar* test results are subject to world copyright and may not be reproduced in whole or part without the Editor's written permission

Twin cam 2-litre *is the same unit as fitted to Thema Turbo but with an overboost facility. Smoothness comes via twin counter-rotating balancer shafts. Intercooler lowers engine temperature*

THE OPPOSITION

AUDI 80 QUATTRO — TBA

The all-new Audi 80 sets a new standard for medium class car aerodynamics, with a Cd figure of 0.29. Underneath the smooth and modern-looking bodyshell, a new gearbox has been introduced with an improved quattro all-wheel-drive arrangement and a Torsen differential. The built-in safety system, Procon Ten, is standard, and requires no maintenance

Tested	N/A
ENGINE	1781cc
Max Power	112bhp at 5800rpm
Torque	118lb ft at 3400rpm
Gearing	20.5mph/1000rpm
WARRANTY	12/UL, 6 anti-rust
Insurance Group	TBA
Automatic	N/A
5-Speed	●
Radio	●
Sunroof	TBA
WEIGHT	2519lb

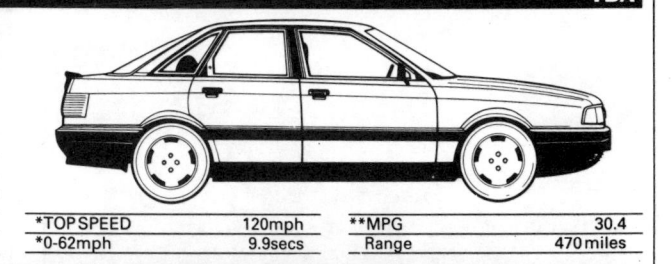

*TOP SPEED	120mph	**MPG	30.4
*0-62mph	9.9secs	Range	470 miles

FORD ESCORT RS TURBO — £10,028

Anyone who drove the original Escort RS Turbo is likely to be disappointed by this latest, softer version. The car is still an impressive contender among hot hatchbacks — mechanical refinement and fuel consumption have been improved, but the increase in overall gearing means that the new RS Turbo has lost some of the edge of the previous car

Tested	27 Aug 1986
ENGINE	1597cc
Max Power	132bhp at 5750rpm
Torque	133lb ft at 2750rpm
Gearing	22.59mph/1000rpm
WARRANTY	12/UL, 6 anti-rust
Insurance Group	5
Automatic	N/A
5-Speed	●
Radio	●
Sunroof	£572 †
WEIGHT	2247lb

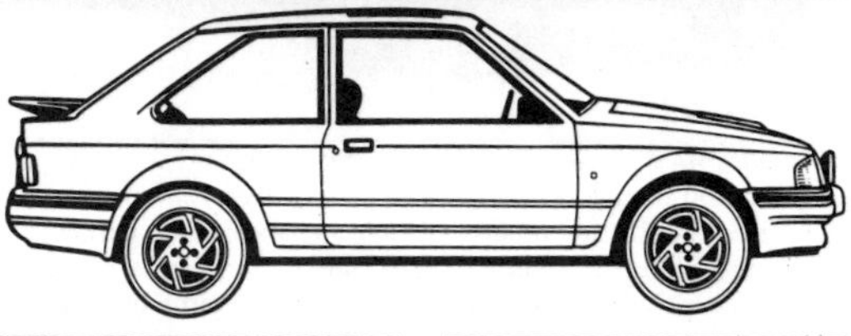

TOP SPEED	124mph	MPG	27.4
0-60mph	9.2secs	Range	290 miles

HONDA PRELUDE 2.0i 16 — £11,900

In common with its predecessor, the latest Prelude offers a willing and extremely flexible engine, a clean, precise gearchange and the usual excellent Honda build quality with a high level of fit and finish throughout. Suspension is biased slightly towards handling rather than ride comfort. It is fitted with Honda's four-wheel, anti-lock braking system as standard

Tested	22 Jan 1986
ENGINE	1958cc
Max Power	137bhp at 6000rpm
Torque	125lb ft at 6000rpm
Gearing	20.7mph/1000rpm
WARRANTY	24/25,000, 6 anti-rust
Insurance Group	OA
Automatic	N/A
5-Speed	●
Radio	●
Sunroof	●
WEIGHT	2315lb

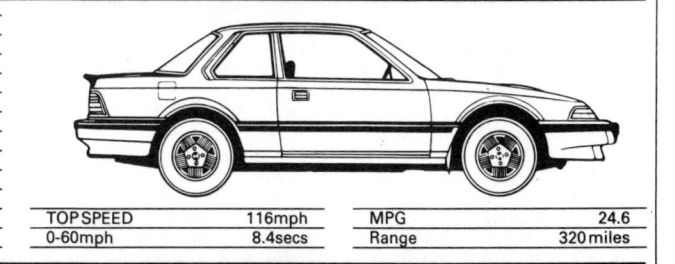

TOP SPEED	116mph	MPG	24.6
0-60mph	8.4secs	Range	320 miles

MAZDA 323 TURBO 4X4 LUX — £11,750

With its sophisticated engine and advanced 4wd system, the Mazda offers good performance and handling, particularly in the rally conditions for which it was primarily intended. But, alongside comparable road cars, the Mazda is less refined, far more thirsty and less attractively packaged. Equipment levels are low for the price being asked by Mazda

Tested	1 Oct 1986
ENGINE	1597cc
Max Power	148bhp at 6000rpm
Torque	144lb ft at 5000rpm
Gearing	20.3mph/1000rpm
WARRANTY	12/UL, 6 anti-rust
Insurance Group	OA
Automatic	N/A
5-Speed	●
Radio	●
Sunroof	●
WEIGHT	2447lb

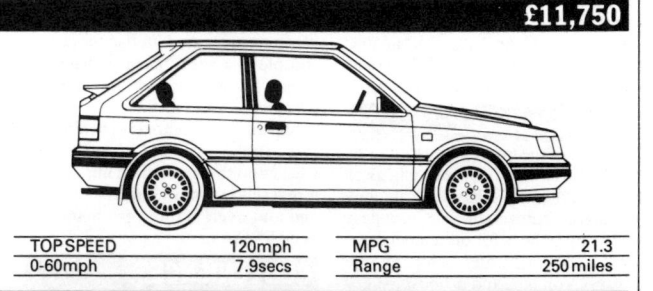

TOP SPEED	120mph	MPG	21.3
0-60mph	7.9secs	Range	250 miles

SUBARU 4WD TURBO COUPE — £10,499

The Subaru is a useful and competent package, but one could argue that it is technically gimmicky. For example, the four-wheel-drive system activated by windscreen wipers, brakes or hard acceleration seems overly complex. Performance is reasonable, however, and the turbocharger installation seems particularly refined. Poor fuel consumption

Tested (Auto)	3 Sep 1986
ENGINE	1781cc
Max Power	134bhp at 5600rpm
Torque	145lb ft at 2800rpm
Gearing	19.34mph/1000rpm
WARRANTY	18/18,000
Insurance group	7
Automatic	£600
5-Speed	●
Radio	●
Sunroof (electric)	£500
WEIGHT	2534lb

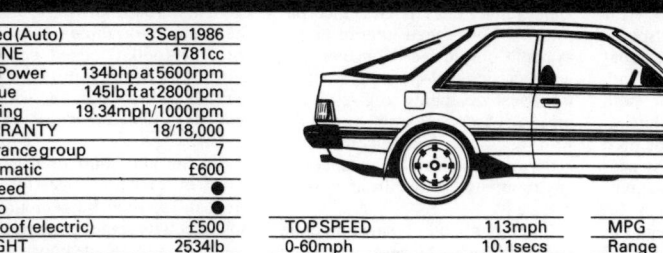

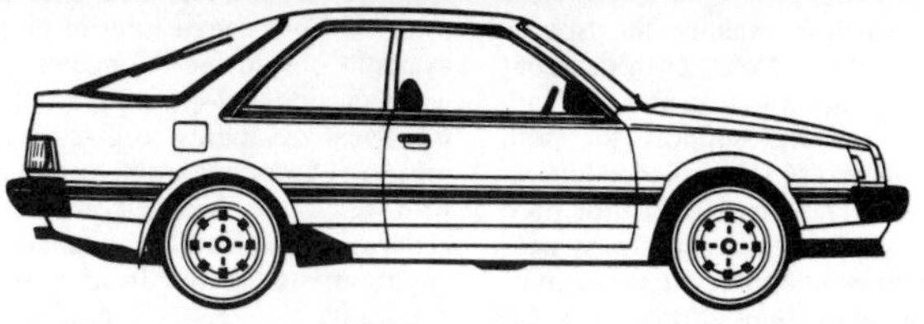

TOP SPEED	113mph	MPG	22.6
0-60mph	10.1secs	Range	295 miles

VOLKSWAGEN GOLF GTI 16V — £10,894

One of the most expensive cars in this class, the 16-valve is the latest version of the trend-setting Golf GTI. Better breathing extends top end performance and keeps the Golf competitive with Japanese rivals. Good levels of refinement, power steering available as an option. Suspension changes are now biased more towards improving handling of the hot hatchback.

Tested	28 May 1986
ENGINE	1781cc
Max Power	139bhp at 6100rpm
Torque	124lb ft at 4600rpm
Gearing	19.8mph/1000rpm
WARRANTY	12/UL, 6 anti-rust
Insurance Group	6
Automatic	N/A
5-Speed	●
Radio	●
Sunroof	●
WEIGHT	2226lb

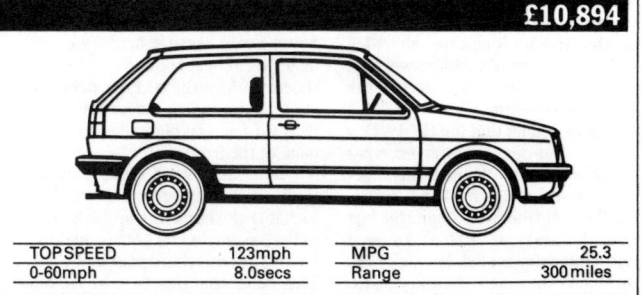

TOP SPEED	123mph	MPG	25.3
0-60mph	8.0secs	Range	300 miles

● Standard OA On application N/A Not applicable † Part of option package * Manufacturer's figures ** European Legislative Average TBA To be announced

ROAD TEST

LANCIA DELTA HF 4WD

HF 4WD is available only in five-door, hatchback form

Wheelarches and spacesaver spare intrude noticeably into boot space

Front seats are Recaro

Grille badge is discreet

that any sensation of speed in fast cornering is masked by its totally unruffled manner. Many four-wheel-drive cars are deceptive in terms of generating high cornering forces, but the Delta is also blessed with a much healthier power to weight ratio than most and feels ideally balanced.

Adding to the overall pleasant feel is the really outstanding power-assisted rack and pinion steering which, unlike that of any other Delta we have yet encountered has tremendous feel, is ideally weighted and highly responsive with just 2.8 turns lock to lock.

The ride is less convincing. It's fine at speed but noticeably busy over sharp, ridged sections at lower speed.

Braking performance is well up to standard with discs all round, ventilated at the front for better heat dissipation. Over 1.0g is attainable with 1.0g produced at 50lb pedal effort without any trace of lock up.

AT THE WHEEL

Another of the impressive features of the Delta HF 4WD is the simplicity of the instrumentation and the amount of driver information it supplies; there are gauges for oil temperature and pressure, water temperature, battery voltage, and a boost gauge, apart from the rev counter and speedometer.

To further enhance the 'sports' feeling of the 4WD, both the front seats are no-nonsense Recaros with plenty of side support for both cushion and squab and good lumbar support. They may feel a little hard initially, but the driver and passenger soon come to appreciate the support and location they afford. They do, however, lack any kind of height or tilt adjustment which is needed because of the slightly awkward wheel-seat-pedal arrangement. It is in the Italian long-arm, short-leg style, but there is a generous wheel rake adjustment to help rectify the situation to a certain extent.

The mere fact that the HF 4WD is an out-and-out performance car should mean that the pedal arrangement is set up for heel-and-toeing, but this is not the case, despite the fact that the test car was an original left-hand-drive model. Braking hard into a corner and then trying to heel-and-toe a neat downchange proved to be almost impossible due to the height differential between the brake and accelerator pedals, and the way the accelerator is mounted too close to the centre bulkhead. This would be one of the first things to rectify in a true competition version of the car.

The gearchange can be smooth and fast, but it takes a certain amount of practice as the 'change is rubbery and a little imprecise; once practised, though, the driver finds no difficulty at all.

Auxiliary controls are located on three stalks around the steering column in classic Lancia/Fiat tradition, with the two on the left controlling side/headlights and full dip beam plus turn indicators, while the right-hand one looks after the intermittent two-speed wipers and windscreen washers. The remainder of the controls are found in the centre console as a line of push-switches set above the ventilation and heating controls.

Slip into the Recaro seat behind the three-spoke, leather rim steering wheel and the immediate impression is that this is a sports car. It feels solid and taut even before the engine has been fired up.

CONVENIENCE

In line with the rest of the revamped Delta range, the HF 4WD incorporates the redesigned interior and is available only in five-door guise. The new facia design slopes away from the front-seat occupants to give an impression of roominess and, combined with adequate head-elbow room, generates a comfortable and spacious environment for the front passengers.

Like many other small-medium size hatchbacks, however, the same cannot be said for the rear passengers. Rear legroom is bearable, but anyone over average height will find insufficient headroom for all but the shortest of journeys. Four doors do make access to the rear good, though, with no need to clamber across the back of the front seat.

Oddment stowage space is adequate but hardly extensive with a lockable glovebox (lift-up lid) in front of the passenger and both front doors have substantial hard plastic pockets. The facia contains two trays, one small one to the left of the steering column and a larger one underneath the glovebox, which are formed as shallow indentations. Corner hard, however, and anything in these trays ends up on the floor. The front seat backs both have elasticated pockets for maps and books.

The gas-strut supported hatchback does not have a remote internal release but it does open from just above bumper height to give good access for heavy loads. The symmetrically split rear seat backs fold down to give a flat load space, but there is a fair amount of wheelarch intrusion. The spacewaver spare wheel is located in the boot vertically, bolted to the nearside rear three-quarter panel with the toolkit lodged behind it. With the rear seats in position the luggage area is 7cu ft, and with both seat backs folded down, it provides 33cu ft.

Heating and ventilation is taken care of by three turnwheels located in the centre console, but ram-air flow is only adequate at the best of times — it is necessary to use the booster fan on its first setting when the weather is warm.

SAFETY

In line with all other production cars, the HF 4WD will have to undergo Type Approval before it is imported into the UK next year to ensure such things as crash durability. The most obvious safety benefit, however, is the four-wheel-drive system.

This ensures maximum grip and traction in all weather and road conditions and in the case of the HF 4WD it does automatically; the driver does not have to worry about selecting four-wheel-drive if it is needed, or about any differential locks. The driver is left to concentrate on driving.

Two front and three rear seat belts are factory fitments, as is a laminated windscreen. The 'sporting' nature of the car means that the engine is likely to be worked hard and to ensure its protection a rev-limiter if fitted — in the case of the test car operating at 7200rpm.

VERDICT

In typical Italian style, Lancia has not gone for any half measures in developing its top of the Delta range high-performance HF 4WD.

The 2-litre turbocharged fuel-injected engine which is also found in the Therma saloon and Station Wagon produces more than enough power and torque for a small/medium sized hatchback with the added advantage of driving all the wheels through a sophisticated four-wheel-drive system.

Apart from producing the 'quickest' hot hatchback, it also provides Lancia with a Group A rally weapon for the 1987 season, and in this guise the HF 4WD will produce a maximum of 230bhp at 6250rpm and develop maximum torque of 240lb ft at 4000rpm. Lancia hopes to have the Delta HF 4WD homologated by the end of this year with the production figure for Group A remaining at 5000 for the 1987 season.

No car is perfect, but considering its size and performance, Lancia has endowed the Delta HF 4WD with admirable refinement levels, stability and handling. The only question marks lie over fuel consumption and price.

Lancia is keen to keep it under £12,000, but by the time it reaches the UK in right-hand-drive form next year it is likely to be show that high performance technology does not come cheap. ■

DRIVING THE NEW CARS

LANCIA DELTA HF INTEGRALE

Would you shell out R80 000 for a car that looks like a body-kitted Golf? Probably you would, concludes Stuart Johnston, once you'd sampled the magic...

WHAT kind of man would spend R80 000 on a car that looks like a body-kitted Mk I Golf? Admittedly a very tastefully kitted hatchback, with finely-wrought 15-inch alloy rims, perfectly crafted steel wheel arch flares and glossy paintwork in a red so bright that it could only come from Italy. And subtle little badges that say "Lancia". But so much money for so little image? Aah, the Lancia Delta Integrale is meant for a very special kind of enthusiast. You won't find the Integrale owner hamming it up outside the disco of a Friday evening. He'll be out in the mountains searching for yet another off-camber bend with a decreasing radius, a loose surface, an approach bumpy enough to challenge the full-house rally-type suspension and full-time four-wheel drive.

Yes, the bodywork of the HF looks like Giorgetto Giugiaro's attempt at designing a Romanic Golf for his mamma an' brother-in-law back home in Turin, and that's basically what the Delta was when it was launched in 1979. The bodywork remains box-like but the underside and underbonnet of the Integrale is a different box of sandwiches.

Let's take a look at the suspension and drive-train first. Borrow somebody else's jacket or beach towel to lie on and start off with the rear end. The first thing you notice is the seriousness of the materials and dimensions of the links and joints. The basic layout is coil over struts, with trailing links and an anti-roll bar of purposeful proportions.

The rear axle contains a fiendishly clever device called a Torsen diff. Using a worm and helical gear principle it is said to be far superior to a limited slip diff, while doing much the same job.

Up front there is lots more hardware to grab your attention. The central diff's torque converter is linked to a Ferguson viscous coupling, which

Cheerful but tasteful Recaro rally seats, a classic leather rim steering and extra gauges turn the Delta's cockpit into an enthusiast's pleasure trip. But the gear lever is placed too far forward.

limits discrepancies in rotational speeds between the front and rear axles, so that axle wind-up is prevented. The front diff is of the floating variety and all this is encased in a common housing. The steering is rack and pinion and somehow Lancia found the underbonnet space to include power assistance in this all-important department. Again, suspension is via some bulldog-like MacPherson struts.

Climb out from under and open the bonnet. If you can find an engine under all the hardware – intercoolers, turbo-plumbing, wastegates, airboxes, electronic ignition an' all – you'll recognise a two-litre, four-cylinder, eight-valve, twin-cam. If it looks like an old friend, don't doubt your memory. Those self same kinky cam boxes (and the basic casting beneath them) first made their appearance back in '69 beneath the bonnet of a Fiat 125. Of course the motor's been up-dated since then, the most notable quantum leap being the insertion of twin balance shafts to jam out the traditional four-cylinder vibes, which become more annoying the bigger the engine size.

In Integrale form the 1 995 cm^3 engine puts 138 kW at the driver's disposal, once the tachometer needle hits 5 300 revs. Maximum torque is produced at a reasonable 3 500 revs and this is achieved with 100 kPa of boost and a compression ratio of 8,0:1.

So much for peeling away the covers but what's it like, boogying with the little red rally rocket? Well, the first thing worth mentioning after you've slammed the left door (the cars

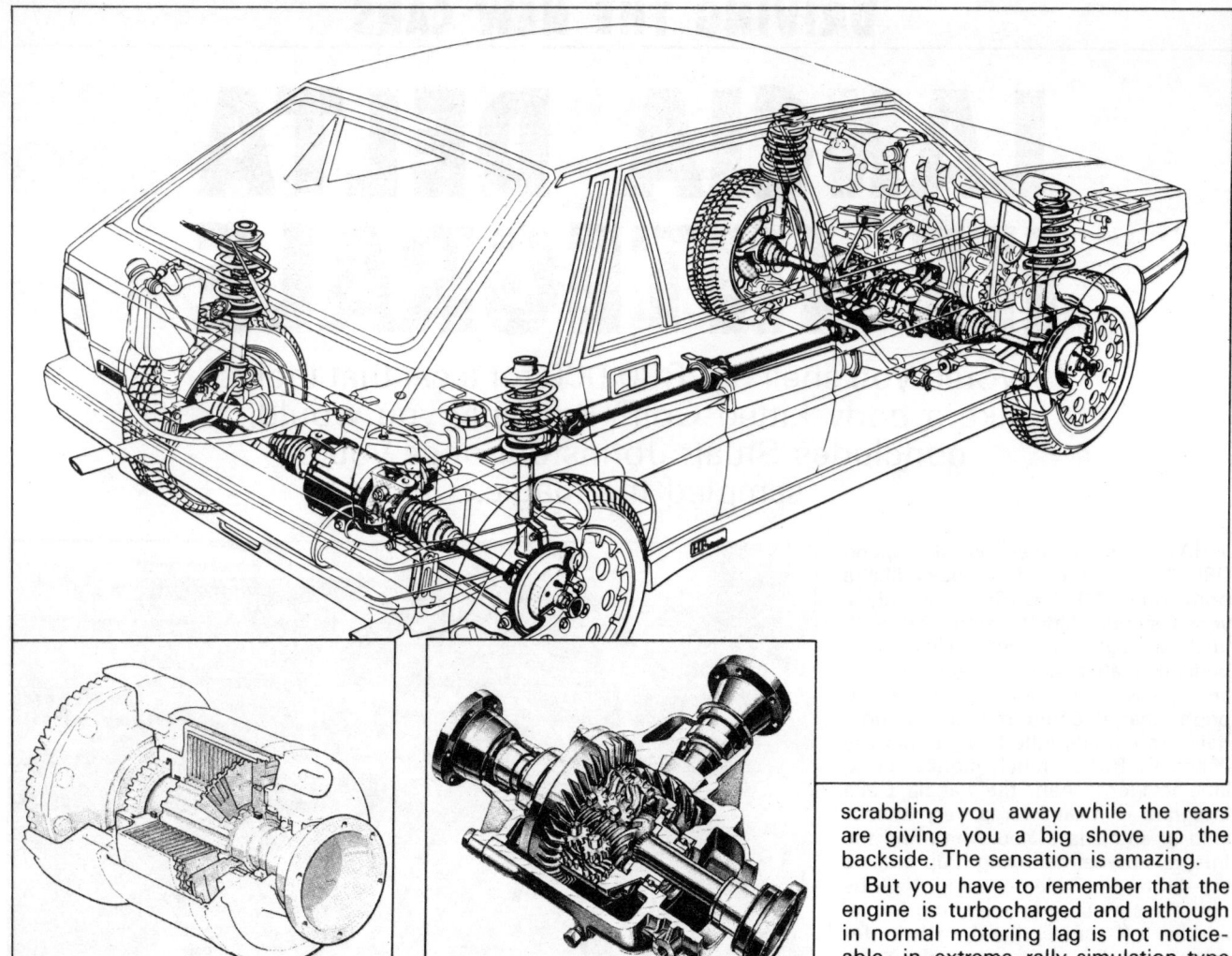

The Ferguson viscous coupling (above, left), is located up front, alongside the central differential. The complex rear Torsen diff (above, right) does a similar job to a limited slip diff. The cutaway drawing (top) shows the rugged struts and the propshaft that passes through the plastic fuel tank.

are only built in l-h-d form, in a limited production run) is that the leather covered three-spoke steering wheel looks and feels totally traditional and wonderful. The seats are made by Recaro, which makes them about the best you can buy and they are covered in a tastefully zany cloth, woven into all sorts of patterns telling you that you are about to have fun.

There may be some people who would be able to get completely comfortable behind the wheel of an Integrale, but they will be of the stump-legged, elongated-armed variety. This is not due to the typical Italian designer approach to the relationship between steering wheel reach and leg space, but because the gear lever is in the wrong place.

To hook fifth gear you have to move away from the seat backrest and stretch your right arm to rib-wrenching lengths, so again you compromise by having your legs a bit cramped and the wheel a bit close. You tend to sit high, which is fortunate, it gives you a good view of the world – and how precious it all is to you. For in such a potent car, the tendency to get hung up on a Miki Biasion impression can be very strong and very dangerous if you overstep the mark.

Not that the Integrale should be considered a dangerous car. In fact its handling is exceptionally safe and forgiving, and short of a Porsche 959 it's hard to imagine any road car that inspires more confidence. Those Michelin 195/55s aren't the widest skins available, but the amount of sheer grip that the Integrale generates is stupefying.

What's more, you can dictate the way the Integrale will handle by the way you set it up for a corner. Enter a bend fast, but in a highish gear with moderate amounts of power on tap and you are looking at a slight understeer on the way out.

So it pays to anticipate the severity of the bend and select the right gear before you commit the car. Drive in hard with lots of revs, keep the power on and you apex right and exit right at terrific speed, the front wheels scrabbling you away while the rears are giving you a big shove up the backside. The sensation is amazing.

But you have to remember that the engine is turbocharged and although in normal motoring lag is not noticeable, in extreme rally-simulation-type conditions it is a factor. If you haven't the right level of power on tap and understeer is developing, to retrieve the situation you can back right off the throttle and tuck the nose back into line. If the situation is urgent you can hit the brakes with the left foot while keeping the power hard on. But this technique shouldn't be practised in the proximity of other animated forms of life.

I discovered a third way of setting the car up for a corner, and this was to lose control of it on the approach. All I'd done, your honour, was to approach the intersection a bit too fast, tugged a bit too energetically on the beautifully-weighted and precise steering, and there we were, driver, passenger and R80 000 worth of rally replica heading sideways for the corner pub, but with a deep canal between us and the swing doors.

In the moment of silence that always descends in such situations, I decided what the hell and floored the throttle. And Vavoom! Front wheels clawed, the rears thrust and we had just executed a perfect flick, point and squirt turn on the absolute limit. Later this passenger, who knows me well, told me she hadn't realised what a good driver I was. It's for moments

When pushed hard, the Pirelli-shod Integrale exhibits phenomenal grip, with slight understeer, a neutral stance or oversteer on offer depending on the way the car is set up for a bend.

like this that you pay all that money...

The Integrale is not perfect. The gear change is mushy and sometimes you are left searching for a cog when you really need it, as in the approach to a bend. This could well be a fault of the particular car loaned to us and not typical of all 5 000 Integrales (the number that Lancia have to build for homologation purposes for Group A rallying).

The ride is not the most refined around, but considering the way this little hatch sticks to the road, the compromise is pretty ingenious. The motor begins to show its ancestry when you rev it past 5 800 but it'll run to 6 000 with just a hint of protest, and 6 200 if you really feel the need.

Oh yes, the Delta Integrale is quick. Quick enough to run 0-100 km/h in about 6,5 seconds and to a top speed of at least 215 km/h. I saw an indicated 235 on a fairly steep downhill, which equated to a genuine 220 or so.

The brakes don't have ABS but they don't need it. They are so powerful and so sensitive, with a minimum of servo assistance, that you are always aware when they are about to lock up.

The nice part is that you get all this performance and wonderful handling with the practicality of a five-door hatchback and just the right mix of exclusivity — there are quite a few people who realise that this car is something special — and anonymity. Driving an exotic can be a drag when you don't feel like posturing.

There are only 12 Integrales scheduled for import here so far, although T.A.K. Motors say that there is a chance they'll bring in a few more. Because, despite the low-profile-high-price ratio, the demand for this little car is certainly there.

Second opinion
by Gordon Wilkins, writing from Italy

A PREVIOUS trip in the competition car with Lancia team driver Michael Eriksson at the wheel suggested that the ultimate limits were higher than I had thought possible, so I tackled snow-bound roads in the production model with all the more confidence. Snug-fitting Recaro seats give support against lateral G and the power steering is superb; instantly responsive and dead accurate, with just the right amount of assistance. An equally precise gear shift controls a close ratio box with overall gearing high enough to make it worthwhile snatching first for a quick spurt away from a tight hairpin, without taking the tachometer needle past the red line at 6 500.

With all those sharp angles, wind noise is high and the engine adds its quota, but this is not the sort of car you buy to listen to Beethoven's fifth, or carry on a conversation about the meaning of existence. It's go! go! and I don't know of a quicker car from point to point, exploiting the gaps in the motorway traffic at up to 200 km/h with the tach showing an easy, smooth 5 500 or so or taking advantage of the compact size and low overhangs to nip past other traffic on ordinary roads.

0-100 km/h comes up in under seven secs. Lancia engineers believe in neutral handling for 4-w-d cars but the production Integrale understeers more than the rally car and this is deliberate.

Putting more of the power through the front wheels helps the driver to feel when he is approaching the limit. A car for sale to the public could get a bad name if drivers of limited skill got no advance warning of breakaway.

There's no ABS, but the grip is so good accelerating or braking that I never felt the need. In snow it is possible to lock the wheels and this is the one condition in which it is an advantage, as the snowplough effect stops the car more quickly. Pedals, with a racy perforated throttle pad are nicely spaced for heel-and-toe downshifts, while braking.

Unlike other cars with pretentions to this kind of performance and road-holding, the Integrale is a four-door four-seater. Agile, nimble and very quick, it is a driver's car to dream about. ●

South African conditions make minimal demands on the phenomenal grip and traction provided by the Integrale four-wheel drive transmission. But in an Italian winter (above) it's quite a different story.

STAGE TEST

Shaken and stirred...

*So far, Lancia's seven victories
With a works pilot about to add the driver's
Jeremy Walton samples*

Comforted by the firm embrace of a full-harness (with crutch straps) and a carbon/Kevlar wrap-around seat, our scribe set about exploiting the Delta's remarkable damp-road traction in the hills above Sanremo

LANCIA DELTA HF 4WD

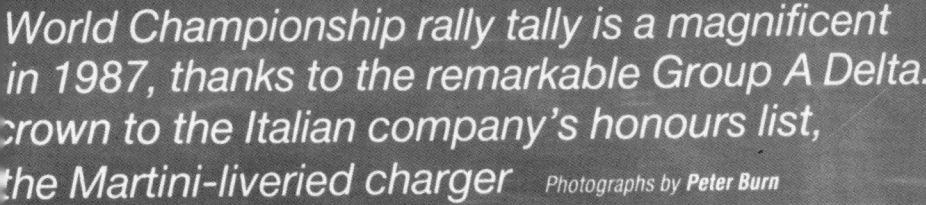

World Championship rally tally is a magnificent
in 1987, thanks to the remarkable Group A Delta.
crown to the Italian company's honours list,
the Martini-liveried charger

Photographs by Peter Burn

Multi-striped and white, the Lancia glides to a standstill in the peaceful Italian village square of Romolo. Yesterday the surrounding cafes and walled streets were packed with wet spectators on Italy's World Championship rally event, witnessing the works Lancias of Biasion and Saby slither to an authoritative 1-2 Sanremo result. It was the seventh 1987 victory for the factory Lancia Deltas in the World Championship of Makes.

Only the Silverstone-based Rothmans BMW (on the tarmac trails of Corsica), the Belgium-based Mazdas 323s (in an icy Sweden) and Ingolstadt's Audi 200 quattro (under a baking East African Safari sun) have interrupted the imperious progress of the Lancia legions this season.

The Lancia HF 4WD was originally hastily converted from the low-volume image-builder of the Delta line to a 5000-plus per annum Group A status World Championship contender. Turin have only recently released the Delta derivative they really would have liked to use from the start — the fattened and uprated Integrale (see page 32).

Today, sunshine and fog sweep in from the Mediterranean coastline below, alternately shrouding and sunning a small white service van. Working from that mobile base are a pair of cheerful Abarth mechanics from the hundreds employed at the *Torinese* Corso Marche factory that has been sending out Fiat-financed winners for decades.

Perched in the hills above Sanremo, Italy's flowery answer to Monaco, the Delta sheds excess heat gratefully. Cooling turbocharged twin-cam engine and dinner-plate-dimensioned disc brakes "tick" away the degrees. I peer reflectively through the blue tint cast by the screen's Lancia Martini banner.

The din of some 250 horsepower, unsynchronised gears and sophisticated 4WD fades into tape-recorded memory. Yet the sense of privilege increases with every recollection of the two outings we have shared.

Lancia's works Delta is the ultimate Group A rallying way to cover such twisty and frequently forested gradients. The power-steered five-door hatch emerges from myriad bumpy and damp hairpins with full power poise preserved, even as another fallen tree is dodged.

You are welcomed by Sparco's Kevlar underpinned seating to a cabin full of roll cage protection. There is everything a visiting driver could want, and clear explanations of the usual works rally car plethora of controls.

Even the instrumentation is dramatic, Abarth-embossed dials flaunting red numerals on a black background. Whilst the speedo is a tiny white and black irrelevance for legal purposes, the major dials speak of 0-10,000 rpm (we're asked to use no more than 7000), the large tachometer matched by a 0-2.5 bar (absolute) boost gauge.

Warning lights suffice for major crisis alerts to low oil pressure and high water temperature, all strategically placed in the sight line beyond Momo's three-spoke wheel and its much-scuffed rim.

To the right in this working cabin are the navigator's electronic assistants and the blue ink trails of hastily stowed kit. Looking down I see the useful Sparco stowage bags that hold my notebook and the drilled foot pedals, massive brace and a brake pedal covered thoughtfully with an anti-slip grip. Still there is more to absorb. A central fuel gauge, Motorola radio communications, 24 fuses and eight relays.

This chunky HF is the car Miki Biasion used to finish second on America's Olympus Rally and for Sanremo recce duties. A 45-55 per cent front to rear power split represents part of the recipe marking Lancia's comeback to winning tarmac form. Abarth Technical Manager Claudio Lombardi confirmed that the biggest engineering challenge with "many, many problems" had been to equal the tarmac pace of Renault's Turbo 11 (front-drive) and BMW's racy (rear-drive) M3.

"The biggest trouble was the size of tyre, the width we could get on our car — 7.75 in front and 7.50 in rear. We are allowed up to 9.0 in by FISA rules, but could not get them under this body shell."

The bespectacled engineer with greying hair explained how they boosted the Delta's tarmac speed. "Engine power is up from 230 bhp on the Monte Carlo to the limit of 260 bhp. Here, on Sanremo, we gained enough ground to turn the boost down from 1.1 to 0.9 quite early in the event, the engine helped a little bit by the wet weather.

"Of course, engine power is only part of the reason for our success. There was much testing work on the Pirelli tyres, suspension and our 4WD system. Plus the weight! That is always a problem on these kind of cars.

"For smoother roads we can have less body protection and welding, plus a titanium roll cage. It makes a minimum of 1130 kg instead of more than 1200 kg for a rough rally like the Greek Acropolis.

"We have come from a special tarmac stage loss of 1 to 1.5 seconds per kilometre to Renault and BMW on the Tour de Corse, to faster times than Renault on Sanremo," said Signor Lombardi diffidently. Looking over the Sanremo result one sees that Lancia really had recovered a lot of tarmac ground, setting fastest times on 11 of 18 such tests versus the best from Renault and Opel.

Pirelli P Zero Corsa prototype road tyres (a full range will be launched in spring 1988) were used on nine stages to win on Sanremo. Abarth's men installed P7 205/50 ZRs with plenty of tread for the varied surfaces ahead of a strange journalist.

"Our" steel-caged HF also lacked the dinky composite air-extracting "hubcaps" that assist brake cooling over tarmac tests like the one we will tackle. TRW Sabelts were cinched good and tight, crotch straps included, a future career in female operatics brutally assured by grinning mechanics.

Abarth dialled in 1 bar/14.2 psi for our test. Claudio Lombardi had already told us that their limit for 260 bhp at 6000 rpm is 1.1 bar, so I am to get a very fair taste of real works power.

At first, I treat the gearbox and clutch too gently. Both are purpose-built for the toughest treatment and respond best to vigorous and rapid use.

The exposed gearchange linkage with its solid joints clunks through five conventionally disposed close ratios. The change quality displayed inspired Formula car speed, one that contrasts notably with the rubbery feel of the road and Group N HFs and Integrales I have been driving recently.

The clutch is not the savage RS200 in/out variety, but it and the competition settings for the multi-plate limited slip diff in

124

the front differential do set up some judder.

More importantly, the 40 per cent torque setting on the front limited slip differential does not upset the steering under hard cornering, full power, conditions. Uncanny how Lancia can do that in a competition car, or the Thema 8-32 front-drive, whilst other manufacturers can't pass 100 bhp through the front wheels without them darting all over a wet road.

The long stroke (84 × 90 mm) fiat ancestry of the generously intercooled twin-cam promises little in high rpm civility. The compensation is a fat 290 lb ft of torque deployed by 3000 rpm. Drop the clutch at 5500 to 6000 rpm and you find the figures mean little against the inspired traction.

The complex all-wheel drive squeezes that hard-won horsepower and torque to astounding effect, the Delta bounding away from an uphill standing start with such prompt energy that you instantly understand one key facet of its success.

The power to weight ratio is not astonishing by current standards (234 bhp per ton for the ultimate works specification on full boost), but most of that power gets to the ground, whatever the works hero is doing with the connivance of the friendliest works chassis I have driven.

After experience with the heavier 180 bhp Integrale, the previous 165 bhp 4WD and this works car I estimate 0-60 mph would occupy under six seconds. More relevantly, much of that sprint speed would be available *regardless* of surface and gradient!

Top speed was irrelevent over the twisty stage mileage, for hitting fourth gear and full boost was a rarity that turned solid scenery to imminent menace. I wold guess no more than 125 mph was available at the works driver's 7500 rpm limit. I can confirm that downhill engagement of fourth gear on a brief straight made 80 mph feel like 180! Lancia's warning of the fate that would befall a crashing journalist became more prominent at each upward gearchange, but I did venture into fifth to confirm its existence!

The transverse twin-cam gargled its Weber Marelli injected and managed way through high-walled streets faultlessly, idling at 1100 rpm to emphasise the old carburetted days of throttle-blipped idle speeds are banished.

In sterner action you wait for the engine and boost to celebrate 3500 rpm before expecting any sympathetic reaction. From 4000 to 6500 rpm Italian horses gallop with thoroughbred exhilaration, deepening large bore exhausts accompanying shrill gear whine and Garrett turbocharger's repertoire of wastegate grunts and full boost whistles. Your conducting job is to provide some punctuation on the gearchange.

Perhaps *the* outstanding driving feature is the power-assisted steering. The rack and pinion is geared to cut about a turn from the production car's already wieldy action. All but full-blooded mountain hairpins can be tackled with one shock-free sweep of the wheel.

Now mix in the ingredients of Formula race car disc brakes (cockpit adjustable for front-rear bias), a ride that will absorb severe bumps without exhibiting road car roll angles, and you can begin to understand the confidence works drivers exhibit.

I pushed on lonely, but public, roads to the point where the car would slide elegantly wide on the exit of second and third gear corners, especially if I used left foot braking to build up the turbo boost in mid-corner. Such mid-corner tricks are not elegant driving technique, but every trace of the initial understeer disappears and the HF bullets away with ego and turbo boost equally sated.

Watching Biasion bashing through the previous day's monsoon underlined just how much capability lies within the Lancia. The 29-year-old home team player would arrive on understeer lock, braking completed during the sliding arc to a tight corner's apex. Even provoked by mid-corner braking (partly for retardation, partly to keep the turbo on boost) the Lancia did not wriggle and twitch in the "nervous" manner of so many rally machines.

The works HF would disappear in a flurry of speedy corrections, front Speedline wheels straightening toward a very modest application of opposite lock and an occasional stab of flame to mark each rapid-fire gearchange.

For a mere mortal the experience of swooping along the Mediterranean hillsides with the engine and transmission noise chasing such speed as I could muster was an emotional education. One akin to a first flight or motor race but, sadly, not so likely to be repeated. Ⓜ

SPECIFICATION

ENGINE
Cylinders	Four in-line
Capacity	1995 (2793 cc turbo rating)
Max power	260 bhp at 6000 rpm
Max torque	292 lb ft at 3000 rpm
Block	Cast iron
Compression	8:1

TRANSMISSION
Layout	Transverse engine, permanent 4-wheel drive, 45-55% front-rear bias
Type	Abarth non-synchro five-speed, Ferguson centre viscous coupling; Torsen limited slip rear differential; ZF multi-plate front

SUSPENSION
Front	MacPherson struts, replacement springs, Bilstein gas damping lower arms and anti-roll bar
Rear	Struts, replacement springs, Bilstein damping, transverse arms, anti-roll bar

BRAKES, WHEELS/TYRES
Front	Brembo 11.8 in dia, 4-piston calipers
Rear	Brembo 11.8 in dia, 2-piston calipers
Front/rear	15 × 7 in Speedline, Pirelli P7 205/50 ZR

WEIGHTS
Road car	1190 kg
Tarmac Rally	1130 kg
Rough Rally	"Over 1200 kg"

Massive roll cage aids bodyshell rigidity, as Kevlar seats support the human frames aboard the Delta

ROAD TEST
By MARCUS PYE

Very together

Above: Despite the most extrovert cornering techniques in damp conditions, the Lanca Delta HF integrale could not be persuaded to cede its leech-like grip, although body roll was surprisingly high. Below: Up front, the potent Garrett-turbocharged 2-litre power unit lurks beneath the distinctive louvred bonnet.

Many motor manufacturers like to claim that competition improves the breed, yet all too often it seems like a glib promise to the consumer who is likely to experience little tangible benefit from the multi-million pound programme for which, ultimately, he or she has helped to pay. Investors in the Lancia Delta HF integrale (with trendy lower case 'i' note) however will be left with no feeling of having been sold short, for the advances made through last year's World Rally Championship-winning performance are available to the public *now*, and have translated well into a very civilised roadster. For under £15,500 on the road, all that technology is competitively priced and packaged too...

The *intergrale* replaces the Delta HF 4WD model, upon which the Group all-conquering Group A rally machine was based. The sporting pedigree of this car is impressive indeed, for it won no fewer than eight of the 11 rounds which comprised the prestigious 1987 World series, and carried Finnish driver Juha Kankkunen to his second successive title. More significantly, perhaps, the GpN class (ostensibly standard production) title fell to Alessandro Fiorio's Delta HF.

When such aces as Kankkunen and his compatriot Markku Alen have had a major hand in the development of your road car, you can bet that there is advantage to be reaped, for both manufacturer and user. And I think one can be fairly sure that their development mileage was a mite more rigorous than the average (?) road driver will manage in a year.

The second generation integrale features a revised version of the famous 2-litre twin-cam engine, the transversely mounted four-cylinder unit pushing out a lusty 185mph at 5300rpm. A larger Garrett T3 turbocharger is fitted, while a lot of attention has been focused on improved cooling, the oil and water radiators, and turbo intercooler being fitted.

Of principal interest, though, is the advanced transmission system, through which its drive is managed and optimised. At the sharp end the Lancia has a robust 5-speed gearbox, but its very healthy torque output (some 224lbs/ft at 3500rpm) is immediately split by the centre differential, the standard bias of which delivers 56 per cent of the power to the front axle, and 44 per cent to the rear.

126

LANCIA DELTA HF INTEGRALE

Depending on road conditions and available grip, however, a Ferguson viscous coupling automatically balances the torque split, front to rear, maximising the adhesion. The rear diff is also torque sensitive for enhanced lateral distribution of power. The front diff is free-floating, to promote traction on indifferent surfaces.

It's a curious car, the integrale. From the moment you climb into the 'wrong' door (lhd only, unless thousands of us send deposits to Turin I guess), adjust that superbly contoured Recaro seat and set the rake of the steering column it feels right, fits like a glove. Fire the free-revving engine and the rorty exhaust note appeals to the driver to cut loose, set the machine free. But should you drive it first in dry conditions, chances are that you will emerge a touch disappointed...

The tweakiest Delta's squat stance, blood red colour (no option here folks!) and sheer Italian *pizazz* stimulate the senses to anticipate more than perhaps it can reasonably be expected to deliver. Its performance actually exceeds preconceptions, and by a heady margin, but the thoroughbred chassis copes so well with everything thrown at it that it feels most unspectacular. Even when a glance at the tachometer reveals that you are travelling 20 per cent more quickly than relaxation levels suggest. Deceptive this – takes some acclimatisation.

Drive the integrale in adverse conditions, though, and its true character emerges. And will blow your mind, unless you are a Kankkunen or Alen, used to such limits of adhesion and control. Now wait for another dry spell and reappraise the vehicle, push it harder and watch it emerge as a truly outstanding all-round machine. Like so many Italian cars, this racy Lancia's company will grow on you. And it is difficult to imagine how this appeal will fade with time.

Despite a comparatively long first gear, which makes the car hesitate on even the fastest starts (there is even turbo lag when one resorts to dumping the clutch at 5500rpm on the loose) it rockets past the 60mph mark in a touch over 6secs. With smooth power delivery right through to the 6500rpm red line, not that much is gained by holding the intermediate gears beyond 5500, progress up towards the 130mph plus top speed is accompanied by a delightful song from the engine bay. Most will revel in its burble, and forgive its noisiness at higher revs as being part of the fun. The gearchange is quick and precise, with a pleasantly weighted action.

It is not the willingness of the engine to rev, but the way in which its prodigious torque is harnessed which sets the integrale apart from other four-wheel drive cars. Aided, no doubt, by the fat Michelin MXV radials its outer limits of roadholding are awesome. Stiffened suspension still allows quite a lot of roll, but does not in any way compromise ride comfort while the suspension set-up and steering geometry permits incredibly fast changes of direction with no loss of feel.

So well engineered is the chassis, and so forgiving, that even such provocations as mid-corner brake-jabbing, throttle lifts and exaggerated steering alterations on the test track failed to upset its inherent balance. Frankly, the Delta barely noticed our taunts, shrugging them off with contempt and a playful squeal of tyres. Directional control remained sharp and instantaneous throughout.

Back on the open road, I was fortunate to encounter an afternoon of very heavy rain in which to put the integrale through its paces. This proved to be the treat of my few days with the car, for the cornering forces attainable in perfect safety are way beyond anything I've previously encountered in comparable conditions. Again, the Lancia transmitted plenty of feel through the chunky driving wheel, and although the rear end did momentarily become unstuck in the deeper puddles, its agility and poise were uncompromised.

The brakes, discs all round naturally, retained a solid, progressive quality regardless of weather, and heel-and-toe downchanges are positively encouraged by the enterprising spacing of pedals. The throttle, by the way, is a sheet metal affair, drilled for lightness. Very boy racer, one of the few real concessions to this idiom.

A bank of circular analogue dials face the driver, in the comfort of his multi-adjustable seat. Functional, with their ochre pointers bold against matt black backgrounds, they are unpretentious, easy to read. The upholstery is, by contrast, garish, its extraordinary diagonal multi-coloured striping um... unusual against the grey base colour. Better than 'drab' black for such a sporting carriage? Certainly less practical.

Externally, the integrale is cleverly understated, a real Q-car. Save for the cooling slots in the bonnet, the very neatly swathed wheel arches and sills, and the pretty 6J wheels, it really looks much like any other Delta. No overt gimmickry, vast spoilers, fancy striping. Nothing to say that this is a stunning little mover. And that is a bonus for which owners will be grateful.

While it is a superstar dynamically, the Lancia's dipped beam headlight performance could be boosted to good effect, and one or two irritating niggles affected 'our' car, which put the spotlight on build quality. Most people could live with a rattle or two, in the short term at least, but I was more perturbed by the ingress of water (from the rather basic sunroof?) which left the rear seat soggy.

These things do, alas, tarnish the image of a car, no matter how good it may otherwise be – and the joys of driving the integrale should make it a great! Even when it is stationary, there is an impulsive satisfaction in blipping the throttle which will gladden the heart of genuine motoring enthusiasts everywhere. And nothing can touch it at the price! ■

The interior can be tailored to fit the driver, from the seat to the steering column.

Integrale – Fun with a cap 'F'!

LANCIA DELTA HF integrale
£15,455

SPECIFICATION
Cylinders/Capacity	4/1995cc
Bore/Stroke	87mm/90mm
Valve gear	Twin cam, 2 valves per cylinder
Power/rpm	185bhp at 5300rpm
Torque/rpm	224.13lb/ft at 3500rpm
Fuel system	IAW Weber fuel injection
Gear ratios	3.50/2.23/1.51/1.13/0.92:1
Final drive	3.11:1
Steering	Power assisted rack and pinion
Brakes	Servo assisted discs all round
Wheels	Light alloy 6J
Tyres	195/55 VR15
Suspension	All indpendent with MacPherson struts, anti-roll bars

DIMENSIONS
Length	153.5ins
Wheelbase	97.6ins
Track (front)	56.1ins
Track (rear)	55.3ins
Width	66.9ins
Weight	1267kgs

PERFORMANCE
Maximum	128mph
0-60mph	6.4s
0-100mph	19.5s
Fuel consumption: overall	24.1mpg

The HF4WD was successful in World rallying – the integrale is currently dominant! Built for the mud, that is where its class shines through.

ROAD TEST

Wolf in Sheep's Clothing

Lancia's task in building the Integrale out of the HF 4×4 was to create a world-beating rallycar, which it has clearly achieved. The road car spin-off is that the agile and rapid HF 4×4 has turned into an automotive Clark Kent — a supercar in plain clothes. Or fairly plain; there are no roof-wings or shovel spoilers on Lancia's super-saloon, but there are brawny wheel-arches looking ready to elbow through traffic, and a new bonnet with large louvres above the radiator.

Summarising the Integrale's improvements, it gains a more powerful version of the 2-litre turbo twin-cam four, the same basic design which is shared with the Thema turbo ie, larger brakes, and wider-track suspension with room for tyres of anything up to 235 width.

More air is scooped into the engine compartment for radiators, intercooler and brakes by a broader and deeper bumper pierced by grilles, and narrow mouldings flare the new arches into the sills. It does look different, but the basic Delta shape is clearly visible, little changed since its arrival in 1979. Whether it is tight purse-strings or deliberate styling policy which have prevented Lancia from fiddling with the Delta shape, it remains a fresh and elegant form and all the better for the lack of interference.

Sales of sporting 4×4 cars have not been up to earlier optimistic forecasts, but Lancia banks on the performance aspect of the Integrale rather than its 4WD abilities. Naturally these are connected to an extent — 185 bhp would be far less manageable through the front wheels only — but the new purchaser of such cars looks for more and more technical sophistication: water-cooled turbos, intercoolers, viscous couplings and Torsen differentials are all good talking points, and the Integrale has them all. It is also restricted to left-hand drive only, which has gained a perverse cachet in motoring circles and has not hindered the sale of the 55 which Lancar ordered for 1988.

In fact, these were sold immediately and only a further few have been obtained, so the car already has a rarity value well beyond its remarkably cheap price of £15,920.

Under the pierced bonnet, the twin-cam engine's parentage goes a long way back, to the Lancia Beta and various Fiats such as 124, 125, 131 and 132. Revised for Delta and Thema and fitted with twin balancer shafts, it later appeared in turbocharged form with an "overboost" feature, allowing the supercharge pressure to jump, for 30 seconds only, from the normal maximum of 1.8 bar to 2 bar. This allowed brief extra-fast spurts for overtaking.

In 185 bhp Integrale form, a Garrett water-cooled T3 turbine of larger capacity forces intake air through a larger vertical intercooler alongside the radiator. Valves and seats, gaskets and water-pump are all uprated, the black boxes of the Weber-Marelli ignition/injection system tweaked, and the time-limit on the boost pressure control removed. Lancia still refers to it as "overboost", with a little light on the dash to excite the driver when maximum pressure is reached, but it is now simply incorporated into the normal boost curve. Larger water and oil radiators and an asbestos-free clutch are also fitted.

Under those arches it is not merely a case of

Balance-shaft twin-cam turbo is sophisticated, flexible and quick to respond.

Lancia Delta HF Integrale

wider wheels: the MacPherson strut suspension system has been equipped with new uprights and links (a form of wishbone at the front and lateral rods plus fore-and-aft links behind) and tougher springs and damping. Wheel size goes up an inch to 15in, and 195/55 VR 15 tyres are standard. To offset the larger wheel diameter a shorter final drive is fitted, though the overall ratio is still slightly higher than before, making cruising a little more relaxed. However, with the increased torque the car accelerates faster (0-62 mph in 6.6 sec) and reaches 135 mph.

Transverse-engined 4WD packages have become very neat, and Lancia's uses an epicyclic box to split the drive 56% front, 44% rear. This is restricted by a viscous coupling when traction is lost at one end, while the rear differential is a Torsen unit with its advantages of no slip and full differential action. The front diff is free. There are no controls for the driver to use, because the system automatically makes the best use possible of the road traction.

Our test route for the Integrale was to Le Mans in hopes of Jaguar's great moment, and very appropriate for this LHD car. Three up, we fitted pretty well into the little hatchback with an assortment of luggage, though the six-foot rear passenger was rather cramped.

Some London acclimatisation had demonstrated the taut feel of the car's suspension and steering, and I wondered whether it would prove over-active on motorways; but as we navigated the traffic cones leading to Dover I found I could relax after all. Despite its short wheelbase, the Integrale is directionally very stable at high speeds, and any pitching is quickly swallowed by the uprated struts.

Taking a Friday afternoon ferry (much more relaxing than the hovercraft, and no doubt more so than the Channel Tunnel trains will be), we ended up sailing an hour and a half late. This meant that our schedule was upset before we even gained the autoroute, and consequently that our speed was to be higher than planned. Yet the Lancia responded without strain, squirting instantly into the gaps in the stream of two-way traffic which joins Calais to the A26 autoroute, and then whistling up to an easy 90 or 100 mph when the motorway arrived.

With some four hours between us and Le Mans, the driver's comfort was going to be important: the nice thick-rimmed steering wheel fits well with the upright driving position, and though the gearshift is a little soft, its intricacies are soon learned, and in fact, the action is relatively good by the standards of most European transverse boxes. Almost all of the various dials can be seen by the driver (it is surprising how rare this asset still is today), while the usual pair of column stalks are joined in normal Italian fashion by one for the lights.

From Paris it is autoroute nearly all the way to Le Mans, but with losing an hour to continental time and then exploring the town

Radical mechanical changes are concealed beneath minimum visible alterations.

for our hotel it was nearly midnight before we parked the car. After a couple of hours at very high speeds, I still have the habit of leaving the engine running for a minute or two to allow the turbo bearings to cool, but with the water-cooled T3 unit there is no longer any need for this.

Over the weekend we sat in a good few queues without any cooling problems, and the middleweight clutch and power-assisted steering eased the strain. There are more subtle power-steering systems, but steering feel is less crucial to a 4WD car with its much reduced chances of breaking traction under power, and the Lancia's is fairly high-geared, making sinuous bends and hairpins alike easy going. In roundabouts it sticks like glue, with little if any adverse response to throttle variations, and it catapults off in the required direction with a low whistle. A torque peak of no less than 224 lb ft at 3500 rpm means that the car will respond promptly even in higher gears.

As we had booked a late ferry on Sunday night, we were able to make more relaxed time using a *Route National* towards Paris, revelling in the memory of those last emotional ten minutes before a Jaguar took the flag, and marvelling at how there seems to be no difficulty in extricating the huge crowds from the Sarthe circuit. Night overtook us, showing up the excellent power of the Integrale's twin-lamp set-up, before our return to Calais (and another hour and a half delay).

Overall the Integrale proved a fine small touring car, turning in fuel consumption of some 21 mpg at high speed, but I had reservations about the braking stability of this particular example. It became very nervous under light pedal pressures, wanting to wag its tail, and of course there is no ABS system; in fact, I found the rear wheels locking under quite mild braking on slippery surfaces, an unpleasant trait which did not afflict the previous one I drove.

For what it offers in performance and handling, this package must be considered very good value. If civilised and rapid cars like these are the spin-off from FISA's attempt to eliminate "rally specials", it is almost tempting to thank them. **GC**

Model: Lancia Delta Integrale.
Importer: Lancar Ltd, Crawley, West Sussex.
Type: four-door 4WD saloon.
Engine: Four in-line, 1995cc (87×90mm), cr 8:1, dohc, belt-driven, 2 valves per cyl. Garrett T3 turbo, air: air intercooler. Weber Marelli electronic ignition/injection, boost and knock control. Power: 185 bhp at 5300 rpm. Torque: 224 lb ft at 3500 rpm.
Transmission: 4WD; front transverse engine, epicyclic centre diff with viscous coupling, Torsen rear diff.
Suspension: (Front): MacPherson strut, antiroll bar. (Rear): MacPherson strut with transverse and longitudinal links, antiroll bar.
Steering: Servo-assisted rack and pinion.
Brakes: Vacuum servo. (Front): ventilated discs. (Rear): Solid discs.
Wheels and tyres: Light alloy 6in × 15 rims with 195/55 VR 15 tyres.
Performance: 0-62 mph: 6.6 sec; Max speed: 134 mph.
Economy: 21.5 mpg overall.
Price: £15,920.

Testarossa corners flat and fast but push hard enough and front will drift, get the power on and it tightens

DAVID AND GOLIATH

In one corner, the £16,000 Lancia Delta Integrale: in the other, the £90,000 Ferrari Testarossa. Can the twice world championship winner hold out against a 180 mph supercar? Howard Lees referees

A final rattle and the Massey tractor was silent, its dusty, overall-clad driver swinging out of the cab and climbing down to Testarossa level. "Beautiful, just beautiful. You don't see too many of these in North Devon." The farmer's eyes roved over the Ferrari's lines with obvious admiration: "I used to have an old 246 Dino — seeing this I wish I'd never let her go."

Go driving in £90,000 worth of Ferrari and that's the sort of reaction you expect, even look for. The contrast between a Testarossa and the other red Italian sports car parked behind it couldn't be more acute — only a few cognoscenti give a Lancia Integrale a second glance. But get behind the wheel and both can deliver a level of usable road performance that's the preserve of very few cars indeed.

The two cars were together to settle a question that had been niggling away at the back of my mind since January. I knew from a winter trip over the Alps (*Autocar*, 10 Feb) that the Integrale's cocktail of overall abilities was outstanding. I knew too, that despite the Testarossa's undeniable performance it had been criticised in some quarters for being just too big and unwieldy to be a serious road car. Was there still justification for owning a mid-engined sports car that cost five times as much as the Lancia?

After heeding the farmer's directions to our overnight halt, we had time to reflect on a day's hard driving that had begun to shed light on two very different cars. The same column switchgear reflects a common Fiat patronage but, other than both being red, the £89,700 Ferrari Testarossa and £15,920 Lancia Delta Integrale approach the serious matter of performance from absolute opposite ends.

Just examining the vital statistics of the two cars is enough to highlight the chasm between them. At 177ins long and 78ins wide, the Testarossa is 23ins and 11ins bigger than the Integrale. That's large — a Transit van is an inch and a half narrower and only five inches longer. The Ferrari's 44.5ins height is another matter though — 10ins lower than the Lancia and only half as high as a Transit.

A good deal of that length and width is taken up by five litres of boxer engine slung in-line behind the seats. The four camshaft, 48-valve flat 12 is a beautiful expanse of high quality alloy die castings, topped by the crackle-finish red cam covers that give the car its name. Oversquare at 82 x 78mm, the 4942cc work of art revs safely to 6800rpm, two separate Bosch fuel injection systems allowing it to develop 390bhp at 6300rpm and 362lb ft of torque at 4500rpm.

With close to 400bhp passing through a five-speed gearbox underneath the engine to the rear wheels, the Testarossa's tyres are surprisingly modest. The 255/50 VR16 Goodyear Eagles hug 10ins wide alloy rims, matched by 225/50 VR 16 rubber on 8ins rims at the front. A Countach has only a little more power but runs 345 section Pirellis at the rear.

Lancia is a strong understeerer and while lifting off helps tighten line it will still drift wide on exit

Lancia sprints to 60 in 6.4 secs, but if you want instant urge off the powerband you'll need a downchange or two

Tiny by comparison, the Integrale's engine bay is still densely-packed. Its alloy head twin cam engine boasts only two valves per cylinder, and a long stroke 90mm crank gives the 87mm bore four cylinder unit a capacity of just 1995cc. But thanks to an intercooled Garrett T3 turbocharger and Weber Marelli fully-mapped control of fuel injection, ignition and boost pressure, the Lancia unit actually betters the Ferrari's specific output.

Lancia's permanent four-wheel-drive system has helped the Delta dominate world championship rallying during the past two years, and it doesn't do any harm on the road either. Drive from the transversely-mounted engine passes through an epicyclic centre diff splitting 56 per cent of the torque to the front wheels and 44 per cent to the rear. A viscous coupling limits slip between front and rear, while across the rear wheels a Torsen diff can progressively feed up to 75 per cent of the torque to the one with the most grip.

You wouldn't expect a Ferrari to sit on anything other than double wishbone suspension and that's what the Testarossa has — coil sprung and Koni gas damped all round with anti-roll bars front and rear. It also has a traditional tubular steel space frame chassis, with a mix of steel, GRP, but mostly alloy, panels hung around it. Rack and pinion steering is resolutely free of power assistance.

Conceived as a family five-door hatchback, the Delta lacks the Ferrari's engineering purity. MacPherson struts sit at each corner of the steel monocoque, and short, fat springs at the front barely clear the tyres. Lower location is by wishbones at the front, transverse and longitudinal links at the rear with an anti-roll bar at each end.

The Integrale needs and gets hydraulic assistance for its steering rack, while disc brakes are ventilated only at the front. A steel shell, iron block engine and all the 4wd hardware means the Integrale weighs only 500lb less than the Testarossa, so its power to weight ratio isn't in the same league as the Ferrari: 146 vs 258 bhp/tonne at the kerb.

Inside the Ferrari, perfectly fitted and stitched beige leather covers everything that isn't glass or carpet. There's enough headroom but, on a wheelbase the same as a Range Rover and with only one row of seats, the Testarossa still falls down badly on legroom.

Drivers taller than six feet are going to have their knees splayed either side of the wheel and will need to use its tilt adjustment to lift it clear of thighs. Firm but well shaped, the seats could do with a bit more support ▶

Testarossa pulls strongly from 1000rpm to 6800 redline: floor the pedal in any gear and it simply takes off

to the base of the spine and the side cushions on the base are going to be too close together for anyone of more than modest girth. At least the pedals are beautifully laid out; ideal for heel and toeing, with the bonus of a substantial left footrest.

Tilt for the base as well as backrest angle is controlled by three tiny switches to the right of the seat, and the rather strange headrests — they look like upholstered tombstones — can be moved up and down. There's a full set of orange on black Veglia instruments — oil temperature and pressure, water temperature and fuel as well as the tacho, speedo and a nasty red digital clock.

Trouble is, like the switchgear, the dials are scattered around the cabin. Why the trip meter needs to be just in front of the gearchange and the heated rear window switch in the roof is beyond me. The overwhelming impression is of a very high quality, but not particularly well planned, cockpit.

Things are still frustratingly Italian in the left-hand-drive only Integrale. Probably the

Ferrari's engine note rises from a deep woofle to a slightly flat-sounding howl as revs climb

nicest steering wheel in the world is set to give a typically long arm, short leg driving position. Even though there's plenty of rear seat movement, taller drivers have to set it a couple of clicks forward with the backrest very vertical in order to grasp the rim.

With four doors, five seats and a hatchback, the Integrale is by far the more practical car. In the Ferrari, there's room only for two inside and a modest set of tailored luggage under the bonnet. The Lancia's excellent Recaro front seats are very firm indeed, but cradle you securely and comfortably in place. They have an extending front cushion for extra thigh support — but I'd willingly trade that for a tilt adjustment.

The Delta's dash is altogether better laid out than the Ferrari, but still far from perfect. It has all the dials of the Testarossa plus a boost gauge and voltmeter — and the same nasty red clock, but this time it is mounted in the roof. With most of the small dials grouped in the centre, the speedo and tacho get pushed to the edge of the panel, and even tilting the wheel up as far as it will go masks the outside top corners.

That thick, leather rimmed 15ins wheel with its centre horn push may be the business, but the column stalks are not. Identical to those in the Ferrari, there's one on the right to handle the wipers but two separate stalks on the left look after lights and dipswitch and the indicators. These are almost the same size and work in the same plane, which can cause problems in the heat of the moment.

To keep in touch with Ferrari, Lancia needs to be kept above 4000rpm, drop off the powerband and car lacks power

Not surprisingly, the Ferrari is better equipped with central locking, electric mirrors and windows, and air conditioning. The Lancia has only electric windows and a manually operated sunroof.

Turn the key in the Testarossa and there's a couple of seconds of mechanical whirring before the flat 12 bursts into life. Responsive to the slightest pressure on the throttle, it sounds much more eager and alive in the cabin than the eight-cylinder Ferraris.

From cold the Testarossa's gearbox is unco-operative. A long, chromed lever and polished alloy gate are no help until the oil temperature has lifted off the stop — the dog-leg from first into second is the most difficult. With all that torque to cope with the clutch is understandably heavy, and even when warm the gearchange needs to be handled firmly. But the great flood of torque from the 5-litre engine means you can be very lazy about gearchanging if you feel like it.

Floor the throttle in any gear and the Testarossa just goes. It will pull strongly with ever-growing urgency from 1000rpm, the engine note rising from a deep woofle to a slightly flat-sounding howl. As you top 3000rpm the car really starts to motor, and by 4000 the Ferrari is flying. Take it round to the 6800rpm redline and the boxer engine is in its element, with not a hint of harshness or protest.

Ferrari claims a top speed of 180mph for the Testarossa. With the standard gearing that's smack on the redline but from the tremendous urge available over 120mph it seems believable. A 0-60 time of 5.9secs is quoted in the owner's manual but, given that we scorched there in 5.5secs in a 328, the Testarossa ought to be a shade faster.

You might just buy a Testarossa for its engine, but not an Integrale. To get 185bhp out of this venerable twin-cam unit, Lancia's engineers have had to forget about the mid-range. Sure, the engine is smooth and docile at low revs, but the turbocharger and complex management system can't disguise its cammyness. There is boost from 2000rpm and thanks to the overboost function you can get an indicated 0.95 bar at just over 3000, but not until the needle hits 4000 does the Integrale really start to fly.

Keep the Lancia above that 4000rpm mark and, although it won't exactly blow the doors off the Testarossa, it can stay in touch. It sprints to 60 in 6.4secs and its standing quarter time of 14.8secs is only just over a second behind.

Drop off the powerband though and you've got turbo lag, and if you want instant urge you'll need a downchange, or two.

With a high first that makes it a serious driving gear, the Integrale at least has a spread of ratios that helps you keep it squarely in the powerband. Change up at the 6500rpm redline in second, for example, and it drops back to 4000, bang on the torque peak. Sensible overall gearing allows the Integrale to nudge 5500rpm at its 128mph maximum, and through the gears it can manage 40, 63, 93 and 124mph.

It takes a while to feel at ease in the Testarossa. Red bodywork seems to stretch in every direction as far as the eye can see. Visibility is exceptionally good for a mid- ▶

FERRARI TESTAROSSA. Limited room and luggage space are compensated for by quality and driveability. Leather upholstery and high equipment levels give feeling of quality. Boxer flat-12 engine gives claimed 180mph

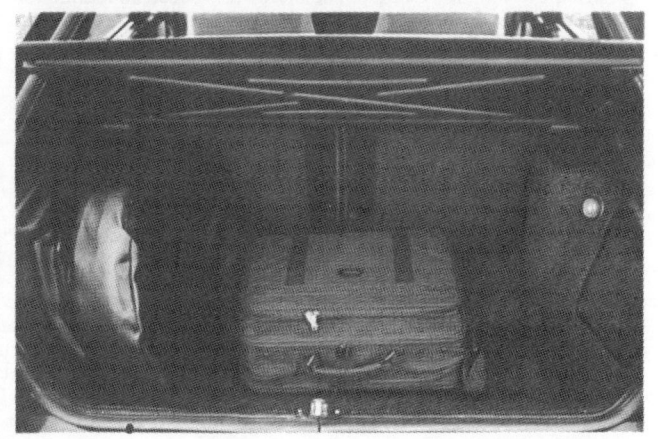

LANCIA INTEGRALE. More practical with five seats and hatch but typical Italian driving position is not ideal. Dash is better laid out and includes boost gauge and voltmeter. Turbo and fuel injection help give 185bhp from 1995cc

engined sports car, even through the rear windows, but those great wide haunches dominate the lower half of the side mirrors.

Trickle around town and the Testarossa is as docile as a kitten. With 225-section tyres, the unassisted steering is stiff and unyielding, the clutch heavy to dip, but the engine will amble happily with the rev counter barely off the stop.

Up the pace and the steering lightens. Above 100mph it comes alive and you feel every bump and camber through the rim. It's wonderfully responsive, but hit a bump under braking and the wheel will snatch and judder. The Testarossa tramlines too, following white lines faithfully on the brakes.

There's plenty of tyre roar and a hardening of the engine note as you ease on the throttle, but even at ridiculous speeds there's not much wind noise. The ride is always firm, but beautifully controlled when you're pressing on, stiff rather than harsh at lower speeds.

With compact dimensions, a higher driving position and better visibility the Lancia is an easier prospect in town. Power steering takes all the effort out, but forget to allow for that tall first gear and cammy engine and the Integrale can be embarrassingly slow off the mark.

The Lancia can't even get close to the Ferrari's immaculate build quality. In this 15,000-mile example there were squeaks and rattles from the dash, doors and trim — and the fit of the add-on skirts and spoilers is average.

Thrashing along the motorway in an Integrale is understandably very different from the Testarossa. That five-door shell makes no pretence at aerodynamics, and generates so much wind noise at high speeds you think you're driving along with the windows open. In contrast, the MXV tyres are almost silent, the engine subdued and the ride more subtle. At speed that is — if anything, around town, the conclusion is that the Lancia is harsher than the Ferrari.

On the tight, twisty lanes leading up over Exmoor we had a chance — at last — to extend the cars. In the Testarossa the steering never really attains the fluidity and response of a GTB, but there's plenty of grip at the front. With almost 3.5 turns from lock to lock there's more arm twirling than there should be, but that's the price to be paid for avoiding power assistance.

Once in the bend the Testarossa corners flat and fast, the Goodyear Eagles providing exceptional levels of grip in the dry. Push the Ferrari harder and the front will drift into the corner; get on the throttle early enough and it tightens, the rear drifting on the exit. With the V-12 engine's tremendous spread of power you can use second gear or even third and there's still enough torque to drift the rear on the throttle alone.

Taking the same series of bends in the Integrale couldn't be more different. Softer, longer travel suspension and a higher centre of gravity mean there's more dive under braking and a good deal more body roll in the corner.

Against that is quite the best power steering around. At 2.8 turns lock to lock it's well-geared, beautifully weighted with pin-point accurate feedback. There's plenty of grip from the 195-section Michelin MXVs so although it's not as crisp as the Testarossa you can turn the Integrale in very sharply.

With a front-biased torque split, the Lancia is a strong understeerer with the power on, and on tight corners you can almost roll the offside front tyre off the rim. Simply lifting off helps tighten the line, but the Lancia will still drift wide on the exit. On a tight corner, the only way to kill the understeer on the way out is to left foot brake to the apex. And on the limit the Ferrari is still ultimately faster round the bend.

It was on all of this that we reflected that first night. The next morning dawned damp and depressing. Around the slippery lanes the Testarossa was controllable but needed care. The Integrale, meanwhile, barely had to slacken pace. In the wet MXVs are probably the best road tyres you can get and, with the Lancia's combination of viscous coupling and Torsen limited slip diffs, it was virtually impossible to get into trouble. Here the Integrale was king.

From our two days of hard driving in the cars it was clear they are, of course, very different animals. But in absolute terms the end result is very close.

On wide open, sweeping roads the Testarossa could haul away from the Lancia. Get among the winding lanes and, if you know them, the Ferrari can still stay ahead. But on blind twisty roads, the Integrale is faster, much more tolerant of mistakes, easier to get back into shape and generally inspires more confidence. In the wet, it's no contest.

The Testarossa emerged as a much better road car than I had been led to expect. It has towering performance (and you can use all of it), perfect manners and very good visibility — all that talk of it being too big is sheer nonsense. It's a car of genuine quality, worth every penny of £90,000.

But the Integrale is as good as it ever was: fast, nimble and with incredible grip. At less than £16,000 it's one of life's bargains. ■

DRIVETRAIN	TESTAROSSA	INTEGRALE
Cylinders	12	4
Capacity	4942cc	1995cc
Bore/Stroke	82x78mm	87x90mm
Valves per cyl	4	2
Valve operation	2 ohc per bank	2 ohc
Compression ratio	9.3:1	8.0:1
Induction	Bosch K-Jetronic fuel injection	Weber mapped fuel injection and ignition
Power/rpm	390bhp/6300	185bhp/5300
Torque/rpm	362lb ft/4500	224lb ft/3500
Gearbox	5-speed	5-speed
Drive	rear	permanent 4wd
Final drive ratio	3.00	3.11
Mph/1000rpm top	26.5	23.4
Tyres:	Goodyear Eagle 225/50 VR 16 front 255/50 VR 16 rear	Michelin MXV 195/55 VR 15
DIMENSIONS		
Length (ins)	176.6	153.5
Width (ins)	77.8	66.9
Height (ins)	44.5	54.3
Wheelbase (ins)	100.4	97.6
Track f/r (ins)	59.8/65.4	56.1/55.3
Kerbweight (lb)	3320	2793
PRICES		
Total in GB	£89,700	£15,920
FUEL CONSUMPTION		
Overall mpg	13.8	20.0
Fuel tank (gals)	25.3	12.5
TOP SPEED		
Mean/rpm	180/6800 (claimed)	128/5450
ACCELERATION (secs)		
0-60mph	5.9secs (claimed)	6.4secs
Standing ¼ mile	13.6secs (claimed)	14.8secs/90mph
Standing km	24.1secs (claimed)	27.9secs/110mph

NEW CARS Lancia Delta HF Integrale 16v

Middle-age spread: a 1.2in slatted bonnet bulge joins the Integrale's familiar wide wheelarches.

Staying One Step Ahead

Lancia has a busy period of reorganisation ahead in Britain. With its models now sold through Fiat (UK) Ltd, instead of Lancar, there will be inevitable dealer changes, but none of the consequent confusion should obscure the new production excellence from Turin.

We have been driving the most exciting example, the third-generation of the four-wheel-drive turbocharged Delta which has consumed the World Rally Championship since 1987. Now equipped with a 16v powerpack of 200 road horsepower (the competition potential is twice that), the HF Integrale 16v remains a supreme driving experience — even at the projected "£19,000 to £20,000" when it arrives in the UK this August, still in left-hand-drive only "for the immediate future".

It was January 1987 before the first Delta 4WD appeared in world-class competition, the necessary 5000 manufactured, whereupon Lancia demolished the opposition to win seven events and the World Championships for both Drivers and Makes. In 1988 the Integrale (its 185 bhp roadgoing base having debuted in October 1987) won all but one qualifying event in the makes series; a second successive double title was never in doubt, and Miki Biasion was created the first official Italian World Rally Champion.

In 1989 the Integrale, easily identified by its Audi quattro-style wheelarch extensions on the original 1978 Delta body, has continued to pummel its rivals from Ford, GM and the Japanese. But the latter are becoming notably more effective through the efforts of Toyota and Mitsubishi, so the 16v derivative, one which also promises to increase the already profitable 15,000 run of Delta 4WD derivatives sold to date, was a commerical and competitive necessity. It should make its World Championship debut any time after August 1.

Fiat/Lancia's familiar iron-block counterbalanced 2-litre now gains 16 valves in vee pattern to complement belt-driven double overhead cams. Pumped intake mixture via a more efficient intercooler and Garrett AiResearch T3 at a maximum 14.2 psi at the usual 8:1 cr assists unusual 16v flexibility.

From 2500 to 4800 revs, over 210 lb ft of torque is supplied (more than the peak of a similarly hefty Sierra RS Cosworth), peaking at 224 lb ft on 3000 rpm. That is 500 rpm down the scale of the previous 8v unit, and maximum 16v power of 200 bhp at 5600 rpm is also highly accessible to the laziest of drivers.

The effect on performance is startling. We did not recall the previous Integrale as lethargic, but the 16v bites 62 mph from rest in 5.7 seconds, a figure which would not disgrace a Porsche 911. The overtaking prowess supplied by the wide power and torque curves should offset considerably the handicap of LHD in Britain.

A 137 mph maximum is almost irrelevant in a machine which continues to employ its strut suspension and sophisticated permanent 4WD to outstanding effect. As before, a Ferguson-patented viscous coupling (VC) and epicyclic gears provide central power apportionment, and the action of the rear differential is further disciplined by Torsen Gleason-patented limited-slip differential. An hydraulic clutch is now specified and the ZF gearbox has been uprated with a new second-gear ratio listed.

For the buying public the good news is not completed at the provision of a flexible 15 bhp bonus in the engine compartment. Lancia and Bosch engineers have worked together upon a four-channel, six-sensor third generation of electronic anti-lock braking for the

> *"Previous Integrales have hardly been lethargic, but the 16v bites 62 mph from rest in just 5.7 seconds, a figure which would not disgrace a Porsche 911"*

quartet of Integrale discs. Listed as an option, along with a sunroof and now 7J × 15in alloy wheels, the ABS has been taught to electronically exchange information with the Marelli-Weber IAW engine-management system.

The effect is to ensure that ABS always intervenes subtly at the consistent brake pedal, even if the car is decelerating sharply over mixed surfaces or has to stop suddenly in mid-corner. It was a 100% success on our experience.

Tyre and suspension settings reflected significantly revised suspension (very few parts interchange with the car's visually similar predecessor). ABS demanded some of these alterations, accompanied by the provision of a slight rearward power-bias. Today the setting is 47% front, 53% rear, almost exactly a reverse of the original. Front-end weight continues to occupy some 63% of the now hefty 1250kg five-door.

Externally and internally it takes a practiced eye to spot the differences between 16v and its rallying champion forerunner, but the 1.2in elevation in slatted bonnet height (the underneath heavily padded to minimise noise levels) is most obvious. Ride height is reduced by almost an inch. I would guess that the intake beneath the spoiler to direct cool air to the gearbox was vitally needed in competition to justify its inclusion.

So the 16v Integrale remains a step ahead in the 4WD "super-hatchback" set, but snags remain. Wind noise from that bluff shape increases sharply above the easy 100 mph and 3900 rpm cruising pace. The interior is a sorry mixture of unsuitable colourings and sprawling switchgear. Recaro seats are the least supportive from that reputable company.

But at the end of a damp but exhilarating day amongst the Torinese lanes, tracks and motorways, my opinion of the latest in Lancia Integrales remains one of unmatched performance versatility. The 16v is the finest all-weather driving experience on sale, one that is only enhanced by the addition of ABS. **JW**

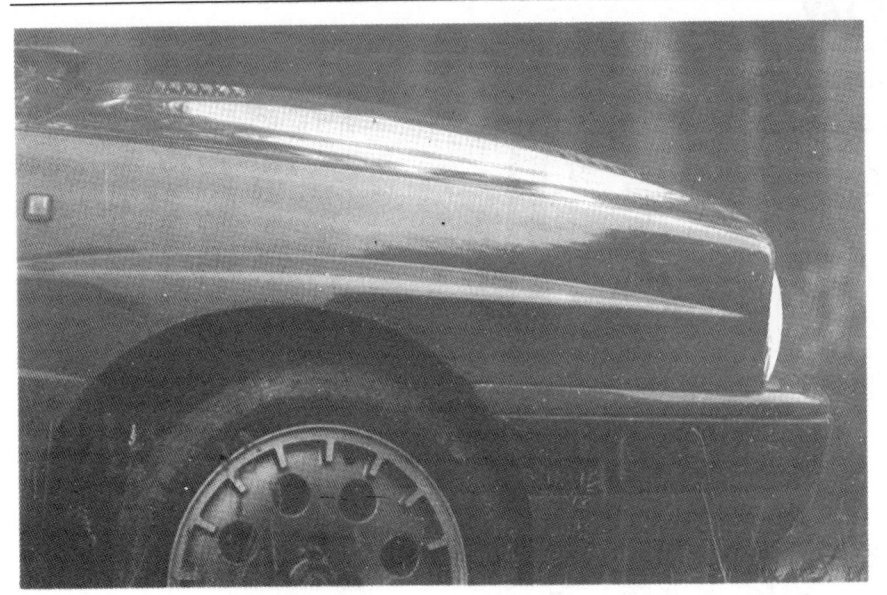

Bonnet bulge, demanded by bigger engine, distinguishes new car. With 53 per cent drive to rear wheels, 16-valve model handles more neutrally

TEST EXTRA
Lancia Delta Integrale 16v

With 16 valves and tauter suspension, the Integrale now delivers more . . . but demands greater commitment

Price *N/A* **Top speed** *129mph* **0-60** *6.3sec* **MPG** *19.1*
For *Performance, handling, steering response*
Against *Left-hand drive only*

NOTHING STANDS STILL IN MOTOR racing for long. Lancia might be sweeping all before it in World Championship rallying, but the Japanese were snapping at the heels of the 4wd Delta Intergrale almost as soon as it was raced. Now, a year later, Lancia has responded to the challenge — a more powerful, 16-valve Integrale will hit the showrooms in August for around £20,000.

The £16,995 8-valve Integrale was always much more than a 5000-off homologation special, as the 7600 people who bought one during 1988 will testify. With the performance to leave any hot hatch for dead, the Lancia's grip and poise on treacherous surfaces made it damn-nigh unbeatable — truly a tough act to follow. On paper the new car is improved in a dozen different ways, but does it add up to such a finely judged package?

Central to the new Integrale is the adoption of four valves per cylinder, the extra bulk of the cylinder head siring the bonnet bulge that provides the main visual clue to the new car. The new powerplant is based on the 16v Thema unit, its twin overhead camshafts opening the eight inlet and eight exhaust valves directly through bucket tappets. Sharing the same 84 × 90mm bore and stroke iron block, the 16v unit gets new, stronger con rods, sodium filled valves and a new version of Garrett's T3 turbocharger with a smaller turbine for better response. The intercooler is no bigger but cools the intake charge more efficiently, while the engine sticks with a compression ratio of 8.0:1.

The new car uses Weber-Marelli mapped control of the ignition, fuel injection and turbocharger wastegate as before, but in road trim, boost now reaches 17psi rather than 14psi. Bigger injectors will allow Lancia to squeeze more power out of the car in rally trim, rather than helping on the road. The bottom line is now 200bhp at 5500rpm, compared with 185bhp at 5300rpm before. The 220lb ft peak torque is only 4lb ft down on the 8v motor but arrives at 3000rpm, a full 500rpm lower.

Shepherding that extra power to the wheels is a beefed-up ZF five-speed box, with a higher second gear closing the gap up to third, and a stronger clutch that gets hydraulic operation. As before, four-wheel drive is provided by an epicyclic centre diff with a viscous coupling lock and a Torsen unit at the rear.

But inside the centre diff lies the biggest single change to the character of the car. With no slip at the rear, 53 per cent of the drive torque is now directed to the rear wheels, compared with 44 per cent before. Lancia claims less understeer and a 'sportier' drive as a result.

Matching the power hike is a host of chassis changes. For a start, 205/50 VR 15 tyres on 7ins rims replace the old 195/55 rubber. Stiffer springs lower the car by three-quarters of an inch and are mated to uprated, progressive action dampers and a stiffer front anti-roll bar. The lower wishbones at the front are reinforced, but the most significant change is the option of anti-lock brakes. At the launch of the original Integrale, Lancia engineers claimed there was no room under the bonnet for anti-lock. Yet, despite the 16-valve engine crowding space still further, Lancia has managed to find room for the complex Bosch anti-lock brake system.

There are very considerable problems in engineering anti-lock braking into a permanent four-wheel-drive system that uses a viscous coupling across the centre differential. If both wheels at one end of the car start to lock, the vc will act to limit the speed difference across the centre diff — either speeding the locked wheels up again or locking up both ends of the car. Since the wheels are not acting independently, an anti-lock system can't act independently on each wheel. ▶

LANCIA DELTA INTEGRALE 16V

PERFORMANCE

MAXIMUM SPEEDS

Gear	mph	km/h	rpm
Top (Mean)	129	208	5550
(Best)	129	208	5550
4th	124	199	6500
3rd	92	149	6500
2nd	64	102	6500
1st	40	64	6500

Standing ¼-mile: 14.9secs, 89mph
Standing km: 28.2secs, 109mph

ACCELERATION FROM REST

True mph	Time (secs)	Speedo mph
30	2.3	32
40	3.5	41
50	4.7	52
60	6.3	63
70	8.7	76
80	11.2	86
90	15.6	91
100	20.5	106
110	28.8	118

ACCELERATION IN EACH GEAR

mph	Top	4th	3rd	2nd
10-30	—	12.5	8.4	4.9
20-40	15.6	9.9	6.1	3.3
30-50	13.4	7.7	4.4	2.6
40-60	11.3	5.8	3.8	2.9
50-70	8.7	5.5	3.9	—
60-80	8.3	5.9	4.6	—
70-90	10.2	6.5	6.1	—
80-100	12.4	8.2	—	—
90-110	16.6	11.8	—	—

FUEL CONSUMPTION

Overall mpg: 19.1 (14.8 litres/100km)
Touring mpg*: 27.7mpg (10.2 litres/100km)
Govt tests mpg: 25.2mpg (urban)
35.8mpg (steady 56mph)
26.9mpg (steady 75mph)
Grade of fuel: 4-star (97RM) or Eurosuper unleaded (95RM)
Tank capacity: 12.5 galls (57 litres)
Max range*: 346 miles
* Based on Government fuel economy figures: 50 per cent of urban cycle, 25 per cent each of 56/75mpg consumptions.

BRAKING

Fade (from 89mph in neutral)
Pedal load (lb) for 0.5g stops

start/end		start/end	
1	29-35	6	35-40
2	29-35	7	35-42
3	29-35	8	37-45
4	29-35	9	40-45
5	30-35	10	40-50

Response (from 30mph in neutral)

Load	g	Distance
10lb	0.22	136ft
20lb	0.37	81ft
30lb	0.55	54ft
50lb	0.68	44ft
70lb	0.85	35ft
90	1.00	30ft
Parking brake	0.33	91ft

WEIGHT

Kerb 2846lb/1292kg
Distribution % F/R 64/36
Test 3227lb/1465kg
Max payload 995lb/450kg
Max towing weight 2562lb/1200kg

TEST CONDITIONS

Wind 5mph
Temperature 15deg C (60deg F)
Barometer 1012mbar
Surface dry asphalt/concrete
Test distance 800 miles

Figures taken at 900 miles by our own staff at the Lotus Group proving ground, Millbrook.
All *Autocar* test results are subject to world copyright and may not be reproduced without the Editor's written permission.

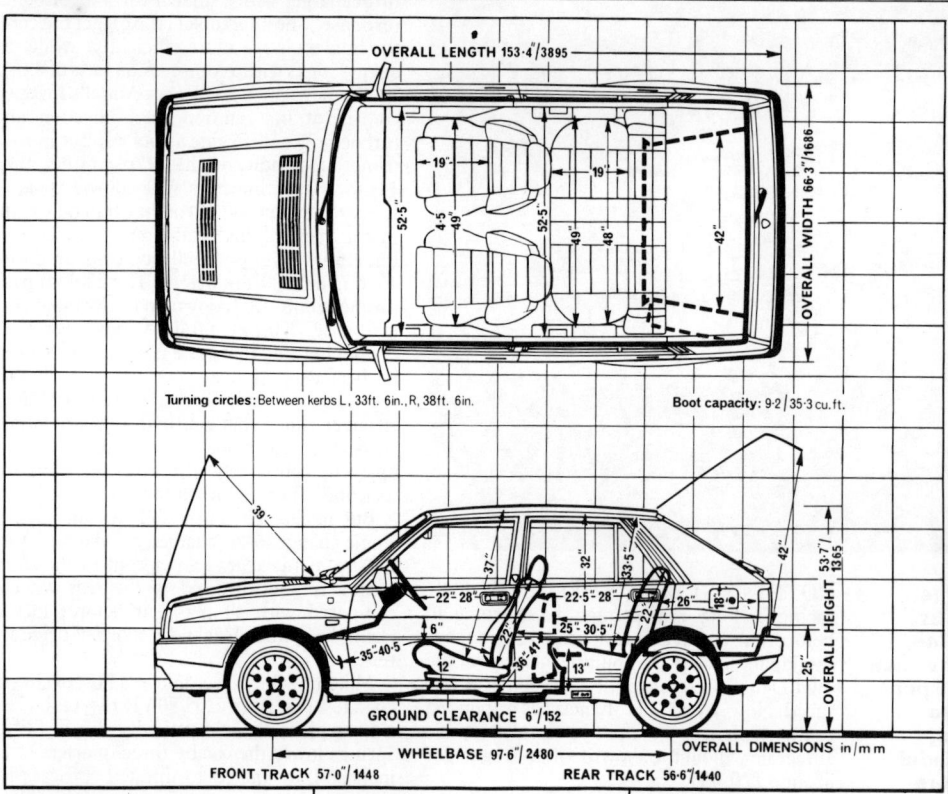

SPECIFICATION

ENGINE
Transverse, front, four-wheel drive.
Capacity 1995cc, 4 cylinders in line.
Bore 84.0mm, **stroke** 90mm.
Compression ratio 8.0 to 1.
Head/block al alloy/cast iron.
Valve gear belt-driven dohc, 4 valves per cylinder.
Ignition and fuel IAW electronic ignition and Multi-point fuel injection management system. Turbocharger.
Max power 200bhp (PS-DIN) (144kW ISO) at 5500rpm. **Max torque** 220lb ft (298 Nm) at 3000rpm.

TRANSMISSION
5-speed manual.

Gear	Ratio	mph/1000rpm
Top	0.929	23.1
4th	1.132	19.0
3rd	1.519	14.2
2nd	2.176	9.8
1st	3.500	6.1

Final drive ratio 3.111 to 1.
Epicyclic centre differential with viscous coupling, 47/53 per cent front/rear, Torsen rear differential.

SUSPENSION
Front, independent, struts, coil springs, lower wishbones, telescopic dampers, anti-roll bar.
Rear, independent, struts, coil springs, lower wishbones, telescopic dampers, anti-roll bar.

STEERING
Rack and pinion, power assisted, 2.8 turns lock to lock.

BRAKES
Front 11.2ins (284mm) dia ventilated discs.
Rear 8.9ins (227mm) dia discs.

WHEELS/TYRES
Cast alloy, 7ins rims. 205/50VR15 Pirelli P700Z tyres.

COSTS

Prices
Total (in UK)	N/A
Delivery, road tax, plates	N/A
On the road price	N/A
Options fitted to test car:	
ABS	N/A
Total as tested	N/A

SERVICE
Major service 12,000 miles — service time N/A hrs. Oil change 6000 miles — service time N/A hrs.

PARTS COST (inc VAT)
Oil filter	N/A
Air filter	N/A
Spark plugs (set)	N/A
Brake pads (2 wheels) front	N/A
Brake pads (2 wheels) rear	N/A
Exhaust complete	N/A
Tyre — each (typical)	£162.33
Windscreen	N/A
Headlamp unit	N/A
Front wing	N/A
Rear bumper	N/A

WARRANTY
12 months/unlimited mileage, 6 years anti-corrosion.

PRODUCED BY
Lancia
27 Via Lancia, 10141 Turin, Italy

Sold in UK by
Lancia
266 Bath Road, Slough
Berkshire SL2 4HJ

EQUIPMENT

Anti-lock brakes	TBA
Alloy wheels	●
Power assisted steering	●
Steering rake adjustment	●
Seat base adjustment	●
Split rear seat	●
Head restraints F	●
Electric mirror adjustment	●
Flick wipe	●
Revcounter	●
Lockable glovebox	●
Radio/cassette player	DO
2 4 6 speakers	DO
Electric windows F	●
Central locking	●
Tailgate wash/wipe	●
Fog lamps	●
Tinted glass	●
Sunroof	●

● Standard DO Dealer option

1 Speedometer, **2** Warning lights, **3** Boost gauge, **4** Rev counter, **5** Oil pressure, temperature, **6** Air vents, **7** Glove box, **8** Stereo, **9** Fog lamp, **10** Heater controls, **11** Wash/wipe, **12** Ignition, **13** Fuel gauge, **14** Indicators, **15** Lights

TEST EXTRA

One way round this is to make the viscous coupling so 'soft' that is doesn't come close to locking the centre diff in the first place. But on slippery surfaces, that means you don't get the full benefit of ABS. Ford, BMW and some other manufacturers take this option, but Lancia wanted no compromise in the Integrale. Developed jointly with Bosch, the Integrale's ABS uses four channels, one to each wheel rather than the more usual common channel to the rear wheels. As well as the individual sensor on each wheel to detect the onset of slip, two more sensors detect longitudinal acceleration, ie the degree of braking, and lateral acceleration, or cornering force.

When the Integrale is braking with different grip on each side, the ABS unit will pulse the brakes on both sides of the car to give more even braking and less yaw, but slightly longer stopping distances. In situations where there is less risk of the car swapping ends — if the sensors detect uniform grip or stable cornering — the unit reverts to pulsing only the wheels close to losing grip, giving the shortest possible stopping distance. Finally, the ABS unit is linked to the engine management system to speed up the idle under braking, preventing excessive engine braking from locking the rear wheels.

From the driving seat you can see the bulge on the bonnet, but everything else is familiar Integrale — well-shaped and supportive, but very firm, Recaro seats, plenty of legroom but a slightly long-arm driving position, and an excellent three-spoke leather-rimmed steering wheel. There's an illuminated check panel to the right of the main instrument panel that is missing on UK-spec 8-valve Integrales, and the carpet over the left wheel arch gets a rubber mat in line with its role as a left footrest. And, at last, you can read the rev counter — zero starts at 3 o'clock rather than 7 o'clock so the working quadrant is not obscured by the rim.

Accelerate away and there's that familiar whistle as the engine comes on boost. But the engine sounds crisper, more eager than before. It certainly boosts from lower revs, building up from 2000rpm to give a full 17psi by 2800rpm and easing back to around 14psi from 3500rpm to the 6500rpm redline, or even on to the 7000rpm rev limiter.

So there's more boost, at lower revs than before — but still as much lag. From a standstill, the tall first gear combines with the delay before the turbo spins up to speed to make it seem slightly sluggish which it isn't.

The 16v is a heavier car, of course, the new cylinder head and the ABS adding an extra 50lb to its 2846lb kerb weight, but the stronger mid-range urge makes the car more usable and marginally quicker at the lower end of the rev range in the gears. There's not much sign of the extra 15bhp at the top end, though. In fact, the Integrale 16v was just 1mph faster on top speed than the 128mph recorded by the 8v. To 60mph, the new car gained a tenth for a time of 6.3 secs, but its standing ¼ mile and kilometre times were fractionally slower.

It could be that the 8-valve car we tested last year was a particularly good example, although our own long-term example was only fractionally slower. There might have been more to come from the 16v with another few thousand miles on the clock but, although it didn't match Lancia's claimed 137mph and 5.7secs to 62mph, the Integrale still delivers very serious performance. And the 16-valve engine scores much better at the pumps, returning 19.1mpg

Integrale's bottom-line is up by 15bhp to 200bhp. 16-valver gets illuminated check panel and new revcounter

overall compared with the old car's 17.6mpg, with the extra benefit of being able to run unleaded without adjustment.

Use that slightly rubbery but quick and accurate gearchange to the full, helped by the perfectly spaced pedals, and there's very little that will stay with an Integrale whether 8-valve or 16-valve.

With considerably more roll stiffness and damping at the front, the Integrale's superb 2.8 turns lock to lock power steering is even more responsive. With the extra bite of 205 section Pirelli P700z tyres, rather than the old 195 Michelin MXVs, the Integrale 16v turns in almost before you've thought of moving the wheel. The trade-off is a slight loss of the old car's arrow-like straight line stability and a tendency to follow the camber under braking.

Once in the corner, the car's rear-biased torque split is glaringly obvious. Gone is the old car's determined understeer on the limit; with 53 per cent of the drive to the rear wheels, the 16-valve stays more neutral. A drifting front end can be eased back in line on the throttle, and even into oversteer if the corner is tight or the road slippery enough. Catching it with a twist of the steering and holding the Integrale in a satisfying four-wheel drift soon becomes an experience that you seek out.

The Pirelli tyres were a disappointment. They were prone to squeal very early in the dry, generating withering glances from passers-by, and although the car was better balanced, in absolute terms there was no more grip available. When the heavens opened, the P700s weren't a patch on the narrower, deep tread MXVs, and the Integrale 16v would enter a four-wheel drift long before the old car would push its nose terminally wide. The 16-valve car comes with both Pirelli P700z and Michelin MXX tyres as original equipment and, based on our experience of the MXX on several bigger, more powerful cars, it would be interesting to sample a set on the Lancia.

Braking was always a strong point of the Integrale; with the 4wd system helping to prevent an individual wheel locking, the old car recorded a full 1.0g deceleration in the wet, with all the power and progression at the pedal, you could want. With ABS, brake feel is slightly reduced and pedal travel increased, but otherwise you would never know it was there. Brake very hard on a slippery surface though, and a surprisingly low frequency pulsing from the pedal lets you know the ABS has cut in, the car coming quickly to a halt under complete control even if you hit the anchors in mid corner.

As you would expect, the stiffer suspension brings with it a coarser ride, still well-controlled and smooth at speed but noticeably harsher around town. There's a noticeable roar from the tyres on concrete surfaces now, with a cat's eye generating quite a thump and just as much wind noise at speed as there ever was from the 10-year old Delta bodyshell.

As a package, the Integrale 16v is more competition biased in almost every respect. It ought to be faster than the 8-valve car that sired it, but extra performance is not at the heart of the new car's appeal. Tauter suspension, better steering response, potentially more grip and very definitely less understeer on the limit bring all the usual trade-offs; a coarser ride, less stability and a slight chance of swapping ends.

Once used to left-hand drive, everyone became a good driver in the old Integrale — it delivered a level of performance, grip and handling out of all proportion to skill behind the wheel. With 16 valves and all the chassis mods that go with them, the Integrale gets serious, delivering more but demanding commitment. Work with it on a winding road, where every bend can conceal a different surface and the Integrale 16v can satisfy like nothing else.

★ *Car supplied by Autohaus, 331-333 Finchley Road, London NW3. Tel: 01-794 5656.*

EVOLVING INTEGRALE

Martin Holmes takes a look at the latest four-wheel drive Lancia Delta Integrale rally car

The long anticipated 16-valve version of the all-conquering Delta Integrale was announced in Italy amid a mixture of curiosity and controversy. International scepticism about the excessive power potential of Lancia's latest rally car, however, was countered by strong assurances from competition chief Claudio Lombardi. With many teams assuming the apparent unenforceability of FISA's 300bhp power limit implies official condonation for more powerful cars, the strength of Lombardi's justifications made the launch of Lancia's latest rally-winner something out of the ordinary. Lombardi affirmed "this year the new car will run at 295bhp. We haven't made tests yet, but we expect next year's 40mm restrictor rule will put us down to 260-270bhp. Physical laws stop us from breaking the 300bhp limit. The 16-valve car has a smaller turbocharger than the 8-valve and the limit of space under the bonnet means the position of the intercooler forces badly shaped curves in the induction pipes." These assurances are contradictory to mathematically projected power figures which suggest outputs of over 350bhp.

Present for the visiting journalists to see at the Mandria test track were not one but two rally versions of the new car, one in asphalt and the other in gravel road specification. Rally drivers Markku Alen and Dario Cerrato were on hand to demonstrate the gravel road version of the new Group A car, and both cars were resplendent in Martini's new red-based colours. It was impossible for passengers to gauge how powerful the car was, although the Finn commented that the new car in any case had much better torque characteristics than its 8-valve predecessor. He added it was too early in development to gain a definitive opinion of the changes of the new car. The 5,000th 16-valve car is to be built in July, so the model should be homologated on 1st August in both Groups A and N. Lombardi added "We don't yet know if we will run the car in the official team immediately but you will soon tell — the red colour is being saved for the 16-valver and the 8 valve cars will remain white!" A relatively insignificant power bulge on the bonnet and discreetly altered name badges are the only other exterior distinctions.

The engine apart, virtually all the improvements of the standard car (for instance the understeer-reducing rear-biased torque split) have already been used in Group A rally cars, but more non-engine development work is still anticipated. This includes (a) the Valeo pedal-less clutch ("Alen's recent clutch problem on the Costa Smeralda Rally was stupid, it was an electronic fault"), (b) direct ignition with separate coils for each cylinder, (c) work with Michelin to allow the ATS expanding foam tyre to be used at speeds of 160kph compared with the present limit of 200bhp, particularly with the Safari Rally in mind, and most importantly (d) a new system of central differential. "The Ferguson is too heat-dependent. We need a system in which the front-to-rear slip coefficient depends on electronically sensed requirements, and not on the temperature of the unit." Nobody said when the Group A car will be seen on World Championship rallies, and for a fact Lancia still don't need it to ward off rival teams. It is possible the new car will even be seen in privateers' hands in Group N form before the crimson racers are brought in to replace the old white cars. The extra power in Group N form plus the improved handling will go a long way to counter the superior power of the Ford Sierra RS Cosworth and the legendary handling characteristics of the Mazda 323 4WD. ●

Delta Force

Lancia's Delta Integrale has decimated its competition in the world rally championship. The new 16-valve road car is equally potent, with performance to match some exotics. **Jesse Crosse** drove one near Turin. **David Gooley** took the photos.

They do say love is better the second time around, but for Lancia's hottest ever road–going property, third time round is definitely a charm. With an indiscreet hood bulge, boxy wheelarches, and fat tires, the four–wheel drive Lancia Delta Integrale 16V is one of the most exciting production cars to come from a European manufacturer in years.

Integrale is the name Lancia gave to a revised Delta HF 4WD once they realized just what a hot little number they'd created. The HF 4WD happened, it seems, almost by accident. Announced at the 1986 Turin show, when everyone's mind was focused on the Thema Ferrari 8.32 (SCI, August 1988), the new HF was powered by the Thema Turbo's two liter four, which developed 165 bhp and 188 lbs. ft. of torque. Coupled to a four–wheel drive system based on that of the Group B Delta S4 with a central epicyclic differential, plus a Torsen torque-sensing rear differential, the driveline was enough to give the Delta spectacular performance, enough to take on anything in its class. The Delta's chassis was always good, with neutral, predictable handling that made it easy to drive flat out. It's no surprise that the Delta HF 4WD cleaned up in 1987, winning both the manufacturers and drivers world rally championships.

Yes, the Delta was definitely a star. Only one thing was wrong: its looks. As often happens with unexpected successes, engineers and marketing men were caught flat-footed by their little car's popularity. Nobody had bothered with the exterior and the only distinguishing feature was a set of big 185 Michelins mounted on standard Delta rims. Because of that, the world remained surprisingly underwhelmed by the Delta.

Enter the first Integrale. It's amazing what a set of 195/55VR-15 Michelin MXVs will do for a car, fitted neatly inside some purposeful arches. Oh, and let's not forget the extra power, up to 185 bhp at 5300 rpm and 224 lbs. ft. at 3500 rpm. Brakes were improved as well, taken from the 8.32; they provided better pedal feel as well as more power. The Integrale, which set Europe alight with its dazzling performance and low cost, was a homologation special. Lancia had to build at least 5,000 (though they built more) to use it in International Group A competition. By now, Lancia had finally realized just what they'd created, and at last decided to go for broke.

No Quarter

The new Delta Integrale 16V takes no prisoners — none at all. With 200 bhp at 5500 rpm and 224 lbs. ft. of torque at 3000, the Integrale now has a power-to-weight ratio of 163 bhp per ton. And although it's only six bhp better than the eight-valve Integrale, maximum torque peaks 500 rpm lower and there's 210 lbs. ft. available between 2500 rpm and 4800 rpm, giving the 16V the extra power edge it needed.

Sitting a full 20 mm lower, with more negative camber on the front wheels and that big hood bulge to make room for the 16-valve cylinder head, the Integrale looks mean where it just looked good before. The basic chassis remains unchanged with four MacPherson struts, lower wishbones, gas shocks, and anti-roll bars front and rear. The Michelins add to the effect, and there's no messing around with skinny wheel rims this time: a 7 x 15 inch alloy hunkers under each fender blister.

Slip into the driver's seat and the car seems generally more muscular than before. Your pulse starts to tick just that little bit faster before you even turn the key; you just know you're going to be in for a treat.

When you do, the effect is surprisingly refined. The engine note is low and promises power, but the sensation is smooth and mellow. The difference is that this one can be revved to 7000, though it's happier making do with 6500 rpm; after that, the power trails off. There are no histrionics when you push the Integrale, and there never have been. The extra power is no match for the fantastic grip offered by four scratching tires.

So opening up the throttle and keeping it there on the twisting, Italian roads near Turin airport sparks no wheel-spinning reprisals from the Delta. It simply sits down a little and gets on with the job. Forty-five mph will arrive rapidly, a quick shift through the rubbery gate of the 5-speed gearbox will take you to 75, third to 105, and fourth to 130 mph. Fifth, as we discovered on the Autostrada, returns an impressive 137 mph — not bad for something with the frontal area of a truck. Lancia claims a 0 to 60 time of 5.7 seconds, putting the Integrale 16-valve right up there with the big boys.

In town, like most modern sports cars, the Integrale is good as gold. Crudely floor the throttle at low revs and you'll find plenty of turbo lag — about half a second's worth — followed by relentless, slingshot acceleration. At sensible rpm there's very little lag, something that's encouraged by a Garrett T3 turbo unit fitted specially with a smaller, lighter rotor producing lower inertia.

Regardless of all that power, handling and road-holding are the Integrale's strongest character traits. Bruno Cena is the charming head of Lancia's experimental department, and he explained just what the differences are between this and the earlier cars. Most significant is the change in torque split between front and rear wheels. The old car was 56/44 percent, front and rear. The new one is 47/53. To modify torque split is not unusual, but to change it completely from front to rear bias is radical.

When using high power in the low gears, says Bruno Cena, the eight-valve Integrale came close to exceeding acceptable limits of understeer in slower corners. Obviously, the extra power of the 16-valve made this problem worse. Cena first investigated other possibilities, like suspension and tire tuning, but in the end it was necessary to alter the torque split drastically.

Pour It In

You might expect that the 16V is a gentle oversteerer, but nothing could be further from the truth. In fact, the characteristics are just about identical to the original. The Integrale is one of those cars you can just pour into a corner and not worry too much about the result: overdo it and the car drifts wide, understeering just a little. Use more power from the apex and it will understeer more, but lift off and it simply tightens line. Cena is right. In slow corners, the power-on understeer is still a little more marked than some would like. The works rally drivers don't mind, though: they simply sling the car sideways before they get to the corner. But for mere mortals, that understeer means safety. Whatever mess you get in, the Integrale will most likely save your bacon. And in very high speed slides, power understeer will be far less marked. It might have been fun to try it with just a little more torque to the rear.

The steering is hard to fault. It has the same 2.8 turns lock-to-lock and despite the power assistance there's plenty of feel at all speeds. The ride is just about right: firm, but surprisingly supple. It has good roll stiffness, but the ride won't loosen your teeth.

Back under the hood, the Integrale's power unit underwent every bit as much careful development as the chassis. Though the engine's basic specification is that of the 16-valve Thema, quite a few changes have been made to find that extra power and torque. The turbocharger has a lighter rotor than the standard version. But more significantly, the ignition and injection systems are different. While the Thema uses Bosch LE Jetronic injection and Marelli Microplex ignition, the Integrale has an IAW–Weber combined ignition/injection system.

Angelo Venturello, head of engine development on the Delta range, admits that they wanted the IAW–Weber set-up on the Thema, but it wasn't ready in time. There are fundamental differences in the way the two systems operate. The Thema turbo engines have always had an overboost capability. So when you open the throttle, there's an instantaneous increase exceeding the normal peak figure. The system is activated by two pneumatic pressure valves. Number one valve trips when the throttle is 60 percent open, and number two when it's fully open. There's no distributor, and the Marelli Microplex ignition system drives two coils which provide direct ignition. The Integrale eight–valve has the same set-up.

The new 16–valve engine is completely different. To begin with, there's a distributor and no overboost capability. There's a valve that acts proportionally, and instead of boost being cut off at a predetermined level at full throttle, it's monitored and continually adjusted according to load, rpm, and manifold pressure. The maximum is 1.2 bar.

Accompanying the revised input side of the engine is an all–new manifold with identical-length intake runners, a bigger intercooler, and a progressive throttle mechanism providing lower gearing and, therefore, better control in the early stages of opening. There are nimonic-tipped, sodium-cooled exhaust valves; and the cast iron block, while retaining its two counter-rotating balancer shafts, has a sturdier crank and much stouter connecting rods.

The IAW system and the ABS braking system are linked, too. So when you lift off to brake, the engine fast–idles for a split second to avoid the instability caused by sudden deceleration of the rear wheels. The ABS system has longitudinal and transverse accelerometers to detect yawing caused by braking hard on split–co-efficient surfaces; it adjusts the braking force on each side of the car accordingly. It can mean that the stopping distances are slightly longer in those instances, but it does produce considerably more stable handling under braking than might normally be expected. The brakes themselves are excellent. The pedal is firm with little travel, and the stopping power won't leave you with any doubts or fears.

Workplace

The Integrale 16V's office is much the same as it was. The original front–wheel drive Delta turbos had unpleasant looking barrel dials with rotating tumblers as the indicators for the secondary instrumentation. All that disappeared in favor of conventional gauges with the first HF 4WD. If you haven't tried a Delta on a trip to Europe, then get ready for the lurid, yellow lettering that leaps at you from the instrument binnacle. The steering wheel is a slotted three-spoke and the seats are by Recaro — again, brightly colored but very supportive. On the floor there's a competition style drilled metal throttle pedal, and the 8.32 brakes now make it far easier to heel–and–toe shift than in the original HF 4WD. In a nutshell, you either like the bright Italian interior or you don't. But you'd have to be easily offended if you didn't; the extensive use of Alcantara trim on the seats and doors makes it pretty sharp.

Now for the bad news. Lancia of course withdrew from the U.S. market many years ago and does not officially import cars here. Worse yet, the 16–valve version won't meet U.S. standards, but a U.S. emissions-spec 177 bhp eight–valve car using the 16V's chassis is available for those wanting to personally import a car. This Integrale is exported to Switzerland, which has emissions laws nearly identical to U.S. laws.

Homologation of the 16–valve car was completed at the end of July this year — in time for the Argentine rally, but five of the 60 per day being produced were equipped with the eight-valve engine and destined for Austria and Switzerland.

Price of the new 16V car in Italy will be around $26,900, and Lancia plans a total production run of 10,000 to 11,000 per year. The U.S. legal car with ABS should sell for just under $26,000 in Switzerland. For the price, it's a substantial chunk of automobile, a car that has been snapped up at every opportunity in most parts of Europe. And for good reason. **SCI**

Vehicle............Lancia Delta Integrale 16V
GENERAL DATA
Vehicle Type...front engine, four-wheel drive, four passenger, four door hatchback
List Price.........................$26,000 (est.)
Body/Chassis.......unitized steel construction
Fuel Economy..................25 city/36 hwy
ENGINE
Type...dohc, 16-valve, transversely mounted, inline four
Displacement..........................1995 cc
Bore/Stroke.......................84 x 90 mm
Horsepower...............200 bhp @ 5500 rpm
Torque...................224 lbs. ft. @ 3000 rpm
Compression..............................8:1
Fuel System.....IAW-Weber injection/ignition
Fuel Required...leaded or unleaded, premium
TRANSMISSION
Type...5-speed manual; viscous center diff; 47/53 torque split; torsen rear diff
Gear Ratios...1st: 3.50; 2nd: 2.176; 3rd: 1.519; 4th: 1.132; 5th: 0.929
Final Drive Ratio........................3.111
DIMENSIONS AND CAPACITIES
Wheelbase...........................97.54 in.
Length.............................153.5 in.
Width...............................66.4 in.
Height..............................53.7 in.
Curb Weight........................2750 lbs.
Fuel Capacity..........................15 gal.
Suspension...(F): MacPherson struts, coil springs, gas shocks (R): Chapman struts, coil springs, gas shocks
Steering Type...rack and pinion, power assisted
Brakes...(F): 11.1 in. vented discs w/ABS (R): 8.9 in. discs w/ABS
Wheels...........................7J x 15 alloy
Tires............................205/50VR-15
PERFORMANCE
0-60................................5.7 sec.
Top Speed..........................137 mph

The choice: Audi v Lancia v Mitsubishi. The driver: Pentti Airikkala

When the Audi quattro first appeared in Britain back in 1980, it created a stir. Its strikingly muscular looks and turbocharged five-cylinder engine were enough to set it apart from the automotive crowd.

Even more fascinating, though, was the car's permanently engaged four-wheel drive system, Audi engineers arguing that the safest and most efficient way of harnessing the car's considerable performance potential was by feeding the power through all four wheels rather than just the front or rear pair.

Audi also boldly predicted that the benefits of four-wheel drive – markedly improved stability, traction and braking compared with an equivalent two-wheel drive car – were such that it would become the norm for high performance cars of the future.

That statement caused more than a few raised eyebrows and the feeling of a number of motor industry pundits was that four-wheel drive was likely to be nothing more than a passing fad. The first sign that perhaps the Audi men were onto something with all-wheel drive came when the quattro entered the motorsport arena.

It quickly proved to be a devastatingly effective rally car in the hands of such drivers as Hannu Mikkola, Walter Röhrl and Stig Blomqvist and swept all before it until the arrival of the shortlived Group B supercar era. Later variants have also enjoyed successful circuit racing careers, specially prepared 200 and 90 quattro models racing with great success in the American Trans-Am and IMSA GTO championships in 1988 and 1989.

More importantly, though, the quattro and subsequent four-wheel drive models from Audi sold in reasonably healthy numbers. Others began to jump on the bandwagon, and now virtually all the world's major car manufacturers boast four-wheel drive vehicles as part of their ranges.

In fact, the concept is now so successful in Europe that between 1979 and 1988, the overall four-wheel drive market (admittedly this figure includes off-road vehicles as well as cars) rose from 65,000 to more than 350,000 units – an increase of more than 500%. It's clear four-wheel drive is here to stay, and that has meant the technology involved has moved ahead at a dizzy pace.

The three cars which Pentti assessed for **CCC** all vary in character, each manufacturer choosing a slightly different approach to the problem of producing an effective four-wheel drive system for the modern passenger car. Significantly, all three have proven to be winning rally cars. As mentioned, the Audi quattro started it all, and it therefore seemed fitting to include the latest version, now fitted with a 220bhp, 20-valve version of the original turbocharged, five-cylinder engine.

No test of road-going four-wheel drive cars would be complete without the inclusion of the hottest hot rod of them all, Lancia's Delta HF Integrale 16-valve. In Group A form, it's provided Juha Kankkunen and Miki Biasion with the weaponry necessary to cart off the drivers and makes titles of the World Rally Championship for the past three years.

Just to add a little spice, we also included Mitsubishi's high-tech 4WD/4WS Galant. The turbocharged VR-4 model with which Pentti won the Group N class of last year's British Open Championship and then, in Group A trim, took him to a nail-biting victory in last November's Lombard RAC Rally isn't yet available in Britain, and until it is, this non-turbocharged version of the four-wheel drive, four-wheel steering Galant was as close as we could get.

There were no worries, though, about this normally aspirated – it's rated at 142bhp at 6500rpm – Galant being shortchanged when compared with its more powerful, turbocharged opponents. It's fair to say Pentti knows the VR-4 intimately after hurling it round the rally stages of England, Wales, Scotland and Ireland over the past 12 months and he's also clocked up a useful amount of mileage with the road-going version of the car.

Just by way of a quick crib, here's a brief description of each of the four-wheel drive systems.

Audi quattro 20V: permanent four-wheel drive with open front differential, self-locking Torsen centre differential and driver-controlled lockable rear differential.

Lancia Delta Integrale 16V: permanent four-wheel drive with open front differential, epicyclic centre differential incorporating Ferguson limited-slip viscous coupling and Torsen rear differential.

Mitsubishi Galant 4WD/4WS: permanent four-wheel drive with open front differential, combined bevel gear centre differential and viscous coupling unit, open rear differential.

4 BY ONE

Take one RAC Rally winner and a group of four-wheel drive road cars that in competition form have all won Britain's World Championship event and you have cause for discussion. Pentti Airikkala joined CCC for some cool high speed driving courtesy of Audi, Mitsubishi and Lancia

Report: Graham Jones

Photography: Norman Hodson

The recipe was comparatively simple: take one Lombard RAC Rally winner, mix with road-going versions of three of the most successful four-wheel drive rally cars in recent times and see what happens.

For our RAC winner, there could be but one choice – Pentti Airikkala. Last November the British domiciled Finn, co-driver Ronan McNamee and their Mitsubishi Galant VR-4 won one of the closest fights in the history of the event. Their eventual margin of victory over the Toyota of Spaniard Carlos Sainz was 1 minute, 28 seconds. In the process Mitsubishi became the first Japanese car manufacturer to win the RAC Rally in its 57-year history.

In his surge to the head of the leader board, Pentti recorded no fewer than 22 fastest stage times, nine of them consecutively on the Wednesday as he began to reel in the leaders.

Those who have watched him on treacherously slippery stages in the past will attest to his uncanny car control, but what may not be quite so well known is that Pentti is also a real thinker and a technician. It's significant, for instance, that his unremitting push to the front on the RAC came after he had requested the rear ride height of his car be dropped to improve traction and that slight adjustments be made to the suspension.

"The Mitsubishi Ralliart team was magnificent when we were setting up the car," he explained afterwards. "They listened to my comments and it turned out to be just as I like it. That's only the second time in my life I've known that."

Pentti's mechanical interests are not limited to rally cars, however, he's also a keen observer of the road car scene and has been known to be quite outspoken on the subject. It was with this in mind that we asked him to assess the road-going versions of three cars which have proven to be ultra-successful rally machines: the Audi quattro 20V, Lancia Delta Integrale 16V and, in non-turbocharged form, the Mitsubishi Galant 4WD/4WS.

All these cars use permanently engaged four-wheel drive systems, but that's where the similarity ends. The quattro is a coupe design featuring a longitudinally mounted, turbocharged, five-cylinder engine mounted ahead of the front axle, the Lancia started life as a competent little front-wheel drive hatchback and has evolved into a mini-supercar with 200bhp on tap from its turbocharged four-cylinder power unit, and the Mitsubishi is an inoffensive-looking, four-door saloon which is a technological tour-de-force with four-wheel drive, four-wheel steering, four-wheel independent suspension and four-wheel anti-lock braking. When in normally aspirated form the Galant's four-cylinder, twin-cam, 2-litre engine produces a healthy 142bhp.

Once we arrived at our test venue – a disused airfield near Thame in Oxfordshire – Pentti outlined what he had in mind for our three cars.

"First we will have a slalom course," he explained while striding along purposefully and setting down marker cones every 25 paces or so. "This is to tell us what the car does in, say, a motorway lane change situation. We can also learn about the steering and how good it is.

"All these cars have power-assisted steering and some of these systems are not that clever. They can suffer from rack catch-up, which means the power steering pump can't cope with quick corrections by the driver and the steering becomes very heavy. We can also see how quickly the car changes direction.

"There will be a long corner which tightens. This will make the cars understeer and then we can assess how badly they understeer and what can be done about it. We can also learn about brake balance in this sort of corner – are the front or rear wheels going to lock, or all four wheels, and what happens when they do lock? In addition, weight transfer and suspension behaviour will come into play here.

"We'll probably have a 'T' junction where it's possible to use the pendulum effect. This isn't something you can use in normal road driving, but it's quite educational all the same. And then perhaps we'll have one long corner where we can just play around generally," he concluded with a wink and a wicked grin.

The first car onto our temporary test track was the Mitsubishi. As on the open road, it felt decidedly undramatic as Pentti hustled it round the cones, stabbed the brakes repeatedly and threw it into muddy four-wheel slides. It handled the abuse with aplomb but certainly didn't seem like the basis for a rally-winning car. It simply felt like a solid, well-sorted and softly suspended family saloon. The impression was deceiving and serves to underline the great strength of the Mitsubishi as a rally car.

"The two main things about this car which need getting used to are the power-assisted steering and the four-wheel steering combined," commented Pentti as we paused momentarily. "At first, it feels like there is far too much power assistance. It's the same as on the rally car, but then that's also fitted with bigger tyres. One of the developments with four-wheel drive cars is that they use tyres which are the same width at the front as at the rear. With rear-wheel drive cars, you might use wider rears.

"In any event, with a rally car you need quite a lot of power assistance, because in emergency situations you have to manoeuvre very quickly and, of course, the steering always gets heavier if you do it quickly. A good level of assistance means not only can you be quicker with the steering but also for the rally car we can go to the maximum tyre size required without worrying about the steering effort required.

"Getting back to the road car, though, I feel quite comfortable in the Galant, it's a car you could live with. Not much effort is required when you're driving it, it's quiet, the heating and ventilation seem to work. It's a car in which you could drive a lot of miles and still have quite a relaxing drive and that, I think, is the clear advantage of the Mitsubishi.

"This car has done 12,000 miles and the basic suspension is still very good. Personally, if I were using this as my own road car, I wouldn't touch the suspension. It's as good as anything I can think of since it's a good compromise with a sporty feel, comfortable ride and it takes jumps remarkably well. Not all Japanese cars are like that, you know. The shock absorbers in Japanese cars are not always geared for fast driving, but the ones in the Galant are fine.

"Earlier today, on the way here, we went round a corner with a slight jump in the middle. This is something you can encounter in a normal road situation and the Galant behaved very nicely. I didn't need to do any corrections with the steering.

"It's well behaved and that shows in the rally car too. On the RAC, nearly everybody

"The nicest feature of the Audi is the engine. It's really impressive with excellent power from low revs, and turbocharger lag is a lot less than the Integrale. It's a fantastic engine – it's just a pity about the location": Airikkala on the Audi quattro 20V

"You can generate very high G-forces in the Mitsubishi and there's no doubt the four wheel steering helps. Some sort of four-wheel steering is needed for all road cars in my experience as well as rally cars. There's no doubt about that": Airikkala on the Mitsubishi Galant

"The Lancia turns into corners more crisply than the Mitsubishi and the Audi. It's also quicker to change direction but feels quite nervous as you exit the last third of a corner. You have to be prepared to correct it": Airikkala on the Lancia Delta 16V

seemed to comment that the Mitsubishi looked very stable and easy to drive. It was, and in a long rally like that, it gives you a good feeling that you don't have to fight the car all the time. And that comes back in normal driving again – you want to have a relaxing car which you don't need to work too hard with."

The big talking point of the Galant, of course, is its four-wheel steering system which is allied to four-wheel drive. Is it a gimmick, or does it work?

"Oh, it works," replied Pentti with conviction. "You can generate very high G-forces with both the road-going and rally versions of the Mitsubishi in cornering and there's no doubt the rear-wheel steering helps. Some sort of four-wheel steering is needed for all road cars in my experience as well as rally cars. There's no doubt about that.

"I don't think anybody has really done it so cleverly," he added with a chuckle, "but it could be done. It just seems that the rear wheels are not yet turning enough. Perhaps the manufacturers aren't yet so confident about their systems and they don't want too much rear-wheel steering effect in case it goes wrong and isn't too nice."

It would have been possible to go on about the Mitsubishi for the rest of the afternoon, but there were two other cars yet to sample. Next up was the Lancia, the diminutive red beast with its bonnet bulges, flared wheel arches and aggressive stance almost looked as though it was daring us to put it through its paces.

It was an impression that was reinforced the moment we headed out onto the course. The car fairly leaped onward as Pentti floored the throttle and it darted between the cones like an enthusiastic puppy.

Pentti then underlined the Lancia's ability to "turn on a coin", as he put it, by yanking the handbrake on to start the car spinning and then balancing it on the throttle as it swung dramatically round on its own axis. "The combination of a Torsen rear differential and viscous coupling in the centre allows you to do this," he explained. "It hasn't got any practical application, but it's great fun to play with.

"This feels so different from the Mitsubishi," he went on. "You can feel all the potholes so it's more like a racing car suspension. I know the Integrale very well because I use it at the Lancia performance driving school and in many ways it's a very enjoyable car to drive on the road. You never get tired of driving it, but if you use it as an everyday car and do a lot of miles, then you'd probably prefer the softer Mitsubishi type of solution."

We headed back down the course a final time, Pentti chucking the little red machine around with unabated enthusiasm. "You know, one of the problems with this latest 16-valve version of the Integrale is the 7in wide wheels which are now fitted with 205-section tyres. They were 6in wide before with 195 tyres and it looks as though when they went to the wider wheel and tyre combination they didn't do anything else with the suspension geometry or steering. It means the car now tends to wander a bit. It's not bad, but it's just a little bit more than it did with the narrower rims on the eight-valve car.

"I think the car looks better with the wider rims, though, so my solution, if you're to own a car like this, would be to run the narrower 195 tyres on the 7in wheels and then you get the best of both worlds – better handling, because it doesn't wander so much, and the improved looks of the wider wheels. You have to change the tyres sooner or later anyway, of course, because this sort of car eats tyres. This type of tyre is not for economy anyway."

Pentti then confirmed our impression that the Integrale has a very "darty" nature. "Oh yes, the Lancia turns into corners more crisply than the Mitsubishi and the Audi. It's also quicker to change direction but feels quite nervous as you exit perhaps the last third of a corner. It's much harder work around this course than the Mitsubishi. You have to be prepared to correct the car and steer it where you want it to go, whereas the Galant remains smooth right through the corner and doesn't ever feel nervous.

His final observation on the Lancia concerned the optional anti-lock brake system with which our car was fitted. "The Integrale is one of the best stoppers in the business and it's hard to understand why it needs anti-lock brakes. I am all for anti-lock brake systems – especially if they are good anti-lock systems – particularly on two-wheel drive cars," he observed.

"Normally, though, I don't think four-wheel drive cars need it because the four-wheel drive system itself distributes the braking effort to all four wheels, or at least to the two axles. The anti-lock brake option isn't one I would choose to take up if I were buying an Integrale."

The last car of the day was the Audi, the car which started the four-wheel drive road car revolution 10 years ago. It was clear from the outset that despite recent technical updates, Audi has paid the price for being a groundbreaker in this area. Specifically, that heavy lump of an engine slung ahead of the front axle just increases the four-wheel drive coupe's tendency to understeer when power is applied.

"In a sense, the Audi is somewhere between the other two cars as far as the suspension is concerned," commented Pentti. "It offers a good mix of ride and sporty handling. It's just obvious with the weight of that engine up front, even on tarmac, that when you put the power on the car runs fairly wide. It feels a little bit clumsy when it comes to change of direction but, on the other hand, it feels a very secure car, so it has to be somewhere between the Lancia and the Mitsubishi.

"Without a doubt, the nicest feature of the Audi is the engine. It's really impressive with excellent power from low revs, and turbocharger lag is a lot less than with the Integrale. It's a really fantastic engine – it's just a pity about the location.

"The car seems to wander quite a bit on the road, possibly the result of those wide, 215-section tyres and wheels which appear to have quite a lot of outset. The Audi also has a small, thick-rimmed steering wheel which probably accentuates the impression.

"There are many lumps and bumps on the wheel which make it quite nice to hold when you are drving on a straight road, but not so good in an emergency situation when you need to manoeuvre quickly – the steering gets quite heavy and your fingers hit all those things on the rim of the steering wheel."

It was the Audi's braking system, though, which came in for the strongest criticism from Pentii.

"The other thing which comes through loud and clear about the Audi is that with the anti-lock system switched on, it's not particularly good in terms of how long it takes to stop the car," he said. "I don't think it's a particularly fast-reacting system, and although I have experienced slower sys-

4 BY ONE

4 BY ONE

tems, you would expect that if they put anti-lock brakes on this sort of car it would be a really good system.

"There is, of course, the option to switch the anti-lock system off, but then the problem is that although the car stops in a straight line, it doesn't want to steer at all. I've never come across such front-biased brakes in my life. The front wheels are locked but the rears aren't and the car just goes straight on regardless of what you do with the steering wheel.

"It seems to show that the Torsen centre differential doesn't work the other way around – it transmits torque but not braking action – and that's probably the reason they had to put anti-lock brakes into the system. If the Torsen did transmit braking action, then when the front wheels are locking, so are the rear wheels. I drove an early quattro without the Torsen diff – it was a locked centre diff, if I recall – and that had really good stopping with all four wheels locking at the same time.

"It has to make you wonder whether the Torsen is a good system. Perhaps the viscous centre differential system that you have in both the Mitsubishi and the Lancia is better since it still allows the car to turn into very tight corners without too much of that binding effect you can get from four-wheel drive systems. It also gives you the benefit of better braking and seems to have good enough traction."

Returning to our original question, how would Pentti rate each of the four-wheel drive systems?

"I must say that I prefer the viscous system," he replied. "I am just a bit suspicious that maybe the Torsen system is a little too 'free', and as a result the car doesn't feel quite so stable. In terms of rating the three systems here, you must remember that the Lancia differs from the Mitsubishi in that it also has a limited slip rear differential.

"When you put the power on with the Lancia it is a slightly oversteering car coming out of corners while the Mitsubishi understeers slightly. Both systems are good and I wouldn't really say that one is better than the other. As I mentioned, the Audi must be marked down for the problem with the brakes and the way it runs wide under power in corners.

And if Pentti had the opportunity to build a composite road car from the best parts of the three test cars, which parts would he choose?

"Perhaps what you would want is the Audi engine, the Mitsubishi suspension and four-wheel steering system, but with a little bit more feel to the steering than it has at the moment, and the Lancia's brakes without anti-lock.

"As far as a good basis for a rally car, I think it would need to be something between the Lancia and the Mitsubishi. As I've already said, the Mitsubishi is very stable and exceedingly fast through high-speed corners in particular, while the Lancia is very, very fast in all the twisty sections. It's nimble, changes direction quickly and stops rapidly, but it's very nervous at high speeds. A combination of the two would perhaps be somewhere close to the ideal rally car."

Pentti paused for a moment, and yet again that impish smile creased his features. "Now dare me to say which manufacturer has those things on its new car."

When it was suggested that such a car might just have a blue and white oval on the bonnet, Pentti simply winked and responded, "Ah, that would be telling." ■

ROAD CAR
BY JONATHAN GILL

LANCIA INTEGRALE

The latest Lancia Integrale sports changes galore under its skin. While the skin itself looks as if it has been pumping iron for some time...

Lancia beefs up the Integrale

The fourth version of the ultra-successful Lancia Delta rally car is on its way, and the road car definitely impresses

Whatever way you may care to judge success, the Lancia Delta homologation specials have been very, very successful in all three guises.

For starters, the HF 4WD, HF Integrale and HF Integrale 16v have notched up four consecutive World Rally Championships for Makes — in other words, a full house of titles since the current Group A regulations were introduced back in 1987.

As if that record isn't impressive enough, Juha Kankkunen's two recent wins in Finland and Australia (victories 39 and 40 for the Delta) have brought a fifth title within reach.

What's more, these very special Deltas have also chalked up amazing showroom sales figures. In all, more than 30,000 roadgoing homologation specials have been built and sold — that's more than double the number required by the FIA rule book and many more than rivals such as BMW and Toyota have managed. For the record, 17,670 old-style M3s came out of Bavaria, while total GT-Four production is 25,981.

Sadly, as a result of left-hand drive only production plus the British public's apathy to all things Lancia, only 493 homologation special Deltas have been sold this side of the Channel.

Much of the above can now be classified under 'history', though, because Lancia has just taken the wraps off a fourth and final generation HF Integrale. This, the ultimate Delta, will be available here (in left-hand drive guise only) around Christmas for a very reasonable price of £22,500.

All being well, the competition version should make its debut next January on the Monte Carlo Rally.

Two thousand of these fabulous new HFs have been built already and the company is confident of reaching the required 5000 mark before the 1992 season starts. Abarth — Lancia's rally team — has been working on the latest offering for some time. Indeed, most of the changes found on the newcomer come directly from the competition department.

The previous evolution concentrated most of the available time and money on replacing the original eight-valve engine with a more potent 16-valve alternative. This heart transplant operation went well and thus, this time, the power plant hasn't needed much attention. A few small tweaks to the management and exhaust systems have boosted power by 5% to 210bhp, but, basically speaking, there's nothing much to report on the turbocharged engine front.

Instead, Lancia has been beavering away on the HF's suspension. To improve stability, front and rear track has been increased by over 2in. Moreover, extensive alterations have been made to the way in which the car is suspended at both ends.

To the fore, you'll now find a new set-up featuring stronger struts, reinforced bushes, bigger diameter and longer travelling dampers plus larger and stiffer springs. There's also an aluminium strut brace linking the turret tops to increase rigidity.

The rear suspension has also been under the microscope. It now includes stronger struts and beefier transverse arms, longer travelling dampers and new geometry. Meanwhile, other very worthwhile revisions have been made to the braking and steering — the latter now features an oil cooling coil on its power assistance unit.

To house all of these mechanical alterations, the latest Delta wears an even more Schwarzeneggerish skin. Wheel arches and bonnet bulges have swelled to still more ridiculous proportions and a small adjustable roof spoiler has sprouted out of the top of the rear hatch. Extra cooling vents are much in evidence, too.

The result is a car which looks 10 times more butch than any *Terminator*, but which behaves with milder manners than a dead Dalek. Well almost...

The chassis changes have come up trumps: the Delta HF Mk4 drives with remarkable refinement and poise. While old versions handled pretty well, this finale is a real cracker. It steers and corners with stunning dexterity.

As before, levels of traction are remarkable and engine refinement should teach Messrs Cosworth a lesson or two. Suffice to say, performance is rarely less than mind-boggling.

One thing's for sure. If the next generation rally car shows similar dynamic improvements, people in rival camps are going to be dieting on dust and weeds next year... ∎

The Delta HF Mk4 steers and corners with stunning dexterity

Problems with fabled 'dubious Italian build quality' didn't materialise. Interior merely suffers from a worn floor mat under the throttle pedal. Alloys are easily bent

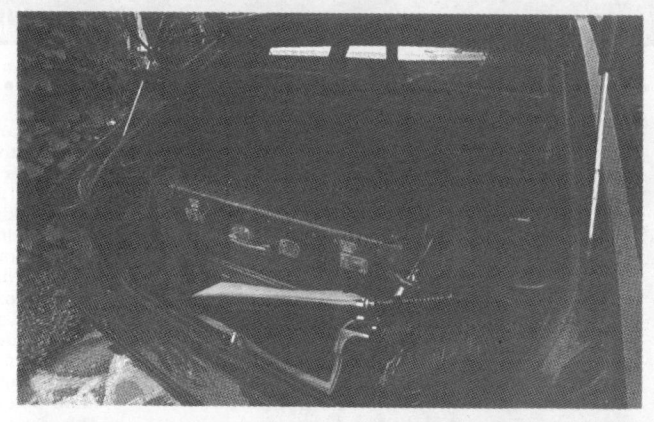

LONG TERM
Lancia Delta HF Integrale

50,000-MILE SPECIAL

They said it wouldn't last: being Italian, it would break. They said we were mad to take on an Integrale long term. But Howard Lees says they were wrong. Photography by Stan Papior

FUNNY HOW YOU FORGET THINGS. One wet, windswept night last month an old flame came to visit, one I hadn't seen for almost two years. She was to stay a couple of days and keep me company during a day-long trip to the Welsh mountains, and I went down the drive to welcome her with no less emotion than when we first met.

It was a clumsy reunion, spoilt by a *faux pas* of such gravity that I felt like calling the whole thing off. F88 JYJ is the left-hand-drive Integrale I fought so hard to get on to the *Autocar & Motor* long-term fleet almost three years ago, a car I thought I knew so well I could recite the owner's manual backwards — yet in an embarrassing attempt to move it into a secure parking spot, I found myself clambering into the passenger seat.

An hour before dawn the morning after our reunion, as I swept the condensation off the windows ready for a breakfast rendezvous in Wales, I sensed I was forgiven. This time I climbed in through the left-hand door and settled into that familiar, brightly upholstered Recaro seat, sliding it almost to the rear of its travel to make room for my legs.

Angling the backrest more vertical than I would have liked to reach the perfectly formed Momo steering wheel, I was also reminded that I could forget about seeing the top halves of either the speedo or revcounter. Some things, and Italian driving positions in particular, never change.

What I could see was the odometer — with over 89,000 kilometres (55,000 miles) showing, the Lancia had travelled the equivalent of one and a half times round the globe since I said farewell to it in January 1990 and in a month's time it would need its first MOT. Apart from a heel-sized hole worn in the protective floor mat under the throttle pedal, the interior looked as good as the day I picked it up.

The turbocharged two-litre twin-cam was a little slow to turn over on the key, a sign that after three years the Integrale's battery was a little weary, but the engine burbled into life after a couple of revolutions. Current custodian of the Integrale, Stan Papior reports that the battery was less successful recently after an extended spell in Gatwick's long-term car park.

Edging out on to the still-deserted M25, *en route* for the M40, a deeper note to the engine backs up Papior's suspicion that the rear silencer box is on its way out, but for now it just makes the Integrale sound a shade more purposeful. The ride is still surprisingly good at speed, there's nothing unpleasent to speak of in the way of tyre noise and the engine sounds smooth and refined. But just as I remembered from driving the very first UK press car back from Italy, at anything above 85mph the Lancia still sounds like it's driving along with its windows open a fraction.

Once past Oxford it's time to forsake the motorway for a favourite B-road through the Cotswolds. The skies darken to a threatening grey and it starts to rain, heavily and without let-up, until we reach the Brecon Beacons. But wet roads never used to worry the Integrale, and they still don't. The chassis feels as ▶

BUYING A USED INTEGRALE

SUCCEEDING THE LESS POWERFUL AND narrow-hipped Delta HF 4WD, the first Integrales arrived on these shores in the spring of 1988. There are a few of these early E-plate eight-valvers around, but most are F-registered. Quite a number were personal imports from the Continent, though there's no reason to fear one of these because the only difference to official UK imports is the lack of mph graduations painted on the speedometer face — the odometer always reads in kilometres, which confuses some people. Beware of right-hand-drive conversions — there are no factory-built right-hand-drive cars. Other weak spots to look out for are dented or buckled wheels, misaligned tracking (look for wear on the inside shoulders of the front tyres) and any signs of clutch slip — fitting a new one is a lengthy, expensive business.

The 16-valve car, with its humped bonnet and wider section tyres, arrived here in mid-1989. It has an extra 15bhp in theory, is firmer sprung, twitchier in the wet and has a peakier power delivery, but ultimately is faster on the road. Stronger front suspension means it is less prone to tracking and the associated tyre wear problems, and anti-lock brakes were introduced as an option. Prices have stayed firm over the past 12 months and you can expect to pay dealer prices of around £10,000 for a decent eight-valver or just over £15,000 for a 16-valver.

◀ taut as ever, and the steering still as sharp as a Silk Cut advertisement. With its 185bhp fed to all four wheels through a sophisticated four-wheel drive system, the Integrale's traction out of soaking wet bends remains outstanding. There's plenty of lateral grip, and it doesn't take much of a revision period before I feel confident to push the Lancia into predictable, controllable and exploitable four-wheel drifts.

Turbo eight-valve two-litre twin-cam: strong as ever

The return journey shows that the Lancia still delivers in the dry. There's still that electrifying surge as the engine comes on boost and the car seems as strong as ever. A set of test figures taken at Millbrook the next day show that 55,000 miles on the bores have done nothing to dull its performance. It equals its own previous top speed of 132mph, 3mph up on our original road test car, and the in-gear figures show that, if anything, the car is fractionally stronger at the top end now.

The Integrale foiled this dealer...

A 30-70mph time of 6.6secs is a couple of tenths faster than F88 JYJ's previous best — the fact that it is a shade slower off the line is due to the fact that the Integrale simply refuses to break traction in the dry (the previous figures were all taken in the wet) and thus can't keep its engine spinning in the meatiest part of its torque curve off the line. Only a rather sloppy gear linkage (which broke with just 6000 miles on the clock) gives any clue to the car's age. Fuel consumption has stayed remarkably consistent, the average having crept up fractionally to 22.6mpg with fewer high-speed Continental blasts in the past 12 months.

Keeping it in such demonstrably fine fettle hasn't been quite as straightforward, though it has fared far, far better than most people in the office predicted when the Integrale arrived. And in truth, at least as much aggravation is down to the quality of Lancia's dealer network as to the Lancia itself. The

LANCIA DELTA HF INTEGRALE

Mileage	55,775
Registration number	F88 JYJ
Date of first registration	December 1988
Date acquired	December 1988

WHAT IT HAS COST
Price new with/without options	£16,995/£16,995
Price now with/without options	£21,451/£21,451
Estimated resale values [1]	£6200/£8750
Depreciation in trade (after 34 months and 55,000 miles)	£10,795

FUEL/OIL
Cost for 55,000 miles
10,980 litres of four-star/super unleaded at 44p/litre average — £4831.20
10 litres of 15W50 grade oil at £4/litre average [2] — £40

TYRES
Type and size	Michelin MXV2 195/55VR15
Cost each	£115
Percentage worn front/rear (current tyres)	20/30 £46/£69
Sixth pair front, fourth pair rear	

FAULTS ON DELIVERY
Turbo boost gauge not working, front tracking not adjusted correctly, nearside door difficult to close

MAJOR SERVICES
Mileage	Date	Cost	Labour/hour
12,198	13.6.89	£253.00	£40
Ivor Hill, 9 Revelstoke Road, Wimbledon Park, SW18 5NJ			
24,472	12.2.90	£233.53	£42
Len Street, 67 Drayton Gardens, Chelsea, SW10 9QZ			
38,718	5.10.90	£170.20	£40
Len Street, 67 Drayton Gardens, Chelsea, SW10 9QZ			
50,772	20.7.91	£81.45	£39
Conduit Cars, 21 Conduit Place, W2 1HS			

REPAIRS/FAULTS
Description	Mileage	Cost
Front tracking adjusted	6000	£62.10
Gear linkage repaired	6000	warranty
Bonnet/wing repair — accident	15,000	£320.85
Windscreen — stone chip	21,000	£161.70
Set tracking	22,000	£118.00
Replace clutch	26,000	£651.05
New wheel bearing and cv gaiter	47,000	£210.44
Cam belt replaced	49,000	£296.28
Number of days off-road [3]		14
Number of breakdowns [4]		3

ANNUAL STANDING COSTS
Road tax (for 34 months)	£283
Insurance premium [5] (for 34 months)	£640/£1025

SPECIFICATION
ENGINE
Transverse, front, four-wheel drive
Capacity 1995cc, 4 cylinders in line
Bore 87mm, **stroke** 90mm
Compression ratio 8:1
Head/block light alloy/cast iron
Valve gear dohc, 2 valves per cylinder
Ignition and fuel Mapped IAW Weber electronic breakerless ignition and fuel injection. Garrett T3 turbocharger with air-to-air intercooler
Max power 185bhp (PS-DIN) (136kW ISO) at 5300rpm
Max torque 224lb ft (303Nm) at 3500rpm
Specific output 93bhp/litre
Power-to-weight ratio 148bhp/ton

TRANSMISSION Five-speed manual

Gear	Ratio	mph/1000rpm
Top	0.928	23.4
4th	1.133	19.2
3rd	1.518	14.3
2nd	2.235	9.7
1st	3.500	6.2

Final drive ratio 3.11:1, Torsen rear, Epicycle centre differential with viscous coupling lock

SUSPENSION
Front independent, MacPherson struts, lower wishbones, telescopic dampers, anti-roll bar
Rear independent, MacPherson struts, transverse links, longitudinal reaction bars, telescopic dampers, anti-roll bar

STEERING Rack and pinion with power assistance

BRAKES
Front 11.2ins (284mm) ventilated discs
Rear 8.9ins (227mm) discs

PERFORMANCE
LT is long-term car tested at 55,000 miles, **RT** is road test car tested (Autocar & Motor, 10 February 1988) at 6050 miles

MAXIMUM SPEEDS
Gear	mph LT	rpm LT	mph RT	rpm RT
Top (mean)	132	5650	128	5450
(best)	135	5800	130	5550
4th	124	6500	124	6500
3rd	93	6500	93	6500
2nd	63	6500	63	6500
1st	40	6500	40	6500

ACCELERATION FROM REST
True mph	Time (secs) LT	Time (secs) RT	Speedo mph RT
30	2.4	2.1	32
40	3.5	3.3	41
50	4.9	4.7	52
60	6.6	6.4	63
70	9.0	8.9	76
80	11.5	11.4	86
90	14.7	14.7	91
100	19.5	19.5	106
110	25.3	26.3	118

Standing quarter mile LT 15.2secs/92mph RT 14.8secs/90mph
Standing km LT 28.1secs/114mph RT 27.9secs/110mph
30-70mph thro' gears LT 6.6secs RT 6.8secs

ACCELERATION IN EACH GEAR
mph	Top LT	Top RT	4th LT	4th RT	3rd LT	3rd RT	2nd LT	2nd RT
10-30	—	—	—	—	8.2	8.1	4.6	4.5
20-40	14.1	14.3	10.3	10.2	6.5	5.9	3.5	3.1
30-50	12.4	13.2	8.6	8.4	5.1	4.5	2.9	2.5
40-60	11.1	11.4	6.9	6.6	4.1	3.8	3.1	2.9
50-70	9.1	8.7	5.7	5.5	4.1	3.7	4.4	—
60-80	7.4	7.3	5.7	5.6	4.8	4.5	—	—
70-90	7.8	8.2	6.5	6.6	6.0	6.0	—	—
80-100	9.6	9.9	7.8	7.8	9.2		—	—
90-110	11.8	12.9	10.1	11.3				
100-120	—	—	—	20.2	—	—	—	—

FUEL CONSUMPTION
Average LT mpg 22.6
Best/worst LT mpg 28.2/15.1
Average RT mpg 17.6

---

[1] First figure for current trade value, second for retail value. [2] Excludes oil changes at services. [3] Includes scheduled services and breakdowns, excludes accident repairs. [4] Based on any occasion when the car needed attention — professional or otherwise — before it could be driven, or when it stopped of its own accord. [5] Fully comprehensive insurance with no excess SDP use and insured as only named driver. Quotations for (1) 35-year-old married male with clean licence, five years' no-claims bonus and car garaged in Oxford (low risk), and (2) 25-year-old single male with clean licence, five years' no-claims bonus and car, not garaged, in Teddington, Middlesex (reasonably high risk). Source: Quotel Motor Insurance Service

... this one failed to get to grips with it

... and this dealer couldn't beat it...

... but Conduit Cars has proved Integrale's equal

Lancia has worked its way through four dealers during our ownership. Only one has managed to deal with its long-lasting problem: apart from routine, if never exactly cheap, servicing, the Integrale has demonstrated an unpredictably voracious appetite for front tyres, getting through three pairs in its first 22,000 miles alone. Tracking that seemed never to be set properly, or dealers who didn't know how to set it, seemed to be at the root of the problem initially, but recent work by Conduit Cars of London W2 pointed the finger at incorrect (and non-adjustable) camber settings. Elongating the mounting holes allowed Conduit to dial the geometry in to what it thought acceptable, but even now the tell-tale signs of severe wear on the inside shoulders of the fifth set of front Michelin MXV2 tyres hint at a problem not entirely solved.

Another problem not completely solved is the mystery of the missing power steering. Sometimes when worked hard during parking, the wheel starts to get very heavy — a change of fluid brought a temporary respite, but a new pump may be the only long-term solution. Other non-routine work has included a new clutch, new clutch cable (which a stranded Papior changed by the roadside in 45mins), a rear wheel bearing, worn-out switchgear restored by Papior with parts from a model shop, preventative maintenance in the form of a new cam belt, a split top hose and, during Michael Harvey's stewardship, new front seats to replace those ruined during an over-enthusiastic valet by a former Lancia dealership.

London's pockmarked roads have put paid to several expensive but easily bent alloy wheels, and stone chips to a windscreen and headlamp unit — the latter being a right-hand dipping item replaced by a UK spec one to give the Integrale cross-eyed night vision. A couple of (£70 each) door mirrors have fallen prey to London's traffic, and a mountain bicyclist dented the bonnet and front wing while the car was parked, resulting in a £320 bill and the only areas of paintwork on the Lancia which do not look as new.

Conceived as a rally car homologation special (and pretty nigh-unbeatable in that role — Lancia has won the 1991 manufacturers' world title, its fifth consecutive year) the Integrale in its various guises has always stacked up as a hell of a road car. Lancia has its latest evolution of the Integrale ready for the 1992 season, but to my mind an original eight-valve car like this is the sweetest of the lot.

That it can be just as sweet and every bit as potent after three hard years, three keepers and over 50,000 miles came, I admit, as a shock to me. With worries about the juxtaposition of Italian build quality and high mileages clearly unfounded, a secondhand Integrale starts to look a like a hell of a proposition. ∎

Steering is still as sharp as ever, the chassis stiff, with plenty of lateral grip. Handling and performance are not dulled by age

ROAD TEST

May 1986: we were privileged to be at the Sardinian launch of a motor car that changed the results sheets of World Championship rallying, and proved to be remarkably able and practical on the road.

At the time, Lancia only expected to make 3000 Delta 4x4s, but when news came through of the abolition of Group B as the formula for World Championship events in 1987, Lancia swiftly (in mid press conference!) announced production of the requisite 5000. Today's test Delta HF integrale is the fifth edition of a line that has sold over 30,500 units, and one which has racked up more than a quarter of all Delta sales in 1990.

The 1986 Lancia Delta HF 4WD had 'only' 165 bhp in road trim, but as Lancia's first 4x4 motor car it gathered prestige via domination of the World Rally Championship. It also perked up interest in the Delta range as a whole, selling 5298 copies in its own right.

Next came the extended wheel arch integrale; 'integrale' is the common Italian designation for 4x4 because Audi swiped, and in 1980 registered, the best Italian label: quattro. This Delta retained the eight-valve, dohc two-litre of '60s Fiat origins, albeit now revised from a Thema base to include twin counter-balancer shafts, turbocharging overboost and 185 bhp. It sold more strongly than ever, 9841 cars being manufactured between November 1987 and 1989's introduction of the 200 bhp/16-valve version, which preceded the test car. The eight-valve unit was cleansed with an exhaust catalyst and remained on sale into the '90s for particularly tricky markets, such as Switzerland (where it's known as the 'kat' variant).

It was the 16v version of the 1979 debutante (the Delta was elected Car of the Year in 1980) that sold best of all, hitting 12,860 examples from spring 1989 to its late 1991 replacement by the current HF integrale, our subject here.

The latter offers a number of vital technical changes which have proved enormously effective in 1992 World Championship Rallying. The (deep breath) Abarth-assembled, Jolly Club-run, Martini Racing Deltas won both opening rounds of the 1992 manufacturers' championship. Frequently we find that what works on the track, or special stage, is a noisy pain in the bottom for the public highway. Yet our 500 miles in the UK with one example and 150 in France were a delight. The Delta has its drawbacks, but you cannot buy more driving pleasure than the latest Delta offers *and* keep a steel roof over your head . . .

UK range

Even by the standards of these distressed times, Lancia sales are at a low ebb, but the Delta no longer has the job of pumping up the volume. Just 150 HF integrales, all in LHD with five-door bodies, are scheduled for the UK in 1992. No other Delta derivatives are currently listed. Post-budget, the HF has dropped £1001 – from £24,250 to £23,249. The only options are metallic paint at a sniff over £182 and the combination of black leather trim and air conditioning which demands £1576.92; neither was fitted to the test car. When we tested the Lancia, its principal British market opponents were the 220 bhp Ford Sierra RS Cosworth

Simply the best?

LANCIA DELTA HF INTEGRALE

four-door at £21,380 and the Toyota Celica GT 4x4 at £24,777, or the 205 bhp Carlos Sainz Limited Edition of the latter, of which only 440 were available. By the time this test reaches your gaze the 227 bhp Escort Cosworth will be more relevant opposition than the Sierra. We have borne in mind the driving characteristics of both RS Fords when writing this, the Sierra RS sharing garage space for a week with our test Lancia.

Technical analysis

The basics of the Delta remain those of 1979's five-door hatchback, one that was originally designed in the conventional transverse engine, front-drive fashion. Beneath those bluff lines, Lancia wrought a 4x4 conversion of exceptional worth that has kept much of the basic hardware in use since 1986. Heart of the system is an epicyclic gear central differential which has its action modified by a Ferguson patented viscous coupling. At the rear, Lancia had a look at what Audi was playing with in Group B competition and adopted the Torsen (torque sensing) limited slip differential, but whereas Audi finally employed the American planetary gear sets and star wheels to act as the central differential power split monitor, Lancia stayed with Ferguson and used the Torsen at the rear, where it remains to this day.

What *has* changed over the years is the basic front-to-rear power split deployed on the roadgoing Lancias. The original eight-valve machines, and today's eight-valve 'kat' emission special had a slight front-drive bias (56 per cent front, 44 per cent rear). The 16v version changed all that to 47/53, a split that is still employed in 1992. The 16v brought the option of Bosch ABS electronic anti-lock braking and this feature – which was adapted with great care to Lancia's requirements and works well on the loose – is standard for the latest HFs in Britain. The 1992 specification also covers enlarged disc brakes, Brembo aluminium twin piston calipers and an eight-inch servo replacing the previous seven-inch hardware.

The biggest engineering changes in the 1992 edition are the significant stretches in front and rear track (54/60 mm respectively), which also increases the overall girth of the flared wheel arch body. In fact these principle dimensions are now similar to more obvious supercars. The Delta is now just 3.6 in thinner than a Lotus Esprit!

The Lancia is startling to behold on the street – rather as if you were seeing it through a fairground mirror – but the fattening process has allowed the factory cars a tremendous reduction in special stage times. It has also significantly reduced cross country times for the production vehicle. (Who else but an Italian car company would specify a saving of four seconds per kilometre over twisty, wet tarmac in its introductory PR spiel? In the dry, the same ground should be covered 1.5 sec more quickly!)

Yet Lancia logic is impeccable. Who needs top speed today? In performance terms it is much more useful to be able to consume cluttered tarmac with phenomenal acceleration and consummate agility.

Accompanying the explosion in width are replacement front suspension components, lower wishbones now offering the kind of box-section construction that most manufacturers homologate in the more radical Group A cars. Lancia struts and bushes are also strengthened by unspecified means. Spring rates are up and the complete strut is attached to the body at a point about half an inch higher than before. Open the forward hinging bonnet and there is a beautifully crafted aluminium bar to stitch both front strut towers together, thus considerably enhancing both front end body strength and the accuracy of suspension geometry under extreme duress. There is more, particularly in the crafty use of anti-roll bar links and larger capacity dampers, but the message is the same. Lancia has uprated this machine in line with the lessons it has learned as the dominant force in World Championship rallying over the past five years.

The rear suspension work is also extensive, embracing replacement transverse arms, fatter struts, uprated springs and dampers (working over a longer travel), fresh geometry for the anti-roll bar and reinforced uprights.

Complementing the suspension moves are detail changes to the power steering pump and rack, an oil cooler added to the system that will be most valuable in Group N (production based) competition.

Wheel rim widths are up just half an inch, but a five-stud location is now necessary and the alloy wheels are of a totally new design, although they support the same tyre dimensions as before. A get-you-home spare is, unfortunately, also necessary.

As is the fashion for the latest evolution or limited edition homologation cars, a modest power bonus is offered. Lancia quotes 210 bhp instead of 200 (although the handbook resolutely quotes the old 16v figure), available some 250 rpm further up a scale that has a limit of 6200 continuous rpm. Maximum torque value remains the same, albeit a further 500 rpm onward.

The latest evolution of the Delta integrale is chunkier than its predecessors. A Lotus Esprit is now only 3.6 inches wider.

Lancia credits the extra power as being sourced via a 6 mm larger diameter exhaust system, one that ends in a single oval rather than the twin exhausts beloved by earlier edition Delta owners. Other key motor statistics, such as the two-litre capacity, 8:1 compression, Garrett T3 turbocharger and Langerer and Reion intercooling are quoted as before, along with one bar (14.2 psi) maximum boost. We found that the 1.2 bar overboost facility was still present, usually reporting at 3000 rpm and gradually absenting itself thereafter.

At maximum speed (shown as 148 mph rather than the honest 132 it was achieving!), boost had slipped back to 0.6 bar at a continuous 5700 rpm (the rev counter was accurate to within 100 rpm) while oil temperature was being maintained at 100 degC and water temperature at 90.

The Lancia remained notably stable at this speed, even though the three-position back wing was on the lowest of three spanner-adjustable settings.

The instrument binnacle remains a riot of colour, with more dials than British Telecom.

Action

Initial impressions of the 3300-mile red demonstrator are mixed. Despite the presence of a former BMW Motorsport manager at Lancia, quality seems to be as mixed as ever. The red paint looked wonderful, the HF galloping elephants emotive, but the door shut gaps were prodigious by current standards.

Unique external features include the wide use of Allen heads in the fuel filler surround and more intake slots (most netted) and vents than any other production car. In fact, the front is just one giant one-way system for cooling air to enter and exit. Any vacant space is set aside for effective quadruple Hella headlamps and under-bumper Carello auxiliaries.

Open the bonnet and there are beautiful castings, some Japanese-style technical boasts ("*Lancia Turbo 16-valve*" shouts the alloy rocker cover), but the rubber pipe feed between intercooler and Weber Marelli induction is perilously clamped by a jubilee clip, a worrying contrast to the purposeful braided lines linking the oil cooler into the Lancia's heartbeats.

The cockpit is still a mess of riotous colour schemes and masses of instruments. An octet of Veglia ('vaguely' seems more apt) dials sport yellow digits and needles on a black background. The cabin is still enlivened by the presence of drilled throttle pedal and droopy ventilation, but the grey roof lining was a comforting touch of class. Standard electrical equipment covers a drowsy steel sunroof panel, four side windows and Grundig stereo system. As a reminder of the late '70s and early '80s, a car check graphic was nostalgic, but not so useful as four-door central locking.

At the Momo three-spoke wheel (adjustable for rake), the Delta overcomes all prejudices about Italian driving positions and its 13 year-old outline. The Delta HF has the sheer ability to consume any winding road, on any surface that will take a modern motor car, more rapidly, and more satisfyingly, than any production car made today, with the possible exception of the Escort RS Cosworth. We say "possible" because our experience of the Escort is limited to overseas, and we know how deceptive first foreign impressions can be. But there is no doubt that Ford has made a major advance with the first production presence of genuine downforce aerodynamics in a production car.

Where the Lancia scores so heavily is in its innate communication skills with the driver. The rapid steering (just under three turns lock-to-lock) is perfectly weighted to inform without chattering and twitching about every dip and camber in the road. Dry road adhesion on P700Zs is so outstanding that the long standing deficiencies of the front seats in occupant location become a scandal. It would be a bold driver who overstepped the enormous limits supplied in dry conditions, but the Lancia proved worryingly less competent both in the wet and on a dry handling circuit. In the latter case we had two drivers check the Lancia's closed road competence and both found that the HF was not so keen to display its prowess in privacy as it was on the public road.

On the queen's highway it is an alert and responsive companion, second to none. Give it a closed track and a Group N Nissan for company and the driver has to battle with very heavy understeer to utilise all the available grip. Our experience with every edition of the Delta has always left us entirely satisfied with the slippery surface margins provided, but the latest HF was notably easy to slip out of line. In power-off situations you have a front-drive car in character; it reverts to a soggy understeer, which can build to such proportions that a definite change in plan is called for. Disappointing.

We have not driven the Escort RS in similar circumstances, so we do not known if they have managed to get their Pirellis to work in these conditions; we do know that the Sierra 4x4 on Bridgestone's ER90 is impressive on slippery surfaces, so Lancia could take a cold look at what Pirelli is providing. Or could such manners originate via significantly stretched front and tracks perhaps upsetting the basic balance of the Delta?

Complementing a chassis that absorbs bumps readily above town speeds (it is awesomely able above 50 mph) are a magnificent set of Bosch-monitored anti-lock disc brakes. The ride is not as amiable as before when below 35 mph but, considering the modest wheelbase, competition intent and 50 per cent aspect ratio tyres, it is perfectly passable.

In the Group A competition variant, extended wheel travel has apparently removed much of its previous skittish 'go-kart' jinks. World class drivers have all commented warmly on its ability to set faster times with less effort. At the test track the Delta did not quite match the factory claims of 137 mph maximum and 0-62 mph in 5.7 sec, but we were very pleased that its performance was at least the equal, or better, than had been recorded by other independents. A 0-60 mph time of less than 6 sec is still impressive to experience, and (allied to the Lancia's stubby outline) makes it an exceptionally wieldy overtaking device for British use. Those with a taste for figures may note that the Lancia is substantially faster than the 330 bhp Aston Martin Virage at sub-70 mph speeds. Lancia and Aston record much the same acceleration times in the 70-110 mph band, showing that the old Lancia body cannot overcome a substantial weight advantage at speeds beyond the British legal limit. The Delta is somewhat the ultimate 'speed limit special', although the aerodynamics are not so poor as you might suppose and it is only beyond 100 mph that sustained cruising becomes downright draughty. We also did rather better than others in the consumption of cheaper unleaded fuels, but it is worth cautioning owners that the mph and mileage recorders are amongst the most inaccurate we have tested in the last two years. A 19-22 mpg band is the true figure for

Under the skin: the best in homologation engineering.

Power is now up to 210 bhp, and torque to 224 lb ft at just 3,500 rpm,

a Delta utilising its considerable capabilities, not the 26-plus mpg indicated by an uncorrected mileage check.

Reliability, or lack of it, is always a key question asked of Lancia drivers. Aside from sundry squeaks and rattles (most from the old dashboard) our UK example had no operational problems in our custody. The French loaned Lancia was smoking heavily on its return, but was running well. We think it had the same sort of problems as the Cosworth Ford breed can get at the test track, oil failing to drain away from the top half of an engine under pressure and making a visual display that is alarming, but not life threatening to its mechanical health.

Verdict

Simply the best in homologation engineering, the dated Delta is the definitive example of a pedigree classic car that just happens to be in production.

The Delta HF integrale in its latest guise has obvious faults, many to do with its age. Yet no admirer of Italian sporting cars should be without one, and it could make converts of us all. **J W**

LANCIA DELTA HF INTEGRALE

ENGINE

Location	transversely front-mounted
Cylinders	four, in-line
Bore × stroke	84 × 90 mm
Capacity	1995 cc
Compression ratio	8 to 1
Valve gear	dohc, four valves per cylinder
Power	210 bhp/5750 rpm
Torque	224 lb ft/3500 rpm
Fuel	unleaded, 95RON

TRANSMISSION

Type	five-speed manual, four-wheel drive

GEARBOX

Gear	ratio	mph 1000 rpm
First	3.500:1	6.22
Second	2.176:1	9.99
Third	1.523:1	14.27
Fourth	1.156:1	18.83
Fifth	0.916:1	23.74
Final drive	3.111:1	

SUSPENSION

Front .... MacPherson struts, lower wishbones, gas-filled dampers, anti-roll bar
Rear .... struts with co-axial coil springs, gas shock absorbers, transverse and longitudinal links, anti-roll bar
Wheels .... aluminium alloy, 7.5Jx15
Tyres .... Pirelli P700Z, 205/50 ZR15

BRAKES

Front/Rear	ventilated discs/discs, ABS

STEERING

Type	rack and pinion, power assisted
Turns, lock to lock	2.8

DIMENSIONS

Wheelbase	2480 mm
Front/Rear track	1502/1500 mm
Overall length	3900 mm
Overall width	1770 mm
Overall height	1365 mm
Kerb weight	1300 kg
Fuel tank	12.54 gallons

PERFORMANCE

0-30 mph	1.89	0-80 mph	10.75
0-40 mph	2.93	0-90 mph	13.56
0-50 mph	4.56	0-100 mph	17.92
0-60 mph	5.91	0-110 mph	23.04
0-70 mph	8.57		

50-70 mph in fourth/fifth gears .... 4.90/7.64 sec
Maximum speed .... 131.77 mph

FUEL CONSUMPTION

Average for test	22.1 mpg
Government figures:	
Urban	25.2
56 mph	35.8
75 mph	26.9

LIST PRICE £23,248.86

NEW CARS TEST ■ Lancia Delta Integrale v Nissan Sunny GTi-R

Red-hot bombers: 200bhp and four-wheel drive may not be strictly necessary – but it spells big-time fun for anyone with more than £20,000 to spend

RALLY REFUGEES

Can Lancia's relatively ancient but artful Integrale match up to the 1990's high-techery of Nissan's hottest-ever hatch? We test the world's fastest supermarket shoppers

So you think the Golf GTi is hot? Go take a peek at the latest rally refugees down at the local Lancia or Nissan showroom, then think again. Packing the sort of power-to-weight ratio more akin to that of a jet fighter than landbound personal transport, and with drop-dead looks certain to create outrage at the Sainsbury car park, the two Exocets-on-wheels tested here are so potent that each should carry a Government health warning.

No ordinary shopping trolleys, these. The Lancia Delta Integrale and the Nissan Sunny GTi-R are homologation cars, built to be eligible for certain classes in international rallying. Just 70 Nissan Sunny GTi-Rs will be headed this way, and even fewer Lancia Integrales. At a list price of £20,552 for the Nissan, £23,249 for the Integrale, and with Ford's aggressively priced Escort Cosworth poised to hit the showrooms about now, the limited supply of Integrales and GTi-Rs is probably wise.

It's hard to believe that the wider, more powerful Integrale is essentially a '70s design. Despite its unprecedented four consecutive world rally titles, the Integrale is now being overshadowed by newer, lighter designs. Its presence alone should ensure a longer shelf life, but Lancia says this will be the last in a classic line of competition Deltas.

ADD-ON BULGES

Place it alongside the fresh-from-the-wrappers Nissan Sunny GTi-R, and the Japanese equivalent appears the seven-stone weakling. The Sunny shares with the Lancia the impression that it has an engine too big to fit, but where the Integrale's flexed muscles endow it with a certain brashness, the Nissan's add-on bulges make you wonder if this was the very car they used for the 30mph concrete splat-crash test.

As for the curious appendage attached to the top of the Sunny's tailgate, no doubt its purpose becomes clear enough when the 140mph top speed is approached. The Integrale, its basic design penned long before drag factors were considered important, manages a mere 134mph. Fast enough, and a sight better looking too.

Powerful brakes, four-wheel drive, and a 2.0-litre turbocharged four-valve-per-cylinder engine are prerequisites of rally dominance. Surprise, surprise, both of these share that specification. But where the Nissan drives its 220bhp to all four wheels through viscous couplings for both the centre and rear differentials, the 210bhp Lancia gains greater sophistication through the use of a Torsen (torque sensing) rear differential, together with a viscous coupling at the centre and a torque split that marginally favours rear traction.

If all this sounds like gobbledygook, suffice to say that each of these is capable of directing its ample power in proportion to the wheels which have the greater grip. None is wasted in wheelspin. Given a bootful of throttle and a few deftly timed gearchanges, the

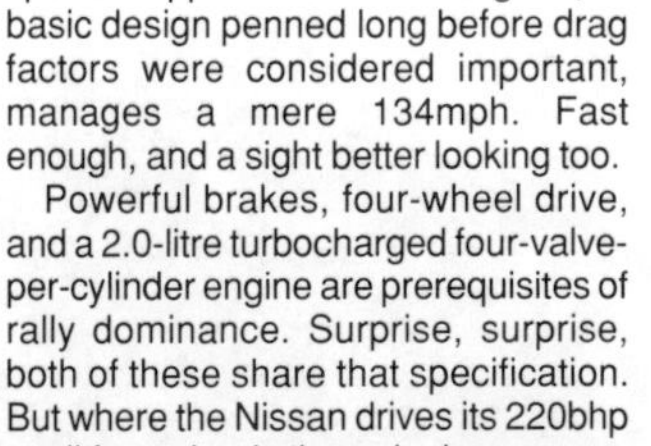

Integrale's Recaros are as good as they look. Not much room in back

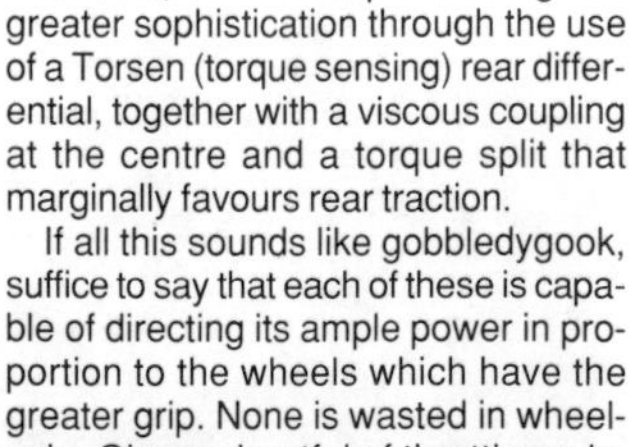

Lights and clocks for every occasion; steering wheel has flat rake

Lancia Delta Integrale v Nissan Sunny GTi-R ■ **NEW CARS TEST**

Massive reserves of grip for slingshot bend exits: keep Lancia's turbo boost needle busy for best results

Brutal bonnet louvres announce presence of Nissan's sledgehammer power. Handling sometimes ragged

Integrale will scorch its way to 60mph in just 5.8sec, to 100mph in a staggering 16.1sec, with 30-70mph – the most meaningful guide to usable performance – accomplished in 5.6sec.

AUDACIOUS PERFORMANCE

That's darned quick, by any standards, yet even the Lancia's masterful showing at the test track can't match that of the Nissan. For the Sunny GTi-R, read 5.0sec to 60, 14.2 to 100, and a mere 5.3 from 30-70mph. Audacious performance from a shopping trolley . . .

Yet for all the Nissan's dominance, you would swear the Integrale is quicker. Perhaps the reason is the more violent power delivery from the Italian car. First there's nothing, then *whooosh*, power arrives all at once. The Nissan's engine is better managed in that the turbo effect is less noticeable, the response electric and, more than anything, it's solidly unstoppable. Faced with a gap opening in the traffic, the Nissan is there, quick as lightning. The Integrale can do the same but only when the engine is kept well spinning.

Cornering further underlines the Nissan's driveability. Unless you happen to be blessed with a Finnish-style left-foot braking technique, powering out of a tight corner in the Integrale can be a frustrating business as you wait for the turbo boost to build.

And that engine behaviour is partly responsible for a disappointing lack of adjustability in the Integrale's cornering poise through the tighter twists. Where the Sunny can be flung into a corner at outrageous speed, with the driver knowing that a mere lift of the throttle followed by a quick burst of power will bring the tail out a touch and the nose heading for the exit, a much more precise and calculated approach is called for with the Integrale.

Go in too fast, and the nose simply pushes wide. Get it right though, and the Latin thoroughbred rewards with a razor precision that the Nissan could never provide. It comprehensively outgrips the Japanese upstart, and provides the driver with a continuous stream of lucid feedback through the wheel. Faster curves merely underline the Integrale's superior finesse. Where the Nissan is a street-fighter, always willing, sometimes ragged, the Integrale is the aristocratic thoroughbred. With either car though, the reserves of cornering and braking ability will leave you speechless with admiration.

SPACE NO PROBLEM

We knew that these thinly-disguised rally cars would be phenomenally quick, but how do they stack up in real life?

Not bad, actually. Each is based on the shell of a mass-market hatch, with five doors in the Lancia's case, so space is not the problem it is in so many high-performance cars. But the Sunny disappoints in its bland, sombre and cheap interior. Just three extra dials and a leather-rim wheel are the only visual clues to distinguish its cabin from that of lesser models at half the GTi-R's price.

The Integrale doesn't share the quality shortfall inside, particularly with the optional leather trim for the wonderfully supportive Recaro seats. Shame it comes in left-hand drive only though. Another not so well hidden cost is the phenomenal thirst of a turbocharged engine driven hard. With either car, don't expect more than 20mpg or 200 miles from a tankful.

While both machines will happily pootle around town sending XR3is scuttling for cover and Golf GTi drivers reassessing their street cred, only one will whisk you over long distances in the manner expected of a proper road car. That's the Integrale. While the occupants are never left in any doubt that its ride is firm, the suspension will smooth off the worst of the lumps and adequately smother the smaller pits to create an acceptable overall ride compromise. The Sunny initially gives the impression that it's more softly sprung, but that doesn't mean it's any more absorbent. There's an underlying harshness and a fidgety quality, even on the motorway, which makes it tiresome for longer journeys.

Equally tiresome is the gruff rasp of the Nissan's engine. It permeates the cabin rather more than is reasonable for a £20,000-plus car, even one so sporting, and the low overall gearing further exacerbates this characteristic. There are high levels of tyre rumble, too. The Integrale is more refined in its noise absorption, although there's still a fair amount of tyre roar on coarse surfaces and more wind noise than the Nissan, but overall, it's acceptable enough.

VERDICT

There's never been a better time to play the rally star. Given the budget for an executive saloon, either of these explosive hot hatches makes it possible to have more fun on tarmac than you ever thought possible. Nowhere this side of a racetrack will you find a more intense expression of dynamic prowess.

Not that such beasts are for anyone; a fair degree of driver skill is an essential ingredient. Given these criteria, the decision lies between the slightly rough and ready but more accessible delights of the Nissan, or the more enduring but harder to master Integrale.

For all the Sunny's ridiculous looks you'll learn to respect its breathtaking abilities, but it's the handsome and distinguished (and more costly) Integrale which gains our vote. A sporting car must possess a certain emotive quality to make the driver feel good, whether it's driving the car or merely washing it on a Sunday afternoon. It's that star quality so exemplified by the thoroughbred Integrale which wins us over to its exotic Latin charms.

RATINGS AND VERDICT		
	LANCIA	NISSAN
PERFORMANCE	●●●●●	●●●●●
HANDLING AND RIDE	●●●●○	●●●○○
BEHIND THE WHEEL	●●●●○	●●●○○
ACCOMMODATION	●●●●○	●●●○○
QUALITY AND EQUIPMENT	●●●○○	●●●○○
SERVICE AND COSTS	●●○○○	●●●○○
VERDICT	●●●●○	●●●●○

HOW THE CARS COMPARE		
	LANCIA DELTA INTEGRALE	NISSAN SUNNY GTi-R
PRICE	£23,249	£20,552
PERFORMANCE		
Max (mph)	134	140
0-60 (sec)	5.8	5.0
30-50 in 4th	7.0	6.1
50-70 in 5th	7.0	8.1
RUNNING COSTS		
Touring mpg[1]	28.3	26.1
Gov't mpg: Urban/56/70	25.2/35.8/26.9	21.4/34.9/26.6
Insurance group	9	9
DIMENSIONS		
Length/width/height (in)	154/70/54	156/67/56
Wheelbase (in)/boot (cu ft)	98/na	96/9.9
Turning circle (ft)/lock turns	34.1/2.7	34.1/2.6
Kerb weight (kg)	1300	1240
MECHANICAL SPECIFICATIONS		
Cyls/cc/fuel system	4/1995/MPFI	4/1998/MPFI
Bore/stroke (mm)	84/90	86/86
Power (bhp/rpm)	**210/5750**	**220/6400**
Torque (lb ft/rpm)	219/3500	197/4800
Brakes F/R	vent disc/disc	vent disc/disc
Suspension front	strut/coil/anti-roll	strut/coil/anti-roll
rear	strut/coil/anti-roll	parallel links/coil/anti-roll
Tyres	205/50 ZR15	195/55 ZR14

[1] Calculated at 'Euromix' mpg (1/2 Urban + 1/4 56mph + 1/4 75mph)

Standard Japanese hatch innards inappropriate for a £20,000 car

GTi-R's dash is functional rather than inspiring; leather-rim wheel

```
  Colour(s)         BLUE
Type of Fuel       PETROL
/Chassis/Frame No. 63H03761
  Engine No.       3164 CC
Cylinder Capacity
Seating Capacity
Taxable Weight    27 05 87
Date of Registration 11 08 88
```

SECOND CHANCES

Lancia Delta HF 4WD

● *Supercar performance and handling, Q-car looks and five-door practicality . . . but what's the catch? John Tipler looks at the market for Lancia's 4wd superstars*

● **Bonnet bulge gives away 16-valve Integrale (left) with HF 4WD forerunner**

The Integrale may be a rare sight on British roads, but there's a very good reason for that — the best possible reason, if you like a car with a sporting pedigree. The Integrale's *raison d'etre* was — and still is — homologation for international rallying; what you buy is essentially the same car that has won three world rally championships.

It's a potent and intoxicating road car, too, and an extremely practical one, being a compact five-door hatchback. It is, however, a car that requires careful and regular servicing and the potential owner should be well aware that this is no ordinary hot hatch . . .

With the groundwork done in the nippy 1600 Delta HF Turbo and the HF 4WD, Lancia introduced the HF Integrale to the UK in February 1988, on the crest of the wave of the HF 4WD's domination of the 1987 World Rally Championship, the first year of the present Group A regulations. The 8-valve Integrale produced 185bhp from its 2-litre water-cooled turbo engine, and it was distinguished from the rare HF 4WD by flared wings and wheel arches plus wraparound bumpers and side-skirts. The car looked as aggressive as its performance proved, given a slightly larger turbocharger. HF, by the way, stands for 'high fidelity', and relates to performance models which in the dim and distant past Lancia would sell only to its 'faithful' customers.

Turbo & Integrale

● **Standard 16-valve engine boasts 200bhp, though some are chipped up to give even more — best to avoid these modified cars**

● *16-valver has even wider low-profile tyres than 8v Integrale*

Lancia's target build for the HF 4WD was 5000 units, and all left the factory in left-hand-drive form; a car with right-hand-drive is a different vehicle as far as homologation is concerned. Ten thousand 8-valve Integrales were made, and there is likely to be a run of just 5000 of the current generation 16-valve cars. The next phase evolution model will undoubtedly use Kevlar panels to reduce weight and stay competitive in international rallying. If you feel the weight of the 16-valve Integrale bonnet, you'll understand why. The evolution model, therefore, will be expensive, and will probably only come to the UK as a personal import — the same way a good ten per cent of Integrales have already entered the country.

Right-hand-drive Integrales remain scarce — perhaps only 40 per cent of UK Integrales have been converted. Mike Spence Ltd did about a hundred, and today there are only a couple of garages in the country which perform the operation. One is John Whalley's in Bishop's Stortford, which has completed about 50 so far. Whalley uses the facia from the regular 1600 HF Delta. Some conversions have used Fiat-derived steering racks, which don't have the precise feel and tight turning circle of the original left-hand-drive set-up. This is well worth bearing in mind, because the rhd conversion lacks some of the pure brilliance of the lhd car and could actually be worth less in years to come.

Coming from the ordinary Delta, the switchgear, positive stalk controls, and the steering wheel and dash may seem a little bit low-rent in a vehicle with the kudos of the Integrale, but they function adequately. The speedo becomes electronic rather than cable-driven, otherwise all components are Delta compatible.

The rhd conversion in the car we drove was near faultless. The Recaro seats are very comfortable and supportive, but there's not a lot of room to rest your left foot; it has to go underneath the clutch pedal. The left-hand-drive cars score much better here.

The rear hatch gives access to a reasonable luggage space, with a space-saver spare wheel on one side. If you've a mind to use the Integrale as a removal van, the rear seats fold down to provide a much larger cargo deck.

The first car we tried at John Whalley Ltd was a red HF 4WD for £11,500, which had done 45,000 miles. Whalley had 'chipped' the 2-litre up to give 210bhp, putting it on a par with the 16-valve Integrale — John thinks that's about the limit for the 4x4 car. With smaller 14-inch wheels and skinnier 185-section tyres, it should have been quicker off the line than the Integrale; the turbo is smaller, so power comes in much lower down the rev band. Performance on

Colour(s)	BLUE
Type of Fuel	PETROL
VIN/Chassis/Frame No.	63H03761
Engine No.	
Cylinder Capacity	3164 cc
Seating Capacity	
Taxable Weight	27 05 87
Date of Registration	11 08 88

SECOND CHANCES

Lancia Delta

the road was little short of sensational. Furthermore, the Delta engine is civilised enough to be driven at a leisurely pace.

The only faults we could find with the HF 4WD were a small stone chip in the windscreen — which you would probably live with — and an inoperative electric window on the driver's side. Otherwise, it was immaculate inside, and the specification included a steel sunroof and Pioneer stereo.

> In a broad, well-sighted sweeper taken flat, the body presses down even harder and the broad tyres form a seemingly unbreakable bond with the tarmac. The lateral 'g' has the odd effect of shaping one's face into a permanent ear-to-ear grin.
> (*Performance Car*, March 1989)

The second of the homologation cars was the 8-valve Integrale, in which the brakes were uprated from the HF 4WD. John Whalley maintains the 8-valve car can be taken to 230bhp for the road and still have sufficient braking and handling reserves. Although he has heard of 285bhp 16-valve cars, his opinion is that the car is so well balanced at 200bhp that it isn't worth going for any more. We would add that unless you are absolutely confident in the work that has been done, avoid 'chipped up' Integrales — they could be more trouble than they're worth.

Official performance figures for the 8-valve Integrale are 0-60mph in 6.2 seconds, and a top speed of 133mph. Hard driving will see the fuel consumption dip below 20mpg, but 25-30mpg should be attainable. Performance figures for the 16-valve car are 0-60mph in 5.5 seconds, with a top speed of 137mph. Knock one or two mpg off the 8-valve's average fuel consumption.

We were impressed by a superb left-hand-drive 8-valve Integrale in Whalley's showroom, priced at £11,500, its one previous owner having logged 35,000kms, or 24,000 miles. The car was first registered in France in 1988, and was a personal import. A rhd conversion performed by Whalley would add a further £3500 to the price.

The 16-valve car has more of a chunky stance about it, fostered by the humped bonnet and wider low-profile P700s on 15-inch alloy rims. Our test car, G-reg 1989, with 19,000 on the odometer and priced at £18,500, had the standard fitment P700s at the rear and Yokohamas on the front, which John Whalley acknowledges are superb dry-weather tyres but is slightly sceptical about in the wet.

John points out that wheels are something the prospective Integrale owner should inspect carefully, for it seems that they are frequently dented or kerbed, often on the inside. Check these thorough-

● **HF 4WD Turbo is a rare beast — this 45,000-mile example was priced at £11,500**

HOW MUCH?

£8000-£9000
If you can find one, a 1987-88 left-hand-drive HF 4WD. Late two-wheel-drive 1600 Delta HF Turbos are cheaper

£10,000-£12,000
Buys you a right-hand-drive 1988 HF 4WD or a good left-hand-drive 1988 8-valve Integrale. Rhd conversion can cost £3500

£13,000-£15,000
The later 8-valve Integrales, possibly with rhd, are to be found here; perhaps someone desperate to offload a 16-valve car in a hurry

£17,000-£19,000
Most of the 16-valve Integrales are priced within this narrow band; they are still under 18 months old, remember

ly, particularly if you are looking at a London car, as a new wheel will cost around £200. The test car, barely 18 months old, had already had one new one.

John has come across cars with as much as 9mm toe-in in the rear suspension geometry, and of course a car so badly out of tune will simply eat tyres. Front suspension tracking is prone to fluctuate, so look for uneven tyre wear on the inner edges, and pulling to the side under braking. Sixteen-valve cars have a stronger suspension, so tracking is less of a problem. Be aware of clutch-slip when pulling away or under acceleration; the clutch may need replacing.

The 16-valve cars handle slightly differently from the HF 4WD and the 8-valve, in that the torque is rear-wheel biased and front-wheel biased on the other two. You have to be going some — or be driving on wet roads — to notice it, however. It's a result of Lancia's deliberate ploy to make the 16-valve Integrale quicker on tarmac, whereas the 8-valve car is faster on loose surfaces. Ultimately the 16-valve is the speedier car, but you are as a consequence closer to the limits of the chassis' capabilities. As Whalley says, 'I'm always more wary of the 16-valve because if you overstep the mark, it says "you did it, you pick it up", whereas the 8-valve car will assist you to get out of the muddle'.

When Whalley took us out in the 16-valve, the heavens opened, but it was difficult to believe the road conditions were tricky. In a standard performance saloon car you'd be sliding all over the road. The Integrale just sticks the power down, and Whalley was hurtling the Integrale round and round a deserted roundabout as if it was completely dry, getting faster and faster. Quite incredible, and thoroughly exhilarating. Just by lifting off and opposite locking, you can feel it turn in, but you can't call it a slide; the car is actually alive on the surface of the road.

It's all a delicate process, which needs learning thoroughly, but is hugely satisfying when you get it right. John Whalley urges his customers to get to know their integrales for a few weeks before they start trying anything like this. This car is a superb all-rounder, with hidden depths.

Because of the relative rarity and sophistication of the HF 4WD and Integrales, not to mention the competition pedigree, John Whalley sees the cars as certain classics. The problem his customers have is that you are supposed not to clock up big mileages with classic cars, but the little Lancias just love to be driven, and hard. They all swear they're not going to use them over much, but it seems most Integrales notch up a relatively high mileage. The 16-valve car we drove had done 19,000 miles in a year, which is a rep's quota.

New cars are usually bought by the young executive type, but as far as secondhand cars are concerned, says John, 'the showroom door opens, and you could see anybody standing there. They are all car enthusiasts, because you don't buy a car like this to impress your next-door neighbour. You buy it because it's one hundred per cent a driver's car, and you're getting some of the latest technology available. It's right out at the front edge of mechanical evolution.'

The values of both 8- and 16-valve Integrales are rising, according to the dealers' bible, CAP; retail values have gone up £1200 since November. Although the 8-valve cars are more plentiful, John says they haven't dropped in value during

THE RIVALS

1989 Peugeot Mi16x4
Styling is sleeker than dated Lancia, plus there's more room inside; excellent chassis but can't match immediacy of Integrale's performance

1987 Audi quattro
Superb build quality, better economy and seamless performance, but heavier and less agile than Integrale; not as nervous, though

1989 Toyota Celica GT-Four
Only two doors, but handsome exterior and sophisticated engine. Graceless interior, and overweight body inhibits performance

1990 RS Cosworth Sapphire
Excellent grip, handling, brakes and balanced steering, combined with towering performance from coarse engine, but repmobile looks and low-rent interior. Rather pricey

> By 4000rpm any sense of lag is firmly dispatched and the Integrale is absolutely on the nail, rocketing forward in response to the drilled steel throttle pedal. As more and more power pours out it is translated into instant speed.
> (*Performance Car*, March 1989)

● Bonnet hump essential on 16v car to clear larger cylinder head

● Recaro seats are superb, but trim is generally cheap and frequently plagued by rattles. Conversion to right-hand drive costs £3500

SECOND CHANCES

Lancia Delta

> For sheer brilliance, it's an impossible car to beat. Fast, frantic and sometimes frustrating; the ultimate under-£20,000 thrill. What's more, the effect shows no sign of wearing off. (*Performance Car*, March 1989)

● Dealer John Whalley says Integrales are bought by enthusiasts — and, in recent years, by collectors

the past year, but it is the 16-valve models which people want at the moment. The HF 4WD cars are more of an unknown quantity, and consequently more difficult to value. They are also a veritable Q-ship. John sees many going into private collections, but invariably they get driven hard and fast too.

If the UK price goes too high, it attracts Integrales in from Germany, and it only takes three or four foreign imports to affect the UK market price. For this reason, John doesn't himself go abroad after the cars, preferring to maintain the status quo, although he's perfectly happy to handle Integrales once they're in the country. However, he says that values are starting to firm up abroad as well, so there soon won't be any price advantage in buying overseas. He cautions anyone thinking of acquiring a foreign import to be absolutely certain that all the relevant taxes have been paid on the car, which means VAT and import duty.

Among the Dedras and Themas, John Whalley Ltd had three second-hand Integrales in stock when we called: two 16-valves, one 8-valve. There was also a brand new 16-valve model, with factory leather upholstery, and a brand new 8-valve car, priced at £18,000. Bearing in mind its potential value as an original classic, he won't convert it to rhd — unless a buyer demands it. ○

● Immaculate G-reg 16v Integrale was advertised for £18,500

OWNER'S VIEW

Jack Thomas is director of an Essex shipping company and lives in Bishop's Stortford. He is a fan of Italian cars 'because they talk to you', and having owned a number of Lancias including a Monte Carlo, he followed the fortunes and development of the Delta S4s and HF 4WD cars in rallying. He had little thought of owning a 4WD car himself, as he certainly wouldn't have bought a left-hand-drive car, until he called at John Whalley Ltd and saw a right-hand-drive conversion on an 8-valve Integrale.

'It was an impressive conversion,' he said, 'using all Lancia parts, pedals and connections and the Tipo steering rack to maintain the original car's 2.8 turns lock-to-lock. John Whalley took me out on one of his demonstrations and tried to scare the pants off me, but really it felt very safe and proved just how good the Integrale is.'

A member of the Lancia Owners' Club, Thomas ordered an 8-valve car in September 1989, and had the Whalley conversion carried out. 'It was about the time they were talking about the 16-valve car,' he recalls, 'but peaky multi-valve engines and complexities like ABS are more appropriate for the rally driver; the 8-valve is much more user-friendly.'

Since then, he's only done about 5000 miles, and the car has not missed a beat. He is in the fortunate position of running an Alfa 164 V6 as a company car, and intends to keep his Integrale 'for a long, long time, so it doesn't get a lot of use.' It suits him to have a car which can run on unleaded petrol, is easy to park in town, and has the potential to out-drag a Porsche.

'It's reliable, starts first time and the heater works in five seconds, and it feels glued to the road, so there's never any complaint from passengers about being thrown around. I've never got anywhere near its handling limits yet, but one of the reasons for owning an Integrale is the potential abilities it has. One of the most interesting runs was last Christmas when we went down to the West Country, and I've never driven in wind and rain like it. The car was absolutely rock-stable.'

Thomas confirms the Integrale's Q-car status. 'Most people don't know what it is. Even the police reaction is neutral. But enthusiasts who do recognise it will flash their lights, stand and stare, or come over and have a chat. It has a sort of self-regulatory speed limit, because it's shaped like a brick, and over 80mph on the motorway the combination of wind and tyre noise gets pretty unbearable!'

Lords of the Forest, Kings of the Road?

ESCORT RS COSWORTH — LANCIA INTEGRALE

Does a rally-winning pedigree breed a better road car? *Mark Hughes* finds out the answer when he drives two rallying success stories with quite different characters

Someone once asked me: 'If rallying success is supposed to sell cars, how come Lancia is not the biggest seller in the world?' Admittedly, at the time said gentleman was trying hard to persuade Ford Motor Company to divert some of its resources away from the forests and on to the race track by way of his bank account, but it's nonetheless difficult to argue with his logic.

Lancia's integrale is an all-time-great rally car, having won six World Rally Championships, yet the marque captures only a minuscule percentage of new car sales. But the question of whether rallying improves the breed has an emphatically more positive answer – at least if the two cars featured here are anything to go by.

Coinciding with the integrale finally bowing out of its top-flight rally career, the UK has seen the release of the final version of the road car it spawned. The integrale Evoluzione 3 differs from its immediate predecessor by having an exhaust catalyst. But in best Italian style, some concurrent changes have been made which result in the catalysed version producing more power than the outgoing 'dirty' model. The smaller Garrett T3 turbo gives better response but the theoretical top-end power penalty has been reversed by recalibrating the electronic engine management, with double the calculation frequency. In all, the 2-litre 16-valve turbo now has 5 percent more power, with a peak of 215bhp. Complementing these changes are bigger wheels and tyres – 205/45/R16s on alloy rims just aching to be kerbed, so small are the super-low-profile tyres' sidewalls – and a rehashed interior which makes full use of Lancia's favourite material, alcantara.

The rest of this classic piece of machinery is left largely untouched. Its power is fed through a very heavy-duty four-wheel-drive system which comprises an epicyclic centre diff, giving a 47-53 percent front-rear torque split under normal running but the viscous coupling attached to this can further vary the split according to conditions. At the rear there's a sophisticated Torsen limited-slip diff. Front MacPherson struts and rear trailing arms take care of the suspension.

Ford may have created the Escort RS Cosworth to challenge the integrale's rallying supremacy – it has come first in five World Championship rallies this season – but the

ESCORT RS COSWORTH – LANCIA INTEGRALE

Its sporty interior looks the part, but the Lancia is not that welcoming: dials are difficult to make out and the side-bolstered seats are strictly for the svelte

Inside the Cosworth gives away little of its true character: apart from the grey Recaros upholstery and flash white on black gauges, the look and feel is standard Ford fare

whole tenor of the road car is more mainstream. You get all the trimmings that might reasonably be expected from a £24,000 car, including a high-quality sound system with automatic volume control. But where the integrale's appearance and stance is pure, raw, purposeful, with sculpted wheel arch flares within an otherwise neatly functional five-door hatchback, the Escort is louder, more flashy, more vulgar in appearance. The dominant feature is that enormous bi-plane rear wing which attracts attention from schoolkids and policemen alike. In fact, with the beautifully understated aggression of the bright red integrale sitting alongside the blue Essex machine, people made a beeline for the Cosworth, brushing past the Italian car without a glance to ask questions about the Ford.

But for all its brash looks, the Escort is just as serious as the Lancia beneath the skin. It's not really an Escort at all, more like a short-wheelbase Sierra dressed in Escort panels. Its engine is mounted longitudinally, not transversely like other Escorts, or the integrale for that matter. Its four-wheel-drive transmission has a similar epicyclic-viscous coupling centre diff to the Lancia and relies on a simple viscous coupling to give a limited-slip effect on the rear diff. It also has more rearward bias than the Lancia at 34-66 percent front-rear. Suspension is relatively simple with MacPherson struts up front and semi-trailing arm rear.

Before outlining the relative strengths and weaknesses of these two cars, it should be emphasised that both offer staggering ability. Their willingness to swallow bends, adverse cambers and wet surfaces while maintaining exceptionally high average speeds in almost complete safety, but with competition-honed,

'For all its brash looks, the Escort is just as serious as the Lancia under the skin. It's not really an Escort at all'

razor-sharp responses, is total. They are not that much more expensive than the best front-wheel-drive hot hatches, but their ability is of a different order.

Neither promises this sort of experience when first started up. Both engines sound mundane at tickover, no exotic multi-cylinder burbles, no racecar snarls; they could be Mondeos or Cavaliers when ticking over. Of course, both are massively potent, with 226 and 215bhp for Ford and Lancia respectively in compact bodies. This translates into 0-60mph times of 5.7sec for both cars, the latest integrale's extra horsepower over the old version shown by slightly brisker acceleration throughout the range. Neither car is as fast all-out as might be imagined with all that power. That upright front and large frontal area make the integrale visibly unaerodynamic; even so we were surprised that it struggled to no more than 124mph – significantly slower than the old model – round Millbrook's bowl. The Escort's rear wing holds its maximum down to a more impressive 141mph.

The Ford's big turbo ensures that its kick-in-the-back acceleration is even more muscular than the Lancia's when fully extended, though its flexibility suffers badly for the same reason. Neither car is truly happy at low revs in a high gear, particularly the Ford in which even fourth gear, never mind fifth, is a waste of time around town. The Lancia's smaller turbo pays dividends in flexibility, as you can see from the in-gear times. Both offer an overboost facility for the first few seconds of full throttle application, though this cuts back rather too abruptly in the Escort. ▷

Ford Escort RS Cosworth Luxury £24,240

RUNNING COSTS
FUEL ECONOMY: Government composite mpg: 27.3. Test mpg: 20.6.
INSURANCE GROUP: 20.
SERVICING/WARRANTY: 6,000-mile service intervals. One-year unlimited mileage warranty, six-year perforation warranty.

ENGINE
Four cylinders inline, front longitudinally-mounted. 1,993cc, 90.8mm bore x 77.0mm stroke, cast-iron block with aluminium-alloy cylinder head. Four valves per cylinder, twin-belt-driven overhead camshafts. Multi-point electronic fuel injection. Single Garrett turbocharger.
PEAK POWER: 227bhp at 6,250rpm.
PEAK TORQUE: 224lb/ft at 3,500rpm.
POWER TO WEIGHT RATIO: (as tested) 162.1bhp per ton.

TRANSMISSION
Five-speed all-synchromesh gearbox, front longitudinally-mounted, driving all four wheels through epicyclic centre differential with viscous coupling. Rear limited-slip viscous coupling differential. 34-66 percent front-rear torque split.

CHASSIS
SUSPENSION: Front MacPherson struts with coil springs, telescopic dampers and anti-roll bar. **Rear** semi-trailing arms with coil springs, telescopic dampers and anti-roll bar.
STEERING: Power-assisted rack and pinion.
BRAKES: Ventilated discs **front** and **rear**. ABS.
WHEELS/TYRES: 16x8in alloy rims with 225/45/ZR16 radials.

DIMENSIONS
EXTERNAL: Length 165.8in, width 68.3in, height 55.3in, wheelbase 100.4in, track front-rear 60.0-62.0in.
INTERNAL: Rear head 37.4in, rear leg 22.2in-29.1in, rear width 51.2in, front head 38.2in, front leg 36.4-42.7in, front width 54.0in.
WEIGHT: 3,123lb (as tested, including full fuel tank and 165lb driver).

EQUIPMENT
Automatic: N/A
Leather upholstery: N/A
Air conditioning: Standard
ABS: Standard
PAS: Standard
Airbags: N/A
Electric seats: N/A

ESCORT RS COSWORTH – LANCIA INTEGRALE

Next to the pure, understated power lines of the integrale, the Cosworth looks brash – even vulgar – but it hogs the limelight

When extracting full performance from the engines, both cars suffer from throttle lag which takes you back to first generation turbo applications. The Ford is worse in this respect – again because of its bigger turbo – but the Lancia is only slightly better. Relative lack of refinement betrays the age of both engines. In each case the muscle is delivered from just under 3,000 to just above 4,000rpm, when the turbos are really doing their stuff. Take them beyond this and neither engine really shines in the way that the performance is delivered. The Lancia's (with contra-rotating balancer shafts) is nowhere near as rough as the Ford's at high revs, but in no way competes for top-end smoothness with more modern engines like the Vauxhall or Rover 2.0-litre twin cams or any number of Japanese equivalents. Although not as musical as the original integrale, there is still a good sporting exhaust note, mixed in with a strong turbo whish when fully extended.

The Cosworth sounds no different at constant revs to any old Pinto engine in a 2.0-litre Cortina or Sierra. It starts to get rattly at around 3,700rpm, is unpleasantly harsh by 4,000 and almost unbearable anywhere near the black line on the white tacho.

White gauges are the strongest distinguishing feature of the Cosworth's interior. Apart from the enveloping grey cloth Recaros, the rest of the interior – switchgear and trim – is standard Ford. The Lancia's is more distinctive and overtly sporting, with a leather Momo wheel, but it's an ergonomic mess. The yellow-on-black dials are attractive but difficult to read, especially the tacho and speedo which, as well as starting from unconventional points on the dial, are counter-rotating. The minor controls above the centre console are unlit and virtually impossible to find at night. The alcantara seats are wonderfully supportive if you fit in them but the side bolsters, particularly those of the squab, demand that the occupants are slim.

The controls of the Italian car feel better than the Ford's and this carries through to their gearboxes. The integrale's change is pleasantly fluid and complemented by an easy clutch; the Escort's box and clutch are slightly heavier. The change is good and positive though – there's never the slightest doubt about the correct selection; it's just not quite the enjoyable precision you get with Lancia.

It's much the same with the steering. The Lancia's is pin-point accurate with a lovely feel: very communicative, not too light or too heavy. The messages still feed back well to the driver despite the new, even lower-profile tyres. Why can't all steering systems be this good? The Cosworth's steering is accurate

Both sound mundane at tickover: no exotic multi-cylinder burbles, no racecar snarls. They could be Mondeos

enough but over-light, masking much of the feel. It also goes curiously extra-light around the straight ahead when you apply or take off lock, giving a false impression that the front tyres are about to lose their grip.

In fact, the Escort's grip levels are enormous. Visibility of the road ahead is the sole determining factor of the speed at which corners can be taken, not the roadholding. It has even more pure roadholding than the Lancia and ultimately a better chassis balance too, though realistically this would only be discovered on the race track. However, it does not flow with quite the same delectable feel as the integrale. The Lancia will turn in with a silky precision that is not too sudden or ▷

Lancia integrale Evoluzione 3 £25,000

FAST LANE Figures

	Ford Escort RS Cosworth	Lancia integrale Evoluzione 3
Acceleration through the gears		
0-30mph	1.9sec	2.0sec
0-40	2.9	3.2
0-50	4.4	4.4
0-60	5.7	5.7
0-70	7.9	8.1
0-80	9.9	10.4
0-90	12.5	13.6
0-100	16.6	17.3
0-110	20.7	22.3
Acceleration in fourth / fifth		
20-40mph	11.3 / ◊	9.0 / 13.9
30-50	9.7 / ◊	6.7 / 11.4
40-60	7.1 / 12.7	5.1 / 9.3
50-70	5.5 / 9.6	4.8 / 7.5
60-80	5.3 / 7.6	5.5 / 7.2
70-90	5.9 / 8.2	6.0 / 8.0
80-100	6.6 / 9.5	7.1 / 8.9
Max Speed:	141.1mph	123.8mph

◊ Car unable to gain these increments before the end of mile straight.
Performance figures taken at Millbrook Proving Ground in dry weather.

RUNNING COSTS
FUEL ECONOMY: Government composite mpg: 26.1. Test mpg: 19.0.
INSURANCE GROUP: 18.
SERVICING/WARRANTY: Service intervals 9,000 miles. One-year/unlimited mileage warranty, eight-year anti-corrosion warranty, one-year recovery service. Optional one- or two-year extra cover at end of first 12 months.

ENGINE
Four cylinders in-line, front transversely mounted, 1,995cc, 84mm bore x 90mm stroke. 8:1 compression ratio. Single Garrett turbocharger. Cast-iron block with aluminium-alloy cylinder head. Counter-rotating balancer shafts. Four valves per cylinder, twin-belt-driven overhead camshafts. Electronic multi-point fuel injection.
PEAK POWER: 215bhp at 5,750rpm.
PEAK TORQUE: 227lb ft at 2,500rpm.
POWER TO WEIGHT RATIO: (as tested) 158.9bhp per ton.

TRANSMISSION
Five-speed, all-synchromesh gearbox driving all four wheels through Epicyclic centre diff with viscous coupling and rear Torsen limited-slip differential. 47-53 percent front-rear torque split.

CHASSIS
SUSPENSION: Front MacPherson struts with coil springs, telescopic dampers and anti-roll bar. **Rear** MacPherson struts with trailing arm, torque reaction bars, coil springs, telescopic dampers and anti-roll bar.
STEERING: Power-assisted rack and pinion.
BRAKES: Front ventilated discs, **rear** solid discs. ABS.
WHEELS/TYRES: 7.5x16in alloy rims with 205/45/ZR16 radials.

DIMENSIONS
EXTERNAL: Length 153.5in, width 69.6in, height 53.7in, wheelbase 97.6in, track front-rear 59.7in-59.2in.
INTERNAL: Rear head 33.3in, rear leg 24.4in-31.1in, rear width 52.8in, front head 38.0in, front leg 33.6-40.9in, front width 52.6in.
WEIGHT: 3,031lb (as tested including full fuel tank and 165lb driver).

EQUIPMENT
Automatic: N/A
Leather upholstery: N/A
Air conditioning: Standard
ABS: Standard
PAS: Standard
Airbag: N/A
Electric seats: N/A

sloppy, but the Escort will go through a brief transition of no man's land where slight body roll begins to build up and the car feels to be edging into the first stages of oversteer. If you then try to encourage the oversteer by planting the throttle, the car will simply pull itself into neutrality and then a shallow understeer. Stay off the throttle and the speed simply comes down without the car ever sliding.

The Lancia has even less body roll than the Ford: once turned in it just digs itself into the corner under power. Ultimately it will understeer quite determinedly in mid-corner, but this will lessen in the second part of the turn. In the wet, it will give a little oversteer flick on the corner's exit. Neither car gripped as impressively as we expected in the wet. We suspect this was down to the tyres, the same make on both cars.

Both have massive ability through the turns. The Ford has even more than the Lancia but the Italian car offers more fluidity.

The downside of the integrale's crisper turn-in is a much firmer ride than the Escort. The Cosworth has excellent ride comfort for such a sporting car and, as a bonus, allows only a limited amount of road noise. The integrale's ride is stiff and jiggly, tending to crash quite heavily over bumps that don't disturb the Cosworth's equilibrium. And road noise is a constant companion for the integrale driver. Both cars have a tendency to tramline – to follow road cambers and white lines – though this is more pronounced in the Escort.

Braking systems on both cars are well up to controlling their performance, but the Ford brakes are especially impressive, with super-short stopping distances and a good, crisp pedal feel. By comparison, and going totally against the otherwise strong tactile appeal of the car, the integrale's pedal lacks feel.

Assuming the enthusiast could get them insured, either one of these cars will thrill. The choice is more likely to be decided on the aesthetic appeal of each to the individual. But for the record, the Escort is faster, has more grip, a better ride and a better handling balance at the limit. What the Lancia has going for it is more difficult to define, but it's most definitely there. It has a much more pleasant engine, a beautiful fluidity in everything it does, fantastic steering and almost unreal body control. But it's more than that. For us, just gazing at the integrale standing there in the forest next to the Cosworth, all taut, aggressive and purposefully beautiful, grabs you where you feel it most. ∎

LANCIA MOTOR CLUB

Founded in 1947, the Lancia Motor Club is the largest and longest-established of the many Clubs devoted to the marque throughout the world. Membership is open to everyone interested in Lancias of all ages from the 1907 Alpha to the latest models.

With 28 area meetings in the UK, a monthly News Sheet and a regular colour magazine, members are kept up to date with all that is happening in the Lancia world. Technical advisers and spares consortia help to keep older cars on the road while local and national rallies, track days and an annual concours provide opportunities to put the cars to good use. An exclusive insurance scheme is available for all Lancias.

For more information and an application form write to the Membership Secretary at the address below:

**Mr. David Baker,
Mount Pleasant, Penhros, Brymbo,
Wrexham, Clwyd LL11 5LY**